COMPACT I

Public Administration in Complexity

COMPACT I

Public Administration in Complexity

Edited by

Lasse Gerrits & Peter Marks

EMERGENT™
PUBLICATIONS
3810 N 188th Ave
Litchfield Park, AZ 85340

COMPACT I: Public Administration in Complexity
Edited by: Lasse Gerrits & Peter Marks

Library of Congress Control Number: 2012951081

ISBN: 978-1-938158-01-8

Printed in the United States of America

CONTENTS

Chapter 1—PUBLIC ADMINISTRATION IN COMPLEXITY

FRANK BOONS, LASSE GERRITS, & PETER MARKS

SECTION I
RESEARCHING COMPLEXITY

Chapter 2—ONTOLOGIES OF MULTIPLICITY AND SIMULATION: POSSIBILITIES FOR PUBLIC AFFAIRS INQUIRY

THOMAS J. CATLAW & YUSHIM KIM

Chapter 3—COMPLEXITY AND NEGOTIATED ORDER

GILL CALLAGHAN

Chapter 4—ADDING A COMPLEX ADAPTIVE SYSTEMS APPROACH TO PUBLIC ADMINISTRATION RESEARCH

ERIK W. JOHNSON, KEVIN C. DESOUZA, QIAN HU, & RASHMI KRISHNAMURTHY

Chapter 5—ASSESSING THE APPLICABILITY OF QUALITATIVE COMPARATIVE ANALYSIS FOR THE EVALUATION OF COMPLEX PROJECTS

STEFAN VERWEIJ & LASSE M. GERRITS

Chapter 6—COMPLEXITY FRIENDLY" MESO-LEVEL FRAMEWORKS FOR MODELING COMPLEX GOVERNANCE SYSTEMS

CHRISTOPHER KOLIBA & ASIM ZIA

Chapter 7—GOVERNANCE NETWORK ANALYSIS: EXPERIMENTAL SIMULATIONS OF ALTERNATE INSTITUTIONAL DESIGNS FOR INTERGOVERNMENTAL PROJECT PRIORITIZATION PROCESSES

ASIM ZIA, CHRISTOPHER KOLIBA & YUAN TIAN

SECTION II
COMPLEXITY IN THE REAL WORLD

Chapter 8—THE APPLICATION OF A COMPLEXITY ANALYTICAL LENS TO EXPLAIN AND UNDERSTAND PUBLIC POLICY PROCESSES

ELIZABETH EPPEL

Chapter 9—PUBLIC SERVICES BETWEEN LAISSEZ FAIRE ECONOMIC AND PREDATORY FINANCIAL CAPITALISM

ALEXANDER DAWOODY & PETER MARKS

Chapter 10—DESIGN, DEVELOPMENT, AND SUSTAINABILITY IN MULTILEVEL GOVERNANCE IN MANAGING EMERGENCIES AND CRISES

NAIM KAPUCU & VENER GARAYEV

Chapter 11—URBAN CLIMATE FRAMEWORK: A SYSTEM APPROACH TOWARDS CLIMATE PROOF CITIES

SONJA DÖPP, FRANSJE HOOIMEIJER & NIENKE MAAS

Chapter 12—ADVANCING PUBLIC SERVICE GOVERNANCE THEORY USING A PERFORMANCE LANDSCAPE MODEL

MARY LEE RHODES

Chapter 13—BALANCE BETWEEN ORDER AND CHAOS AS EMERGING VALUE IN COMPLEXITY LEADERSHIP

GEERT TEISMAN, SIBOUT NOOTEBOOM, & YTSEN DEELSTRA

SECTION III
HAPPY MARRIAGE?

Chapter 14—THE MICRO-MACRO PROBLEM AS THE CORE PROBLEM OF COMPLEXITY THEORY OF POLICY PROCESSES

GÖKTUĞ MORÇÖL

Chapter 15—WHY TAKING COMPLEXITY SERIOUSLY IMPLIES A PARADIGM SHIFT FOR POLICY STUDIES

DERK LOORBACH & NIKI FRANTZESKAKI

Chapter 16—PROPOSITIONS FOR COMPLEXITY THEORY IN APPLICATION TO THE STUDY AND PRACTICE OF PUBLIC ADMINISTRATION

LASSE GERRITS & JACK W. MEEK

Chapter 17—COLLECTIVE DECISION-MAKING AS AN EVOLUTIONARY PROCESS

FRANK BOONS, LASSE GERRITS, & PETER MARKS

Chapter 1

PUBLIC ADMINISTRATION IN COMPLEXITY

Frank Boons (Erasmus University Rotterdam, NLD)
Lasse Gerrits (Erasmus University Rotterdam, NLD)
Peter Marks (Erasmus University Rotterdam, NLD)

INTRODUCTION

There is an argument that says that research in Public Administration deals with complexity by definition. This argument is true. There is also an argument that says that Public Administration is actually very little informed by complexity. This is equally true. The differences lie in the different takes on complexity. The first statement builds on the premise of complexity as an expression of the complicatedness of public administration; the second statement builds on a very particular and articulate conception of complexity that is inspired by the loose collection of complexity sciences. As documented elsewhere (e.g., Meek, 2010), the latter has increasingly gained attention in the realm of Public Administration and related domains. As befits a new perspective, it has led to both claims of 'doing something completely new' and claims of 'doing something old using fancy jargon' (Pollitt, 2009). It shows that complexity-thinking has the potential to do things in a different way but it also shows that one should not get carried away and should keep an eye on what is being done elsewhere.

Much of the mixed response can be traced back to the fact that the concepts grouped under the header of 'complexity theory' in Public Administration itself is neither one coherent research program nor a coherent research community. While there are common themes and common concepts in this realm, there is also a wild variety of topics, approaches, conceptualizations and methodologies. No wonder then that the community of people seeking to make use of complexity theory is still working out where it stands and what its work comprises. An international research community is beginning to emerge, fuelled by regular meetings in both Europe and the United States of America. One such meeting took place in Rotterdam, the Netherlands, in June 2011. Under the header of COMPACT Work (Challenges Of Making Public Administration and Complexity Theory Work), researchers from around the globe convened to discuss their take on complexity, how it informs their research and, if possible, how that could translate into a more coherent research program within the field of Public Administration. This volume contains revised papers that were presented at the workshop and additional comments from the editors. Also discussed during the workshop is a white paper that outlines the research program which is being developed separately from this volume.

IN APPRAISAL OF CONFUSION...

One of the main questions that has been around since the emergence and diffusion of complexity sciences has been whether it is possible to develop a formal universal theory of complexity. The social sciences, of which Public Administration is a sub-domain, has seen similar attempts at such unifying frameworks, such as in the works of Talcott Parsons (1951) or Niklas Luhmann (1995). But the ongoing discussions in the complexity community lead us to believe that a unified theory of the complexity of public administration is unattainable. Indeed, the papers presented here confirm the observation by Dennard, Richardson and Morçöl (2008) that the complexity community in our domain is very diverse. We can trace this back to divergence of perspectives in three categories: complexity, epistemology and ontology, and public administration. These three deserve some discussion here in order to make more sense of the debates going in within and between the contributions in this volume.

What Complexity?

The word 'complexity' equals 'difficult' or 'complicated' in everyday parlance. It is true that complexity is difficult but not everything difficult is complex. Then there are simplicity, order and chaos. Simplicity could be the opposite of complexity, and order could be that too. But there can be some degree of order and simplicity within complexity. Chaos is sometimes also used as shorthand for complexity, or to indicate that the perceived complexity is experienced as something negative. To those that look into it, this simple word has as many meanings as the word 'snow' has for an Eskimo. There are many ways to think about complexity, which lead to many ways of doing complexity research in Public Administration (Gerrits, 2012). Rescher's work (1998) is useful to position the many approaches to complexity presented in this volume. He notes that there is no agreed definition of 'complexity' and that there are multiple ways to approach it. Here the complexity of the notion of 'complexity' becomes important. While 'complexity' can be categorized in three main modes, these categories are neither hierarchically related, nor mutually exclusive. In fact, much thinking and theorizing about complexity intersects and builds on different modes but still approach complexity from different angles. It is necessary to revisit the three main modes to better understand the approaches to complexity underlying research, without of course fully reproducing Rescher's taxonomy.

The first category of epistemic modes contains three subtypes of complexity. Descriptive complexity concerns the minimum length of a description of complexity that is accurate. Generative complexity concerns the minimum length of set of instructions necessary to produce a complex system. Computational complexity concerns the effort necessary to solve a complex issue. The more complex the issue, the more effort it takes to solve it.

The second category of ontological modes contains four subtypes of complexity. Constitutional complexity concerns the number of components making up a system, whereas taxonomical complexity concerns the heterogeneity of said components. Organizational complexity concerns the arrangement of the components, and hierarchical complexity concerns the elaborateness of hierarchical relationships between levels, which points at the thesis that lower-level components are less complex than higher-level components.

The third category concerns the functional complexity and has two subtypes. Operational complexity concerns the variety of modes of operation, i.e. the intricateness of the functions. Nomic complexity, as the final subtype, concerns the elaborateness of the laws governing a system, i.e. more complex systems have more complex law structures.

Each mode and its subtypes attempt to capture some part or dimension of complexity in its view and analysis. Complexity thinking postulates that the complexity of the real world is always more extended than can be captured in mental models. In other words, each representation of complexity in for instance a theoretical framework or a computational model will necessarily focus on certain aspects whilst other aspects are pushed to the background or left out altogether. Complex systems in the real world are open by definition and each researcher has to define system boundaries in order to keep the research manageable (Cilliers, 2001; 2005; Rescher, 1998). Cilliers then states that the real world keeps evolving and transforming and that each model or description of complexity is not only partial at best but also temporal. Consequently, any analysis of complexity fails to fully match the actual complexity. In fact, we can only arrive at knowledge *because* we draw boundaries (Cilliers, 2005). That inevitably leads to the question: what are we looking at with our models and within our boundaries?

The different perspectives on complexity as discussed above may suggest that there is so much variety in research that there is very little that is in common in complexity thinking. But there is one powerful common denominator: complexity is understood to be real. Complexity-thinking disagrees with nihilist postmodernist accounts where research is reduced to story-telling, especially where it concerns Public Administration (e.g., Frissen, 1998). In that sense, the emergence of complexity thinking reaffirms a realist approach to the world we investigate and talk about.

Most authors in complexity adhere to the point that attempts can be undertaken to understand this reality. It echoes Kant's stance on empirical realism that there is a reality that exists independently outside our knowledge and perception (Losch, 2009) but goes beyond that because it assumes that research is a method that can approach reality and that has the tools to come closer to reality (Sayer, 1992). Yet this does not imply a positivist stance where science delivers true, objective knowledge and that the world can be wholly and unambiguously understood through the deployment of said research tools (Byrne, 1998). Each demarcation in research bears the researchers' normative stances, thus determining what is being researched. Moreover, the continuous evolution of the world means that research is contingent; complexity prevents researchers from delivering universal knowledge (Rescher, 1998). Thus the golden standards of positivism, with its focus on controllable conditions and repeatability do not hold much value in the face of complexity (e.g., Callaghan, 2008).

What Public Administration?

Now that we have a first grasp of the complexity of complexity and the worldviews behind it, it is time to discuss the position of Public Administration within this complexity. As can be glanced from the title of this volume, we think that complexity comes first and that the domain of Public Administration is a specific highlighted section of this complexity. But that begs the question: what exactly is Public Administration? We adhere to Waldo's classic distinction between Public Administration as the scientific domain and public administration as its empirical focus. It is striking to see how all authors concern themselves with issues in the public domain and how much confusion results from the ambiguity surrounding core concepts. For example, there is the difference between government and governance. But what one calls 'governance' is called 'public administration' by the other. 'Collective' is regarded by some as a neutral term to indicate that groups of people coordinate issues among each other, whereas others think of it as a normative ideal. There is more confusion about e.g., public policy, public management, project, process and networks, etc.

Much of this variety is explained by the cultural backgrounds. Our conference had people from all across the globe. Concepts, while technically often translatable, still carry specific cultural backgrounds that are very hard to translate. 'Governance' is one such example. It is a concept that is well-known for its ambiguity (e.g., Blatter, 2003). Much literature suggests that the concept of steering concerns attempts by the state to structure and guide society but that this concept has to be abandoned in favor of a perspective where the concept is toned down to a conceptualization where multiple actors have mutual influence (e.g., Kooiman, 1993; 2003) because of the impossibility of a central authority determining everything (Kickert, Klijn & Koppenjan, 1997). Thus there is a major tendency in Public Administration to focus on governance as a process of steering attempts by many. It focuses on the fact that many actors hold some degree of influence and that the outcomes of a policy process are aggregated results of those attempts. This approach is interpreted and used in many ways. To some, governance is a specific type of government, others think that governance is the only thing that governments do. There are many more interpretations.

These differences intersect with different attitudes of research vis-à-vis its public. Some researchers consider themselves policy analysts in an almost

Easton-ian sense, informing policy-makers about the possible choices and consequences and implementation. Other researchers are more geared towards theory development. And then there are those who attempt to bridge the gap between science and practice by trying to make complexity meaningful to practitioners who find perhaps more utility in metaphorical or allegorical stories than in refined theories or data crunching. On top of that come the cultural and national differences. Each country has its specific arrangement of public organizations, state-structure and history of state delivery and state reform. There is no Dutch translation for the word 'governance', whilst the term 'process management' as used by Dutch authors on network governance is associated with manufacturing in English. And you, the reader, reinterpret those terms, concepts, and meaning in your context, informed by your own knowledge. Acknowledging this, it becomes clear that even the name of the conference could be interpreted in many ways.

... AND IN SEARCH FOR A DIALOGUE

More specifically, we see the relevance of organizing a meeting like the COMPACT workshop as a way of balancing the evolutionary process of science with something that at least seeks to approach something like a dialogue among the members of the community. Science can be conceived as an evolutionary process through which human beings seek to gain an understanding of themselves and the world around them (Popper, 1972; Toulmin, 1972; Hull, 1988). The variation that is presented in these proceedings can be seen as the necessary fuel for this evolutionary process. The question then is: what can be learned from this particular collection of ideas? One could argue that this is a set of papers that coincidentally has ended up in the same volume, and seeking for any coherence would be beside the point. This position might be construed as legitimating the publication of the proceedings without further introduction or conclusion.

We adopt a different position, based on a closer examination of the evolutionary mechanisms of retention and selection at work (Campbell 1969). An evolutionary perspective is often equated with the analysis of a static situation that is an equilibrium outcome of the interplay between variation, retention and selection. But the snapshot that this collection presents of the ongoing evolution of ideas about the complexity of public administration need not be such an equilibrium. It may be a product of a process akin to genetic drift. In social systems ideas may diffuse in a fashion-like way (Abrahamson

& Fairchild, 1999), and the natural and social sciences are no exception (Bell, 1960; Abrahamson, 2009). This implies that reflection on the direction taken by the evolutionary process is useful.

Luckily, the mechanisms of retention and selection have specific qualities in social systems that are a result of such reflection; while biological evolution proceeds through a succession of local maxima, scientists may follow the promise of an idea and thus 'jump' through the fitness landscape (Sterelny, 1994). Also, scientists can, and do, reflect on the way in which peer review, editorial policies, and communication among scientific groups, affects the development of knowledge.

In addition, the boundary of the system in which the evolutionary process takes place needs to be considered. This is especially relevant for our situation, as the workshop sought to bring together people that drew on at least two domains of scientific inquiry: complexity theory or complexity sciences and Public Administration. Evolution takes place within the bounds of a community of research, the interacting individuals that constitute a recognized area of scientific inquiry (Toulmin, 1972). While the ideas that were central to COMPACT are the product of the coevolution of two of such communities (the community of complexity science and of Public Administration), in fact it brought together a small number of individuals that constitute a sub community within the public administration field. As noted above, we found that even within this group different definitions of the community manifested themselves. Thus, the proceedings contain the traces of an evolutionary process rather than the product of a mature community with established ways of retention and selection.

The evolutionary perspective provides us with an adequate representation of the socio-historic process in which scientists are engaged, but it does not give us any confidence that collectively they are moving towards a better understanding; the evolutionary perspective does not provide any standards to determine progress. This may lead to a position where the status quo is accepted (Lakatos, 1976). In fact, the evolutionary success of a theory (its survival in a scientific field) may be a source of stagnation. As Munevar (1986: 122) puts it: "The more successful a view is, the greater the tendency, at least among certain groups, to take it for granted. If consensus spreads throughout the entire discipline, the level of criticism aimed at that view may be seriously diminished." Such a mechanism can occur in small groups (where it is referred to as groupthink; Janis, 1971, 1982) and in organizational fields (see the theory of institutional isomorphism; DiMaggio and Powell 1983), and there is no reason why scientific communities should be exempt from them. For this reason, fostering reflexivity (or embracing new variety) is important.

There are important reasons for including human interaction as part of our assessment of the confusion. For one thing, scientists as well as those that look at scientific development, can be reflexive about that development, and seek to guide it in another direction. As a part of this, they can organize conferences that bring together members of different fields (see above) in an attempt to cross-fertilization. To move beyond the pure evolutionary nature of the scientific process, there is a need to truly try and connect to the variety that is presented.

For this, the practice of dialogue is crucial. Isaacs (1993: 25) defines dialogue as "a sustained collective inquiry into the processes, assumptions, and certainties that compose everyday experience (…) -the experience of the meaning embodied in a community of people. Relating it to organizational learning, he sees it as a way to uncover root definitions and underlying assumptions, enabling collective learning. While a scientific community is more loosely coupled than most other forms of organization, we feel it is useful to pick up on this promise to use reflexive dialogues to learn from the evolutionary process. For this reason, it may be good to look at three principles of a dialogue as identified by Yankelovich (1999).

A first requirement is equality among participants. The participants in a dialogue should not base themselves in any way on the positions and roles they have as scientists, whether in formal terms or in the informal structure. If they are unable to do so, the exchange will be clouded by the status and power attached to roles and positions, and a dialogue will not occur.

A second principle is *listening with empathy*, which Yankelovich (1999: 43) describes as 'the ability to think someone else's thoughts and feel someone else's feelings'. This amounts to suspending ones judgment, and instead try to understand from the inside what a participant to a dialogue brings in.

The third principle is to aim for *bringing out assumptions into the open*. What normally remains implicit in an exchange should be uncovered: the underlying judgments about what is good and bad, important and unimportant, as well as ways of seeing the world. This requires a safe context, which to a great extent is provided by the other participants ability to listen empathically and refraining from adhering to their status and power.

From the description of these principles it is clear that dialogue is primarily understood as a face-to-face exchange among persons who can experience each other's presence. To some extent, the possibility of being engaged in such dialogue has drawn participants to our workshop. We therefore organized the workshop in a specific way. Drawing on earlier experiences, we informed participants well ahead of the actual workshop that we wanted them to present each other's, rather than their own, paper. This simple rule has a number of advantages, which include that it makes it clear that papers need to be finished well ahead of the workshop, and need to be read before entering the conference room. Most importantly, the presentation of a paper by another participant requires that an empathic stance is taken by the presenter. At the same time, the presenter will have distance towards the paper, so the presentation tends to uncover some potential basic assumptions of the author(s). Also, it serves as a conduit between statement and interpretation by others. We found that this rule worked well in getting quickly into exchanges where we talked about the boundaries of our inquiries, the nature of disciplines (and the fact that there are differences within the same discipline in different parts of the world). Bringing this out in the open was catalyzed by the fact that each day we concluded with a session where we wanted to draw out implications of the discussion for a collective paper that should come out of the workshop. In all, the COMPACT workshop dug deep into the minds and hearts of the participants.

For these proceedings, we have chosen to ask participants to review two of the papers presented. This served to improve the quality of the work, and extends the dialogue among authors. In fact, the review in several cases gave rise to a conversation between author(s) and reviewer(s). Below we present our impression of some of these exchanges.

In most of the exchanges, we find that both sides of the exchange tend towards empathy. Several feedback points are formulated in terms of questions for clarification and elaboration. No doubt, this is also a function of the fact that participants are to some extent in the same boat: they all seek to advance a complexity-informed Public Administration. What is most clear however is the fact that assumptions are uncovered and being questioned.

Several papers in the section on *researching complexity* focus on the way in which a specific methodology (actor-based modeling, QCA) could help to study complexity in public administration. These papers make clear how a focus on methodology directly leads to questions about core assumptions that underlie the complexity perspective on public administration. As each of these methods can be used with other theoretical frameworks, each time the question arises: To what extent does this method enable an analysis of complexity? The papers that bring in specific *empirical material* (often case studies) have the same effect: they beg the question to what extent an analysis drawing from complexity theory brings new insights. In both sections, papers invited a discussion on the extent to which the complexity perspective enables prediction of future events, and thus of the success of policy interventions. This is an ontological and epistemological issue that is one of the fault lines through the community of complexity scientists, and it is no surprise that it surfaced both during the workshop and in several of the exchanges among authors and reviewers.

The papers in the *'happy marriage'* section display a property that is shared by most, if not all, of the participants of the workshop: they express the uncertainty about bringing complexity theory to bear on Public Administration. While all of us have the intuitive feeling that there 'is something there', it is difficult to really bring that point home even for like-minded persons. This community is not putting forward a strong claim based on a rigid framework; instead, it is searching for ways in which to conceptually and empirically take seriously the fact that public administration is (part of) complex social systems.

Of course, proceedings can never take the place of being present at a workshop. Nevertheless, we hope to have given some flavor of the exchanges among participants, which we hope will evolve towards a research community rich in ideas and contribution to the study and practice of public administration.

This volume shows that researchers undertake different, diverging and again converging attempts to grasp complexity. There are attempts to model complexity meticulously to understand the patterns or drivers underlying complex, emerging systems. There are attempts to understand particular cases qualitatively and to compare different cases where the richness of the data informs the comparison instead of being regarded as white noise. There are also metaphorical accounts where concepts from complexity sciences are used to transfer the gestalt behind the experience without mapping its causality or its details. The undertaking proves elusive. Research is not a linear progression where each step uncovers more of reality. Rather, it is shifting our veil of ignorance. Each next step doesn't bring us closer to the horizon of what we know but shifts the horizon accordingly. It means that we learn more but also that we understand that we are never done learning.

We admit that we felt the temptation to synchronize all the differences in an effort to offer a more streamlined, less ambiguous volume. But then we decided against it. Variety in approaches is the main property of a theory or a group of theories under development and conceptual purity is not always useful. Restrictive use of concepts could cut off the potential of added explanatory power and may in turn, frustrate further theoretical development. Williams (2000, in Haynes, 2003) posits a thesis that while purity itself is a good thing, it should not serve to restrict further development within the realm of the social sciences or Public Administration. We subscribe to the perspective.

REFERENCES

Abrahamson, E. and Fairchild, G. (1999). Management fashion: Lifecycles, triggers, and collective learning processes, *Administrative Science Quarterly* 44(4): 708-740.

Blatter, J. (2003). Beyond hierarchies and networks: Institutional logics and change in transboundary spaces. Governance, 16(4): 503-526.

Byrne, D. (1998). *Complexity Theory and the Social Sciences*. London, New York: Routledge.

Callaghan, G. (2008). Evaluation and negotiated order. *Evaluation*, 14(4): 399-411.

Campbell, D. (1969). Variation and selective retention in socio-cultural evolution. *General Systems,* 16: 69-85.

Cilliers, P. (1998). *Complexity & Postmodernism*. New York: Routledge.

Cilliers, P. (2001). Boundaries, Hierarchies and Networks in Complex Systems. *International Journal of Innovation Management, 5*(2): 135-147.

Cilliers, P. (2005). Complexity, Deconstruction and Relativism. *Theory, Culture & Society*, 22(5): 255-267.

Dennard, L.F., Richardson, K.A. & Morçöl, G. (2008). *Complexity and Policy Analysis. Tools and Methods for Designing Robust Policies in a Complex World.* Goodyear, AZ: ISCE Publishing.

DiMaggio, P. & Powell, W. (1983). The iron cage revisited: institutional isomorphism and collective rationality in organizational fields. American Sociological Review, 48: 147-160.

Fell, H. (1960). Fashion in Cell Biology. *Science,* 132(3440): 1625-1627.

Frissen, P. H. A. (1999). *De Lege Staat*. Amsterdam: Uitgeverij Nieuwezijds.

Gerrits, L.M. (2012). *Punching Clouds. An Introduction to the Complexity of Public Decision-making*. Emergent Publications: forthcoming.

Haynes, P. (2003). *Managing Complexity in the Public Services*. Maidenhead: Open University Press.

Isaacs, W. (1993).Taking flight: Dialogue, collective thinking, and organizational learning, *Organizational Dynamics,* 22(2): 24-39.

Janis, I. (1971). Groupthink.Psychology Today, November: 43-46, 74-76.

Janis, I. (1982). Groupthink (2nd ed.). Boston: Houghton-Mifflin.

Kickert, W. J. M., Klijn, E. H., & Koppenjan, J. F. M. (eds.). (1997). Managing complex networks: Strategies for the public sector. London: Sage Publications Ltd.

Lakatos, I. (1976). Understanding Toulmin, Minerva 14: 126-43.

Losch, A. (2009). On the origins of critical realism. Theology and Science, 7(1): 85-106.

Luhmann, N. (1995). Social Systems. Stanford: Stanford University Press

Meek, J.W. (2010) Complexity Theory for Public Administration and Policy. *Emergence: Complexity and Organization,* 12(1).

Munevar, G. (1986). Consensus and evolution in science, *Proceedings of the Biennial Meeting of the Philosophy of Science Association* 2: 120-129.

Parsons, T. (1951). *The Social System*. Glenco: The Free Press.

Pollitt, C. (2009). Complexity Theory and Evolutionary Public Administration: A Sceptical Afterword. In: Teisman, G.R., Van Buuren, A. & Gerrits, L. *Managing Complex Governance Systems; Dynamics, Self-organization and Coevolution in Public Investments*. New York: Routledge.

Popper, K. (1972). Objective Knowledge. Oxford: Clarendon.

Rescher, N. (1998). *Complexity; a Philosophical Overview*. London: Transaction Publishers.

Sayer, A. (2000). Realism and social science. London: Sage.

Toulmin, S. (1972). Human Understanding. Oxford: Clarendon.

Yankelovich, D. (1999). *The magic of dialogue. Transforming conflict into cooperation.* London: Nicholas Brealy Publishing.

Dr. **Frank Boons** currently works as researcher and teacher at the Department of Public Administration of Erasmus University Rotterdam, The Netherlands. As an organizational & economic sociologist, his interest is in the dynamics of inter-organizational processes that relate to sustainable development. He has published widely on the ecological strategies of firms in global and regional networks of firms and policy actors and issues of complex governance.

Lasse Gerrits, Ph.D. studied Public Administration and Urban Planning at the Erasmus University Rotterdam and now works as associate professor at the research group Governance of Complex Systems. His research interests concern the complexity of public decision-making and the development of the urban. He also teaches courses in complexity and systems.

Dr. **Peter Marks** works at the department of Public Administration at the Erasmus University Rotterdam. Having studied economics and philosophy his research mainly focuses on complex decision making processes from an evolutionary (economics) perspective. The target domain in his research is public safety policy issues. He also teaches courses in systems and evolutionary economics.

INTRODUCTION TO THE CHAPTERS

The collection of papers in this volume covers three areas; researching complexity, complexity in the real world, and 'happy marriage'.

Researching complexity covers 6 chapters that deal with various methods and perspectives on how to study complexity. In Chapter 2 Thomas Catlaw and Yushim Kim introduce the philosophers Gilles Deleuze and Alain Badiou who propose—in distinct but complementary, impressive and rigorous ways—ontologies of multiplicity that hold the potential to illuminate the intersection of contemporary public affairs and an analytical instrument of complexity sciences at the problem of ontology. They discuss how multiplicity and the concept of assemblages help to clarify the ontology of simulations, an approach that has been used in the study of networks, governance, and sectoral-boundary blurring. Specifically, they argue that the strength of simulations, such as agent-based modeling, lay less in its capability to model or represent the world, but in its capacity to embrace an ontology of multiplicity and its flexibility in combining with and creating other assemblages. They conclude the article with a brief discussion of the implication of an ontology of multiplicity and the simulation tool for the study of networks and governance, specifically as it concerns the problem of agency.

Gill Gallaghan in Chapter 3 considers an approach to researching complex social settings in ways that allows space for agency to emerge. It incorporates insights from realist based complexity theory that highlight the nested nature of reality, giving rise to relatively permanent social structures but also to emergence and change. The starting point is that such explanations must be derived through observation of human agents as active and reflexive beings. The chapter begins by briefly identifying the problem posed in combining structure and agency to explain the sedimentation of practices in places. It goes on to outline the potential of negotiated order theory to provide a practice theory that can help in operationalizing public policy research. The empirical work drawn on to illustrate the utility of this approach reflects on a policy implementation process that explicitly used systems notions.

The changing contour in public administration and public policy only rarely reflects the traditional hierarchical form of government apparatus; instead a complex web of interactions among diverse stakeholders within the context of policy systems is increasingly common. Even with the emerging organizational forms and behavioral understandings, the field of public administration remains predominated by the "rationality" assumptions and positivist epistemology. This may have created structural blind research spots and allowed researchers to focus on management and policy solutions that do not fully take into consideration complexity of human behaviors and social systems. Thus, to advance systems thinking that is cognizant of these dimensions Erik Johnston, Kevin Desouza and Qian Hu in Chapter 4 complement existing research with methodologies that explore and capture the underlying complexities in policy and human interactions. Methods traditionally associated with complex adaptive systems, like computational modeling, have great potential and provide tools to researchers to explore complex web of interactions that are seldom linear. In this paper, we first briefly review existing research foci and epistemology in public administration and discuss challenges for policymaking and government management practice. Next, we introduce a complex adaptive systems approach to public administration research. And finally, we conclude that, rather than replacing one method with the other, contemporary management and policy challenges requires the use of multiple methods to understand the emergent complexities.

In Chapter 5 Stefan Verweij and Lasse Gerrits introduce and assess Qualitative Comparative Analysis (QCA) as a complexity-informed framework for evaluating complex infrastructure projects coupled with area development. This is done in four steps. First, the properties of infrastructure and area development are discussed. When combined with a complexity perspective, it follows that the context of a given project is explanatory for its outcomes. Secondly, prerequisites for a methodological framework following this point of departure are developed. Thirdly, common infrastructure evaluation methods are evaluated against these requisites. It follows that a comparative case-based approach is most suitable to study the relationship between context and outcomes in projects. Fourthly, fuzzy set Qualitative Comparative Analysis (fsQCA) is introduced and assessed the developed requisites. The paper concludes with a discussion of the further development of fsQCA.

In the 6th chapter five theoretical frameworks of policy and governance systems including the multiple policy streams, punctuated equilibrium, institutional analysis and development, advocacy coalition, and governance network frameworks are presented by Christopher Koliba and Asim Zia as some of the meso-level, "complexity friendly" frameworks that incorporate whole systems properties, tend to avoid reductionism, allow for emergence and path dependencies, and accommodate dynamic systems characteristics. The extent to which these frameworks are accommodating to complex adaptive systems modeling and how they may be used to generate lines of inquiry that may be tested, within limits, using complex adaptive systems modeling is assessed. Further, the potential utilization of complex systems modeling tools to inform a meta-theoretical research program to compare and refine alternate theoretical frameworks is discussed. Implications for a meta-theoretical research and modelling program are explored.

Increasingly, governance networks are comprised of actors from a diverse array of public agencies at multiple scales and private and non-profit sectors to facilitate decision making on complex public policy problems, such as building sustainable transportation infrastructure, reducing greenhouse gas emissions and improving water quality in freshwater lakes. In Chapter 7 Asim Zia, Christopher Koliba and Yuan Tian introduce "governance network analysis" as a policy analytical tool to compare alternate institutional designs of intergovernmental decision-making processes that are aimed at addressing complex public policy problems. A prototypical Agent Based Model (ABM) of transportation governance network in the state of Vermont is presented to demonstrate an application of governance network analysis in real world public policy settings. This ABM simulates the dynamics of transportation project prioritization processes under alternate inter-governmental institutional rule structures and assesses their impacts on financial investment flows from federal to state, regional and local scale governments. The current version of the ABM is limited to simulating roadway projects that are primarily funded through US Surface Transportation Program (STP) and Interstate Maintenance Program (IMP). Multiple focus groups, individual interviews, and analysis of federal, state and MPO transportation project and program data informed the development of this prototypical ABM. This paper presents the results from experimental simulations to test system-wide effects of alternate institutional designs on the differential emergence of roadway project prioritization patterns and funding allocations across regions and local towns. Implications are drawn for the application of governance network analysis to study alternate institutional designs of inter-organizational governance networks for addressing complex public policy problems.

Complexity in the real world deals with six chapter that in one way or another are about the empirical complexity of public administration. In the 8th chapter Elizabeth Eppel uses an analytical lens developed from complexity concepts to create a useful explanation of policy processes. This lens is demonstrated through an empirical example based on a qualitative, interpretative study of tertiary education policy processes in New Zealand (Eppel, 2010). The lens, incorporating the concepts of interdependent actors, interacting complex systems, feedback loops, adaptation, self-organization and emergence is used to create a holistic understanding of the socially interactive, reflexive dynamism of policy processes. Implications for the development and implementation of public policy and the practice of policy analysis are identified. Rather than the conventional notion of thinking of policy processes as achieving optimal or best solutions to problems, complexity theory indicates that we need to think of them as facilitating the achievement of self-consistent, stable patterns which allow for adaptation, coevolution and emergence to occur in local contexts.

With globalization and the morphing of multi-national corporations, Laissez-Faire economic is abandoning its impartial spectator and gradually disappearing for the sake of the newly-emerging predatory financial capitalism. The result is the shrinking of the middle class, the undermining of governmental authorities and the rule of public administrators, and the increase in threats to economic security. In Chapter 9 Alexander Dawoody and Peter marks address a complexity response to the ill-logic of artificial engineering in bailing out predatory financial capitalism by tax-dollars and emphasizes on complexity solutions in allowing for Pareto-optimal move through self-organization and emergence of new dynamic system that is better capable of dealing with changes in interconnected global economy without artificial engineering. They illustrate the importance of coordination, autonomy, networking and decentralization in the materialization of expertise and shared-resources. Understanding public policy's role in the economy through complexity lens enables us to capsulate the bigger picture in change dynamics and have better appreciation of the multilayered dynamics that interplay during their display

Organizations are restructured into collaborative systems in order to address complex problems by combining expertise distributed across functions, knowledge specialties, and geographic locations. Often times these systems face complex and multifaceted goals requiring distinct teams of public managers to coordinate their efforts and compile crucial information distributed across a network. In Chapter 10, by Naim Kapucu and Vener Garayev, renders an understanding of the requisite structural and relational dynamics that enable network organizations to function effectively in preparedness, mitigation, response to and recovery from disasters. A network of public, private, and nonprofit organizations with stronger and denser ties is critical for the success of disaster preparedness, response, and, recovery networks. Networks not only need to be designed, but also developed and sustained effectively. Information communication technologies are a cornerstone to achieve these goals. Findings from this study contribute insights to the literature on networks and multi-team systems (e.g., intra-organizational and inter-organizational level). In particular, this knowledge informs scholars about needed alterations to the design, leadership, training, policy, and feedback systems in such organizations.

The urgency of climate proofing urban areas is increasingly recognized and various adaptation and mitigation measures are being developed. However, to combine these measures in comprehensive strategies and actually implement them is challenging. The incorporation of climate change issues in day to day urban development requires better understanding of the urban system and the relations between climate and non-climate urban issues. In Chapter 11 Sonja Döpp, Fransje Hooimeijer and Nienke Maas describe a theoretical framework and practical method to get a grip on the urban complexity and climate change effects: the Urban Climate Framework (UCF). The UCF approach is based on different system approaches and aims to expose underlying relations in the urban system and takes the complexity perspective. This perspective takes the unpredictable behavior and unforeseen consequences as starting point and enables better understanding of second order effects of climate change impact and the identification of win-win solutions to balance between mitigation and adaptation. The framework is developed in the context of adaptive management and can be used to structure information of different fields of knowledge and stakeholder groups. The UCF serves as a knowledge brokerage method to incorporate climate change in the complex processes of urban development and can be used by researcher, policy makers, urban planners and designers.

In Chapter 12 Mary Lee Rhodes assess assesses the fit of empirical data relating to the Irish housing system to a complex adaptive system framework with the aim of contributing to the development of governance theory in relation to public services. The 'performance landscape' (PL) model (Kauffman 1993, Siggelkow & Levinthal 2003) was selected as the most appropriate complex adaptive systems framework to integrate existing public management theory and to facilitate the development of governance theory of public services. The PL model is mapped against data on the decision making processes of actors in the Irish housing system to assess the fit of the model to the data and to determine what, if any, public administration behaviour could not be addressed based on this model. Implications of this mapping for theory and practice are highlighted in the conclusion.

Leaders in modern societies are confronted with a paradox: simplicity in terms of goals and results are desired; they have to be achieved in a reality that is complex. In Chapter 13 Geert Teisman, Sibout Nooteboom and Ytsen Deelstra elaborate on the question how three 'professionals in charge' reflect on the impact of their leadership actions on interaction patterns and on the change that results from actions and interactions. The hypothesis that they are balancing between the 'self' and 'network' drivers and between 'self-referential' (conservative) and 'dissipative' (adaptive) communication seems to fit their reality. They are balancing between the order of self and the chaos in interactions? Combining self-interest and network needs and as well as self-referential knowledge and external knowledge seems to be the most effective style, but impossible to achieve for a longer period. Complexity leadership combined daily home based action and temporary uphill interactions that can create innovations.

The happy marriage section deals with four chapter in which the authors are searching for ways in which to conceptually and empirically take seriously the fact that public administration is (part of) complex social systems. Göktuğ Morçöl makes the case in Chapter 14 for a complexity theory of public policy. He argues that the micro–macro problem should be the core of this theory. The three sub-problem areas of the micro–macro problem—assumption about the nature of individual actors, how macro structures and processes emerge from the interactions of these actors, and how macro structures and processes affect individual actors—are discussed. The implications of complexity theory in these three areas are compared with those of the institutional analysis and development framework. These comparisons show that complexity theorists still need to conceptually separate their theory from others and further refine and empirically verify their conceptualizations.

In Chapter 15 Derk Loorbach and Niki Frantzeskaki argue that Public Administration should draw more fundamental lessons from complex system sciences than it has so far. Taking societal complexity seriously in the context of persistent societal challenges like the transitions in energy, food production, health care and the economy requires a fundamental rethink of the role of policy in society. They argue that in complex, uncertain and ambiguous contexts, policy is one of many factors influencing long-term change. So far, policy studies have mainly used complexity perspective to improve the existing policy paradigm and processes. They explore how a complexity perspective could help to develop a new policy perspective as well as new policy practices to influence the speed and direction of transitions.

Complexity theory has potential to inform the field of public administration in both theory and practice. For almost two decades, researchers have constructed research efforts that draw upon concepts derived from complexity theory to highlight characteristics and patterns in public affairs and administration. Critics of this intellectual project note the lack of analytic clarity in conceptualization and question the epistemological assumptions of complexity theory in application to social phenomenon. In Chapter 16 Lasse Gerrits and Jack Meek view these constructive criticisms as translation errors that often hinder metaphorical application and assert three propositions regarding the promise of complexity theory for the study and practice of public affairs and administration. While these propositions to do not necessarily invalidate critical comments in regard to the application of complexity theory, they do attempt to establish the basis for encouraging continued examination of how concepts derived from complexity theory. Finally, these propositions do not diminish the promise of other avenues of inquiry in the field; they are intended to establish complexity theory as one avenue of that also has promise.

In the final chapter of this volume Frank Boons, Lasse Gerrits and Peter Marks argue that that collective decision making processes can be considered as a process where populations of ideas, consisting of problem framing and solution framing, evolve over time. Actions and perceptions of actors within networks, material changes induced by decisions, and chance events drive the evolution in decision making. The creation and selection of problem ideas and solution ideas within populations are intertwined and are part of the same feedback loop. The variation, selection and retention of problem and solution ideas take place within a system of heterogeneous actors and this takes place in multilevel systems. Hence, the problem-solution space evolves. Not only are actors connected to other actors through dynamic boundaries, they are also connected to other physical systems, which is not constant and exogenous to the individual species' evolution, but influenced by it. Social and physical systems can exert selection pressures on each other, influencing the variation and connected selections in a continuous process. This continuous process is not necessarily progress, but at best a local optimum or selective adaptation. The authors aim to conceptualize collective decision making as an evolution of process and content and to develop a research agenda to investigate the explanatory power of this conceptualization.

SECTION I

RESEARCHING COMPLEXITY

Chapter 2

ONTOLOGIES OF MULTIPLICITY AND SIMULATION: POSSIBILITIES FOR PUBLIC AFFAIRS INQUIRY

Thomas J. Catlaw (Arizona State University, USA)
Yushim Kim (Arizona State University, USA)

Philosophers Gilles Deleuze and Alain Badiou propose—in distinct but complementary, impressive and rigorous ways—ontologies of *multiplicity* that hold the potential to illuminate the intersection of contemporary public affairs and an analytical instrument of complexity sciences at the problem of ontology. We discuss how multiplicity and the concept of *assemblages* help to clarify the ontology of simulations, an approach that has been used in the study of networks, governance, and sectoral-boundary blurring. Specifically, we argue that the strength of simulations, such as agent-based modeling, lay less in its capability to model or represent the world, but in its capacity to embrace an ontology of multiplicity and its flexibility in combining with and creating other assemblages. We conclude the article with a brief discussion of the implication of an ontology of multiplicity and the simulation tool for the study of networks and governance, specifically as it concerns the problem of agency.

INTRODUCTION

There is a pervasive sense among contemporary scientists and publics that the world of the early twenty-first century is highly *complex* and that the tools (theories, concepts, methods, etc.) that humans have used previously to understand and interact with their environment are insufficient for making sense of this complexity. In the realm of public affairs research, this recognition of societal complexity is evident in the scholarship on networks, governance, and sectoral-boundary blurring. Not surprisingly, some scholars of public affairs have productively turned to the science of complexity in order to better study and understand the dynamics of governance and, in particular, have shown that the study of networks may be usefully integrated with the concepts (e.g., emergence and nonlinearity) and methods (e.g., agent-based modeling) of complexity science (Dennard, Richardson, & Morçöl, 2008; Koliba, Meek, & Zia, 2010; Morçöl & Wachhaus, 2009; Rhodes, Murphy, Muir, & Murphy, 2010; Teisman, van Buuren, & Gerrits, 2009).

We do not contest this impulse to seek new theories and concepts and use new technologies in the face of our demanding conditions but—to paraphrase Dwight Waldo's assessment on the final page of *The Administrative State* (1948)—we think that public administration and policy may still be seeking to address the problems of the complex human world on too low of a plane. But what would a "higher" or more fundamental plane be?

Waldo argues that any political theory rests upon a "metaphysic, a concept of the ultimate nature of reality" (1948: 21) and contends that public administration's metaphysic rested on the "verdict of science." Recently, however, Catlaw (2007a) critiques and extends Waldo's analysis to show that this verdict is a subset of a more specific and basic metaphysics or ontology called "representation" (described below). It is Catlaw's contention that this ontological commitment is failing and that this conditions our experience of complexity today. A different ontology is needed, though public administrative thought continues to operate within the ontological confines of representation (see Stout & Staton, 2011, for a possible exception). Network and governance thought, similarly, remains embedded in representational ontology (see Catlaw, 2006a, 2007b, 2008 for extended analyses). The focus there is on matters of methodological instruments and their utility in studying various phenomena in nature and society (or both) rather than on the being of those phenomena as such. However our understanding and appreciation of the theoretical underpinning and, in turn, the practical implications of those methods and tools is constrained by the general lack of attention to matters of ontology.

To address this lacuna in ontological inquiry, we present the thinking of two French philosophers, Alain Badiou and Gilles Deleuze. We consider Deleuze and Badiou because they both propose—in distinct but complementary, impressive, and rigorous ways—ontologies of *multiplicity* that hold the potential to illuminate the intersection of contemporary public affairs and a central methodological instrument of complexity sciences, simulation, on the basis of a *nonrepresentational* ontology. The first section reviews the philosophical meaning of ontology, which in our view has narrowed in social scientific thinking. We then sketch some core elements of Badiou's and Deleuze's nonrepresentational ontologies as they concern the basic proposition that Being is fundamentally a *multiplicity* rather the One, as held in representation. Next we discuss how multiplicity and its attendant concept of *assemblage* help us to think differently about the nature of objects and, as such, to identify different sets of theoretical and method-specific questions. Specifically, we argue that the strength of simulations, such as agent-based models, lay less in its capability to represent the world than in its theoretical and analytical fidelity to an ontology of multiplicity and its capability to combine with other entities and to create new assemblages. Moreover, this ontological foundation helps to address some of the epistemological concerns that have been expressed about simulations. We conclude the article with a brief discussion of the implications of these ontologies and the simulation tool for the inquiry into networks, governance, and sectoral-boundary blurring.

We note at the outset that given space constraints our exposition will be introductory by necessity and we will elide some relevant philosophical problems and opportunities for debate. We hope, however, that the presentation will be sufficiently clear and interesting to stimulate further discussion of these important matters with scholars of differing and competing ontological views.

ONTOLOGIES OF MULTIPLICITY

Badiou and Deleuze view ontology as the science of being qua being, what *is* it *to be*. Ontology in this sense concerns itself with the "be-ing (in the verbal sense) of beings (in the substantial sense)" (Hallward, 2003:

50). (We will retain this terminological distinction throughout the article.) This is distinguished from more common usages of the term ontology in social science. First, ontology as the science of be-ing is different from the notion that ontology names the fundamental or "paradigmatic" propositions about reality; it is more than a conceptual frame. Second, ontology as the science of be-ing is to be distinguished from the *ontic*, or the "real" objects that exist in the world (cf. Byrne, 1998: 35-53) and the things we can say about them and their relationships with one another. Ontology is both more and less than a set of assumptions, objects (beings), and their predicates. It is the inquiry into what is sayable about be-ing *irrespective* of the specificity of those objects and predicates. Badiou and Deleuze offer ontologies in this sense, both of which are articulated from the concept of multiplicity.

Badiou's Mathematical Ontology of the Multiple

For Badiou, ontological inquiry does not purport to ground itself in anything "empirical" or ontic and thus is, by definition, an effort in speculative metaphysics that rests upon what he says is a "decision." This decision is grounded in answers to two questions. First, Number or Things? The choice between initiating ontology from numbers or things reflects a priority given to concrete entities or to the rigors of mathematical formalization. In other words, do we begin ontological thinking from a speculative inquiry of the world as it appears to be presented to the senses or, rather, from the nonphenomenal domain of mathematical symbolization? Badiou choses the latter course and, provocatively, declares, "mathematics *is* ontology." We will unpack this enigmatic claim below.

The second question, is be-ing One or Multiple? Badiou (2006: 23) writes, "Since its Parmendian organization, ontology has built the portico of its ruined temple out of the following experience: what *presents* itself is essentially multiple; *what* presents itself is essentially one." Elsewhere he (2004) writes, "We can therefore define metaphysics as the commandeering of be[-]ing by the one" (40). What Badiou is getting at here is the predilection of philosophical thinking to reduce the phenomena we see and experience to some underlying or more fundamental reality, unity, or substance. This reality has been given various names: God, the Idea, Substance, Structure, matter, energy, etc. Thus, to the question "One or Many?" traditional ontology choses the One. The many, which is the diverse world of objects in which we live, is essentially epiphemenonal and it, ultimately, reduces to one of these underlying elements which is—and this is critical—*more real* than the objects of experience.

Philosophy, further, tends to reduce the objects we encounter in the world to the "oneness" of entities and objects themselves. So, for example, say that we are sitting at our desks, looking out the window at a group of sparrows. Typically, we will group these entities together under the term "sparrow" or "bird" because we presume that there is some property common to those entities. But while that property allows us to group entities with another, it also gives a unity to the single bird and tells us what that entity *is*. The concept organizes it for us as an object. So, not only is be-ing one, but in order to be a being (an entity), one must be *one* kind of object. Both aspects are key dimensions of a *representational* ontology.

In opposition to this representational ontology of the One, Badiou "hold[s] that what is thinkable of be[-]ing takes the form of radical multiplicity, a multiplicity that is not subordinated to the power of the one" (2004: 41). What does this mean? To begin, Badiou does not contest the notion that the world is *presented* to us *as one*. We encounter beings like birds and people as unified coherent entities rather than as disconnected parts. Yet Badiou argues that these unities are not in-themselves one but are the result of an *operation*, called the *counts-as-one*, that produces a particular *situation*. Think again of our sparrow. We encounter the sparrow as a single entity but this is the result of an operation that obscures the multiplicity that the bird is: its distinct body parts, muscles, and nerves; its biological systems; its cells, hormones; all the way down to the molecular and sub-molecular domains and the physical forces at work there. As Badiou contends, even when we shift our attention to those other dimensions we still encounter situations, to which he gives the term *consistent multiplicities*. Badiou calls the particular way of counting in a given situation the "structure of the situation." Given this structure, only certain entities appear with each other and only in certain gradations of identity (sameness) and difference.

So far we have largely remained on the familiar terrain of objects and entities, though Badiou has reconceived them as consistent multiplicities. Often this is where "ontological" inquiry stops: with a list of real entities and their characteristics, and a principle for differentiating them from un- or less-real entities. Badiou, however, wants to know what the *be-ing* of these multiplicities is. Recall that the be-ing of multiplicities is what is sayable about them once all the properties and descriptions we might attribute to them are stripped away. Badiou's reply is that the be-ing of these consistent multiplicities is to be as an *inconsistent* multiplicity and ontology can only be a "*theory of inconsistent multiplicities as such*" (Badiou, 2006: 28).

This is, admittedly, an impenetrable statement on its face. The entre to making sense of it is to keep in mind that Badiou seeks to avoid two basic problems of representational ontology. The first is the reduction of the diversity of the world to an *underlying* unity or substance; the second is sort of the inverse: to remain only on the surface "after the count," that is, in the domain of unified objects, or (in the Kantian sense of detailing limits) consistent multiplicities. This is no easy trick. Badiou must find a way to critique the appearing world of objects without reducing them to a more fundamental, describable reality. Not only would this *reduce* the world of objects but also by specifying traits of that fundamental reality, we would simply make be-ing into another object.

It is to avoid these problems that Badiou claims that the be-ing of consistent multiplicities is to be as an *inconsistent* or pure multiplicity. An inconsistent multiplicity is what exists *prior* to the count, but is not and cannot be fully *re-presented* in the integrating operation. Still, Badiou argues, that left-out part does not go away; it persists as an excess to every situation or object. It is precisely because all counts are incomplete or not-whole that there is instability or unpredictability to all our encounters with objects in the world. There is always something more to the situation that unsettles our representations. Because they exceed the count that creates the situation, inconsistent multiplicities are *ontologically* indescribable as such and, thus, the be-ing of objects, counter-intuitively, is uncountable (for such a description, again, would subject be-ing to the count-as-one and generate a consistent multiple).

But how can we have a rigorous scientific inquiry into what is ineffable? It is here that Badiou's decision for Number over Things and his statement that "mathematics is ontology" becomes critical. He asks, "But what kind of thinking never defines what it thinks and never expounds it as an object?" (2004: 43). His answer is *axiomatic* thinking which "grasps the disposition of undefined terms [and] . . . never encounters either a definition of its terms or a serviceable explanation of they are not" (43). He sees this kind of thinking most vitally in post-Cantorian set theory. This is not the place to rehearse the complex details of set theory and how Badiou puts them to work, but suffice it here to say that Badiou sees in set theory a formalized system of thought that demonstrates the unavoidable incompleteness of the count and so the be-ing of beings as inconsistent multiplicities.

To sum up, we return to our sparrow. We can call this entity a consistent multiplicity and also a *set* that gathers together many diverse elements. Each of those elements, in turn, is also a multiplicity, a set of other elements; those elements, too, are sets, and so on. Because each element is a set there is, *ontologically*, no difference or hierarchy of "realness" between a set and its elements. The genus and species are, so to speak, equivalent yet different. In constructing each and every set, every consistent multiplicity is marked by an inconsistency that is, paradoxically, its very be-ing. Badiou names be-ing qua be-ing *the void* or *empty-set*.

Deleuze, Multiplicity, and the Univocity of Being

Whereas Badiou thinks multiplicity via mathematics and formalization, Deleuze thinks it via the concrete and processes of becoming. If Badiou choses Number, Deleuze choses Things (Hallward, 2003). Yet both are bound by their shared concern for the concept of multiplicity and the problems posed by representational ontologies.

One approach to Deleuze's challenge to representation is to consider it as an effort to overturn a specific "image of thought." Many expressions of this image recur in Deleuze's work but this passage describes his philosophical target clearly: "The popular and technical images of the philosopher seem to have been set by Platonism: the philosopher is a being of ascents; he is the one who leaves the cave and rises up. The more he rises the more he is purified... [E]verything happens in the heights..." (1990: 127). As Deleuze sees it, this ascension from low to high is not simply for its own sake but establishes a hierarchy in be-ing and the basic practice of judgment, insofar as judgment entails the power to categorize entities. As we will see, though, for Deleuze such judgment will always be normative insofar as it will turn on determining "good" and "bad" representations of some posited ideal or determined set of criteria. This is a view he will reject entirely.

From this image of thought follows a specific way of posing the *problem of difference* called *representation*. Arguably the fundamental concern in Deleuze's philosophy is to show that representation *cannot* think difference *in itself*. He (1994/1968) writes, "Difference is not and cannot be thought in itself... It seems that it can become thinkable [within representation] only when tamed" (262). According to Deleuze, in representation we think of difference in terms of categorization, the different buckets that we put objects into, such as those we described in the discussion of Badiou.

For example, when we think about birds, we might make a preliminary distinction that birds are different than plants and minerals. Then we might say that birds themselves are different than the many kinds of other animals and insects we can observe; we differentiate birds from, say, bees. Next, we will notice that there is great variation among these birds themselves; there are many species. We will segregate birds, then, according to the category of species. When we reach the species level, these categories are considered to be "natural kinds"; there is an essence or an ideal form associated with each (DeLanda, 2006). Along the way we will create other categories to refine our categorization and will do our best to put each thing into one *and only one* category.

In this simple, familiar illustration of classification, we can see how representation's ontology of the One tames difference. Differences are related to one another or mediated by the concept. So, specific birds are connected via "sparrow" but birds themselves are related across the concept of species. Differences are defined via opposition in the predicate insofar as the predicates distributed positively (has X) but also negatively (does not have X). Judgment works by analogy since we reason from a set of predicates, which we apply or extend to specific objects, which are "like X." Finally, we are guided by our perception to identify resemblances based upon the "ideal" laid out in the categories' predicates. There *must* be an ideal because, at some point, categories must differentiate themselves without overlap. So while there is "overlap" within the species of birds, we must define an ideal or model between species.

For Deleuze, the dilemma is that representation does not allow us to think difference prior to the concept or without predicates (Williams, 2003: 61-62). There is no proper notion of difference because difference is actually a function of *identity* (sameness) within or across concepts. *To be* essentially means to possess a specific set of predicates. The sparrow exists insofar as it exhibits a set of characteristics; if it does not demonstrate them it means that it exists in another classifiable way. Regardless, the entities themselves exist only insofar as they have an identity within the class, category, or concept; and it is, by implication, precisely because the concept determines the entity's being as identity that the concept is a *higher* reality than the entity itself.

This matters because it means that we define the be-ing of beings in terms of what *they are* and, by implication, what they *are not*. It is a formulation of be-ing that rests on separation and opposition (Williams, 2003: 61-62). Furthermore, the necessity of judgment in the movement from depth (entities) to height (models, ideals) entails the creation not only of hierarchies in terms of categories and subcategories but also, as suggested above, onto-*normative* distributions of *better to worse* because ideal forms or models are required to make the representational system run. We ask, which is in and which is out? Which one is a better copy, which one is worse (see Catlaw, 2007a)? All relations among entities are, in the end, mediated by these judgments.

Deleuze seeks an alternative to this by positing a distinction between the *virtual* and the *actual*. Contrary to the model-copy, or representation-represented, relation neither the virtual nor actual is more "real" than the other; they are both real and ontologically equivalent but they name an important distinction. Deleuze considers the virtual as a problem or a problem-space. As an illustration, Ronald Bogue (1989) asks that we imagine a

> *U-curve that is symmetrically bisected along its vertical axis, which is rotated 360 degrees (hence generating a three-dimensional, hemispherical shape). All the points on the curve will move, except for the point at which the axis bisects the curve—the singular point of the curve. One cannot specify the value of that singular point without determining the values of all the other points. However one can determine that such point exists without specifying its value; in doing so, one delineates the general topology of the problem, the general abstract domain of the possible curves that can pass through the singular point.* (59)

In this example, there is a kind of "structure" (Deleuze, 1974) that bounds what the U-curve can do and there is a set of solutions to the problems of the values of the points along the curve. But any *actual* U-curve will differ from this virtual structure in its specificity. These curves *actualize* the virtual. They generate something unique, rather than representing an ideal form. Living animals can be considered in this fashion, too. DNA and genes name a virtual structure for an organism and its species. Like the singular points on the curve, chromosomes are differential elements *in relation with one another* that define a virtual structure. But the way in which the virtual is actualized will differ actually from animal to animal, and context to context.

In be-ing, there are not better or worse—normal—actualizations. There is only the production of differences in the unfolding continuum from the virtual to the actual. This gets to perhaps the core idea in Deleuze's philosophy: that be-ing is *univocal*: "There has only ever been one ontological proposition: Being is univocal" (1994/1968: 35). In other words, be-ing is the *same* and said in the same way for all differences but how it is said is precisely the *declaration of differences*. *Ontologically*, everything is equal because everything is, in itself, different. *To be* (be-ing) is to individuate and to differ.

ASSEMBLAGES AND SOCIAL INQUIRY

Badiou's and Deleuze's philosophies of multiplicity are committed to what may be called a *flat ontology* (DeLanda, 2002). Both posit the generation of unique entities that may differ in many empirical, observable ways but not in ontological status. One is not more "real" than the other. As we discuss below, these flat ontologies have some notable "payoffs" in terms of addressing traditional problems concerning the foundations of social research. Before discussing those, however, it is useful to couple new social-scientific terminology to these new philosophical ideas in order to register the shift in the domain of inquiry.

So, to mark the movement away from the previous abstract philosophical argumentation to entities of the world that interest social researchers, we can call these unique combinations or compositions of multiplicities *assemblages*. We do this, further, because the term assemblage already has currency in some areas of social inquiry (e.g., Ong, 1999; Sassen, 2008) and has been given rigorous and systematic formulation in the work of Manual DeLanda (2006), whose thinking is rooted Deleuze's philosophy and his collaborations with Felix Guattari (1983, 1987).

Broadly speaking, just as "natural" objects like sparrows and human beings are assemblages of distinct body parts, muscles, nerves, biological systems, cells, hormones, and so on, we can say that there are assemblages of people, such as interpersonal networks, institutions, organizations, and communities. Specific organizations, such as public administrative agencies, are assemblages of public officials, citizens, documents, management practices, and physical spaces. Economies are assemblages of individuals, organizations, flows of labor and capital, physical infrastructure, natural resources, and so on. Ecosystems can be seen as assemblages of environmental policies, engineers, deep ecologists, rain forests, animals, and the Jet Stream.

But beyond a superficial terminological change, to consider these familiar topics of social research as assemblages within a flat ontology of multiplicity helps to reground and redirect core aspects of social scientific research. The most obvious implication is that a flat ontology dissolves the contest about the ontological status of individuals viz. groups and societies. From Badiou's vantage, all appearing assemblages are consistent multiplicities, composed themselves of multiplicities. The material question is to specify the structure of the local situation, or set, and count in question. From Deleuze's vantage, various assemblages can be viewed as unfolding along a continuum from the virtual to the actual. In both cases, since all "levels of analysis" are *ontologically* equivalent, the specific (cosmological, evolutionary, or human) historical process or operation by which the set or actualized assemblage emerges is of interest.

An additional virtue of the ontology of the assemblage is that it allows us to combine and, thus, investigate an array of entities without running headlong into either the natural-social or organic-inorganic ontological divides. That is, the ontology of multiplicity accounts for the be-ing of *all* phenomena and does not posit different ontological *kinds* of objects, such as natural or social ones. At the same time, though, avoiding these divides is accomplished without reducing these different entities in the assemblage to a common substance or viewing the assemblage itself as the mere aggregation of its parts.

As entities, assemblages are *irreducible* because they are the result of the exercise of the parts' capacity rather than an aggregation of parts' properties, but they can still be *decomposable* because each part has its own autonomy as an entity—each is itself a multiplicity after all. Depending on the specific composition of the assemblage, parts may be detached and plugged in, which influences the interactions among the component parts through feedback, not by generating a fused new totality. Nevertheless, constituent parts in stable assemblages must be redundant so that the replacement of a part does not alter the functional capacity of the assemblage as a whole.

Given the ontological status of assemblages, from the vantage of social inquiry two issues bear emphasizing. The first is methodological. Since what we experience in daily life are simply differently composed assemblages of varying scales, it becomes imperative to be clear and precise in scientific inquiry about how one combines entities and decides the ways in which one will "count" or produce an assemblage for investigation or terrain of action. The second issue speaks to a possible core focus for inquiry. As suggested above, the assemblage's flat ontology allows us, in principle, to combine what conventionally are viewed as different kinds of objects from the natural and social worlds (e.g., mountains, public employees, endangered species, and algorithms). However, understanding which situated objects *actually* do combine, why and how they do so, and the dynamics of the assemblage become central matters for empirical investigation.

SIMULATIONS AS ASSEMBLAGES

Though assemblages and multiplicities illuminate some ontological dilemmas, they also beg some practical questions. How can we study these dynamic, heterogeneous entities at different scales? Which analytical mode comports with the ontology of the assemblage?

We explore these questions by scrutinizing *computational simulation models* (specifically, agent-based models) and tools advanced in complexity sciences so that we can shed light on the linkage between these ontologies and complexity science-inspired research in public affairs. As we suggest, not only do simulations seem especially compatible ontologically and methodologically with the concept of assemblages, but, in turn, these ontologies help address some theoretical questions that have dogged simulations.

Agent-based modelings (ABM) are nothing more than a name for Object-Oriented Programming (OOP) languages, such as C^{++} and Java. That is, ABM can be built based on unique functionalities that these programming languages provide. In working with OOP, its flexibility, adaptability, and scalability are well acknowledged (Barrett *et al.*, 2011). These modelings are widely used in many areas such as artificial intelligence, business management, geography, computer sciences, and so on for various purposes.

In OOP, one can certainly work with a primitive data type—i.e., numbers and characters—as any programming language does; but more importantly, objects can embody heterogeneous, individuated entities as a fundamental decision-making unit (e.g., individuals, birds, schools, ideas, or cells). In the example of the classical predator-prey model, grass, sheep, and wolves are coupled to form an ecosystem. Others have built an artificial social system that shows how an environmentally unjust society emerges from interactions among land, pollutants, residents who seek to settle there, and firms who provide jobs and pollutants (Eckerd, Kim, & Campbell, 2011). Land, people, and firms are distinct objects that are linked. Another artificial systems may be designed to understand public service delivery programs, including program recipients, contracting stores that deliver benefits to recipients, and public agencies as distinct decision-making entities (Kim & Xiao, 2008).

It is also well known that in OOP control can be decentralized to autonomous objects (i.e., agents). Agents are populated with their own attributes (i.e., variables or data) and methods (i.e., functions, rules or schema). Attributes are designated, used, and updated when agents engage in decision making using local and global information collected from the environment and exercise their discretion based on their internal state at the moment of decision given specific conditions. An object's relation to other objects is specified through schemas that are designed to elaborate, experiment with, and deliberate on rules, theories, or stories. For example, in the classical predator-prey simulation model, what governs the behavior of sheep can be the level of its energy and the amount of grass present at each moment of simulation; whereas the wolves' behavior can be governed by its own changing energy levels and the number of sheep in the system. Each agent controls and decides their actions given their internal states and the external environment within the context modeled. Key attributes, capacities, and relations among agents in these examples need to be *contextualized* through the interpretive process of making sense of the world or problem-space one wants to explore and also must be *parameterized* in order to be operational as a tool.

Ontology of the Simulation

Simply, an ABM simulation *itself* is an assemblage. ABMs combine an array of *distinct and different* entities (e.g., material and immaterial, natural and social) that are combined as parts of an assemblage for simulation. Each is an element that gathers other elements that are sets as well. Each set is distinct, but ontologically equivalent. Seen from an ontology of multiplicity, in these artificial systems, individualized entities are multiplicities and remain heterogeneous even if they are interdependent and interact with each other through causal mechanisms.

From the perspective of a flat ontology of multiplicities, there is no reason to reject this new reality that appears in computer simulations as pejoratively "artificial." As such, this *ontological* perspective also allows us to reconsider a persistent *epistemological* problem in ABM and simulation, namely the relationship between the simulation and the reality purportedly simulated in the model. The recent approach to ABM has been to seek to attain some empirical validity of simulated realities against arguably existing, ontologically more real phenomena (i.e., the systems being modeled or represented in models and simulations) (Grimm *et al.*, 2005; Janssen & Ostrom, 2006). From this vantage, there are well-known, practical difficulties. Synchronizing two dynamic, multidimensional systems in detail is highly unlikely, as it is evident in weather forecasting simulations. At the same time, a (impossibly) complete replica of the simulated system is not a simulation at all!

DeLanda (2011) proposes the examination of what he calls "diagrams" that may be shared or overlapped by the two systems because "the isomorphism between models and what they model can be explained as *coactualization* of the same diagram, or of different but overlapping diagrams" (202). In this sense, even well thought out primitive prototype simulation models can shed light on understanding different multiplicities that present themselves to us. Perhaps, then, the focus should be less on improving the resolution of the simulation model, and more on identifying possible diagrams of the system that the simulation aims to understand.

The simulation itself is also a crucial part of other assemblages: 1) a simulation and a simulated-system and; 2) a simulation, a system simulated, and a system that consists of the simulation, the system simulated, and simulators. It is worth noting, more particularly, that many simulations for social inquiry are part of assemblages that consist of people (e.g., researchers, practitioners, and modelers) and computers as tools in the context of modeling. The simulation-assemblage can be formed for specific purposes and can be an especially useful location for combining different knowledge and professional capacities to explore concrete problems of public administration and policy. This often allows us to enable and enhance that knowledge and those capacities in ways that would have not been possible had the process of modeling not brought them together.

But even if simulation models are developed from such an open collaborative process, the model still cannot guarantee precise answers about the world. The simulation model/assemblage is a newly created world where human and nonhumans engage and experiment in order to form intuitions and guide actions. What is provided from the process of building simulation models is, ultimately, a learning opportunity. This assemblage of people and computers certainly provides a site for actors to affect and affected by another. But as DeLanda (2011) says, how we can creatively exploit the capacities of simulations in research is up to social scientists and the assemblages they inhabit and the new ones they help to create.

POSSIBILITIES FOR PUBLIC AFFAIRS RESEARCH

Our presentation of the ontologies of Deleuze and Badiou is preliminary and does not explore differences between them (see Badiou, 2000/1997; May, 2004; Smith, 2004). Still, in concluding, we can begin to see how these ontologies offer opportunities to consider networks and governance in public affairs research in a new light. The broad approach to these topics, generically, might be framed around the problem of how, why, and whether entities within a world become connected; how assemblages are created or *not* created. In public administration, Catlaw (2008) calls this an inquiry into the dynamics of "code sharing." Many existing network governance questions may be productively folded into this, such as: the roles of metagovernors, democratic anchorage, capacity building, quality of social "ties," power dynamics, and so on (Sorensen & Torfing, 2007, 2009; Koliba, Meek, & Zia, 2010). However our lexicon would be deepened through incorporation of the concepts of ontological multiplicity and assemblages, and the scope of our interest widened by a concern for the dynamics and conditions of (dis) connection in a flat ontology. Our thinking about political institutions and practices would similarly be challenged by the expanded spaces and agents involved. As we discussed, simulations appear to be especially useful tools for examining the dynamics of combining and for including a wide range of heterogeneous participants—both in the simulation models *and* the modeling process.

This ontological vantage also may help to enrich our thinking about some central questions in public affairs research, such as managerial control and individual agency. We note two possible directions for future development.

First, the philosophical position presented here ontologically rules out the possibility of total managerial control. However unlike Simon's notion of bounded rationality or the postmodern emphasis on the constitutive and limiting affects of language (see Catlaw, 2006), the basis for this conclusion rests not in human finitude or logistical hurdles but rather on the ontological nature of real objects themselves. This ontology establishes epistemological limits. At the same time, this does not mean that all scientific explanation or prediction is impossible; only that such explanations are inherently unstable (given some time horizon) and local to specified processes, systems, or assemblages. Second and as suggested in the discussion of simulations, the rich heterogeneity of objects ontologically permitted in assemblages invites us to reconsider agency beyond the confines of intentional human action. That is, since we need not bound inquiry to only human "actors" we can also expand our thinking to consider the agency of nonhuman entities (Catlaw, 2008; see also Latour 2004). And that would be something new indeed.

ACKNOWLEDGEMENTS

The paper was presented at the Challenges of Making Public Administration and Complexity Theory (COMPACT) Work conference, Rotterdam, The Netherlands, June 23-25, 2011. This work was partially supported by the National Research Foundation of Korea Grant funded by the Korean Government (NRF-2010-330-B00262) and the Decision Center for a Desert City at Arizona State University. Many thanks to Goktug Morçöl and Derk Loorbach for their generous criticism and commentary on previous drafts of the paper.

REFERENCES

Badiou, A. (2004). The question of being today (R. B. A. Toscano, Trans.). In R. Brassier & A. Toscano (Eds.), *Theoretical Writings* (pp. 39-58), ISBN 9780826493248.

Badiou, A. (2006). *Being and Event*, O. Feltham (trans.), ISBN 0826494706.

Barrett, C. L., Eubank, S., Marathe, A., Marathe, M. V., Pan, Z., & Swarup, S. (2011). Information integration to support model-based policy informatics. *The Innovation Journal: The Public Sector Innovation Journal,* ISBN 1715-3816, article 2.

Bedau, M.A. (2008). Downward causation and autonomy in weak emergence. In M.A. Bedau and P. Humphreys. (Eds.). Emergence, ISBN 978-0-262-02621-5.

Bogue, R. (1989). *Deleuze and Guatarri*, ISBN 978-0415024433.

Byrne, D. (1998). *Complexity Theory and the Social Sciences: An Introduction*, ISBN 0-415-16296-3.

Catlaw, T. J. (2006). Discourse, decision and the doublet: An essay on the crisis of modern authority. In G. Morçol (Ed.), *The handbook on decision-making* (pp. 99-118), ISBN 978-1574445480.

Catlaw, T. J. (2007a). *Fabricating the People: Politics and administration in the biopolitical state*, ISBN 978-0-8173-1572-6.

Catlaw, T. J. (2007b). "From representations to compositions: Governance beyond the three-sector society." *Administrative Theory & Praxis*, ISSN 1084-1806, 29: 225-259.

Catlaw, T. J. (2008). "Governance and networks at the limits of representation." *American Review of Public Administration*, ISSN 0275-0740, *39*: 478-498.

DeLanda, M. (2002). *Intensive Science & Virtual Philosophy*, ISBN 0-8264-5623-5.

DeLanda, M. (2006). *A New Philosophy of Society: Assemblage Theory and Social Complexity*, ISBN 0-8264-9169-3.

DeLanda, M. (2011). *Philosophy and Simulation: The Emergence of Synthetic Reason*, ISBN 978-1-4411-7028-6.

Deleuze, G. (1974). "How Do We Recognize Structuralism? *Desert Islands and Other Texts, 1953-1974* (pp. 170-192), ISBN 978-1584350187.

Deleuze, G. (1990). *The Logic of Sense*, ISBN 0231059825.

Deleuze, G. (1994/1968). *Difference and Repetition*, ISBN 0231081596.

Deleuze, G., & Guattari, F. (1983). *Anti-Oedipus: Capitalism and Schizophrenia*, ISBN 978-0-8166-1225-3.

Deleuze, G., & Guattari, F. (1987). *A Thousand Plateaus: Capitalism and Schizophrenia*, ISBN 0816614024.

Dennard, L., Richardson, K., & Morçöl, G. (Eds.) (2008). *Complexity and Policy Analysis: Tools and Concepts for Designing of Robust Policies in a Complex World*, ISBN 978-0981703220.

Feltham, O. and Clemens, J. (2005). "An introduction to Alain Badiou's philosophy," O. Feltham and J. Clemens. *Infinite Thought: Truth and The Return to Philosophy [by Alain Badiou]* (pp. 1-38), ISBN 9780826479297.

Eckerd, A., Kim, Y., & Campbell, H. (2011). Using ABM to illuminate social processes leading to environmental injustice. *Computational Social Science Society of America*. [Online] http://computationalsocialscience.org/publication/eckerd

Grimm, V., Revilla, E., Berger, U., Jeltsch, F., Mooij, W.M., Railsback, S.F., Thulke, H., Weiner, J., Wiegand, T., & DeAngellis, D.L. (2005). Pattern-oriented modeling of agent-based complex systems: Lessons from ecology. *Science*, ISSN 0036-8075, *310*

Hallward, P. (2003). *Badiou: A Subject to Truth*, ISBN 978-0-8166-3461-3.

Janssen, M.A., & Ostrom, E. (2006). Empirically based, agent-based models, *Ecology and Society*, ISSN: 1708-3087, 11(2): 37.

Kim, Y., & Xiao, N. (2008). FraudSim: Simulating fraud in a public delivery program. In L. Liu and J. Eck (eds.), *Artificial Crime Analysis Systems: Using Computer Simulations and Geographic Information Systems* (pp. 319-338), ISBN 1599045915.

Koliba, C., Meek, J.W., and Zia, A. (2011). *Governance Networks in Public Administration and Public Policy*, ISBN 978-1-4200-7126-9.

Latour, B. (2004). *Politics of Nature: How to Bring the Sciences into Democracy*, C. Porter (trans.), ISBN 0-674-01347-6.

May, T. (2004). Badiou and Deleuze on the one and the many. In P. Hallward (ed.), *Think Again: Alain Badiou and the Future of Philosophy* (pp. 65-76), ISBN 9780826459077.

Mitchell, M. (2009). *Complexity: A Guided Tour,* ISBN 0195124413.

Morçöl, G., & Wachhaus, A. (2009). "Network and complexity theories: A comparison and prospects for a synthesis." *Administrative Theory & Praxis,* ISSN 1084-1806, *31*: 44-58.

Ong, A. (1999). *Flexible Citizenship: The Cultural Logics of Transnationality*, ISBN 9780822322696

Rhodes, M. L., Murphy, J., Muir, J., & Murray, J. A. (2010). *Public Management and Complexity Theory: Richer Decision-making in Public Services*, ISBN 978-0-415-45753-8.

Sassen, S. (2008). *Territory, Authority, Rights: From Medieval to Global Assemblages* (Updated ed.), ISBN 9780691136455.

Simon, H. A. (1996). *The Science of The Artificial*, ISBN 0-262-19374-4.

Smith, D. W. (2004). "Badiou and Deleuze on ontology of mathematics." In P. Hallward (ed.), *Think Again: Alain Badiou and the Future of Philosophy* (pp. 77-94), ISBN 9780980668346.

Sørensen, E., & Torfing, J. (Eds.). (2007). *Theoretical Approaches to Democratic Network Governance*, ISBN 9780230624535.

Stout, M., & Staton, C. M. (2011). The ontology of process philosophy in Follett's administrative theory. *Administrative Theory & Praxis,* ISSN 1084-1806, *33*: 266–290.

Teisman, G., van Buuren, A., & Gerrits, L. M. (Eds.). (2009). *Managing Complex Governance Systems*, ISBN 0415459737.

Waldo, D. (1948). *The Administrative State: A Study of the Political Theory of American Public Administration*, ISBN 1-4128-0597-X.

Williams, J. (2003). *Gilles Deleuze's Difference and Repetition: A Critical Introduction and Guide*, ISBN 9780748618187.

Thomas J. Catlaw, PhD., is Associate Professor in the School of Public Affairs at Arizona State University in Phoenix. His research centers on the application of poststructural and postmodern theory to problems of governance, authority, legitimacy, and political change. His work has appeared in *American Behavioral Scientist, Administration & Society, Administrative Theory & Praxis, Public Administration Review,* and the *American Review of Public Administration,* among other venues. His first book, *Fabricating the People: Politics and Administration in the Biopolitical State,* was published by the University of Alabama Press in 2007. Dr. Catlaw is also the former editor of *Administrative Theory & Praxis,* an international journal dedicated to the theoretical and critical analysis of governmental practice. He worked previously for the U.S. Office of Management and Budget in the Executive Office of the President, and the Center for Excellence in Municipal Management in Washington, DC.

Yushim Kim, Ph.D. is an Assistant Professor at the School of Public Affairs, a founding faculty of the Center for Policy Informatics, and a Senior Sustainability Scholar at the Global Institute of Sustainability at Arizona State University. Her research program examines how crises, emergencies, and unintended consequences occur and progress in complex social and policy systems, as well as how these systems can recover. She has modeled social interdependencies and interactions in order to understand systemic patterns emerging from humans' and organizational entities' behaviors and decisions in the context of fraud in public service delivery, environmental justice, and pandemic influenza outbreak.

Chapter 3

COMPLEXITY AND NEGOTIATED ORDER

Gill Callaghan (Durham University, UK)

This paper considers an approach to researching complex social settings in way that allows space for agency to emerge. It incorporates insights from realist based complexity theory that highlight the nested nature of reality, giving rise to relatively permanent social structures but also to emergence and change. The starting point of the paper is that such explanations must be derived through observation of human agents as active and reflexive beings. The paper begins by briefly identifying the problem posed in combining structure and agency to explain the sedimentation of practices in places. It goes on to outline the potential of Negotiated Order theory for framing public policy research. The chapter draws on empirical work in a health and social care service in the UK to illustrate the utility of this approach, reflecting on a policy implementation process that explicitly used systems notions.

STRUCTURE VERSUS AGENCY

Structure and agency have been established as two comprehensive, but alternative, bases of explanation. The problem for each lies in incorporating the other in a meaningful way (Dawe, 1971). We can see cause in the social world in terms of the structures which create observable continuities and regularities. Human life is not random but is shaped by experiences of social class, ethnicity, gender etc and it is possible to derive explanations of how people understand and behave in the world as a result of those shaping experiences. This was a dominant mode of explanation in mid 20th century sociology, taken to its logical extreme among structural functionalists using a particular interpretation of the work of Parsons (1937). Critics of this approach pointed out that the model of human action that is implied is a very limited and consequential one, seeing agency as stimulated behavior, rather than an expression of motivations based in desires and constraints, both current and anticipated. They argued that agency has transformative power that cannot be read straightforwardly from the structural account. In such explanations the place for structure often disappeared and was replaced by human agents creating and recreating their world. Those (such as Garfinkel, 1967) who espoused this view focused on structures internal to discourse but largely ignored the role of external social structures (Giddens, 1979). Each approach offered a comprehensive explanation, but taken to its logical extreme, each account excluded the other.

Structure and Agency

Yet few social theorists would seek to sustain either side as the whole explanation. For those, particularly, who have engaged with empirical work, the problem is how can we explain the relationship between the continuities and constraints of structures alongside the intentional and non-intentional action of people (individual and collective). The aim is then to achieve an explanation based in mutual immanence in which structures are embodied in agents in the sense that individuals live their lives in contexts and shape their expectations according to what is known and what is possible, but that these are structures, at the same time, created, sustained and repudiated by agents. A crucial feature of this process is the fact that the relationship to the social is embodied. What we perceive and how we interpret the world is shaped by our experience of it. This applies even to the most personal sphere of emotion, having both interactive and organismic dimensions (Hochschild, 1979). Moreover the interaction of these facets of experience mean that social structures are dispositional structures, we experience 'objective chances as expectations' (Bourdieu, 1994). This has consequences for empirical research because it suggests that accounts that simply identify structural factors as contributing to a way of being through generalized aggregation miss important elements of the interactive process. To develop an explanation that encompasses both structure and agency we need to understand the interaction between embodied knowledge, conscious agency and a structural field that is experienced as relatively permanent.

Appreciating the need to derive explanation in both the tacit knowledge of agents and the more frequently accessed propositional knowledge, (described by Gerrans, 2005, as 'knowing how' rather than 'knowing that') has implications for researching human agency. The model of agency that is implied is neither legalistic (rule following) nor wholly voluntaristic (conscious choice). Rather in Bourdieu's terms it represents 'a feel for the game'. This poses a problem for deriving a frame for analysing social process that does justice to complexity. While undoubtedly analysis must be in some way reductive, the danger of relying on such models is particularly significant when they achieve an unreflective, hegemonic status. Rules that seek to represent human agency, such as those deployed in rational choice theory, represent pale imitations of the relationship between the agent and the social order. Rational action theory is pervasive across social (LeGrand, 2007, as proponent) and economic policy (Skidelsky, 2008, as critique) as well as in the rational/hierarchical models on which the 'dominant management discourse' is founded (Stacey 2010). It continues to frame ways of knowing, even in the face of evidence of the disjuncture between action in accordance with formal organizational design and the role of informal interactive processes (Lipsky, 1980). The strange persistence of such models despite the absence of evidence to support them has been persuasively documented (Stacey, 2010).

This could not be more central for an applied discipline. It reminds us that empirical observation is central to the development of theory if it is not to become simple, meaningless abstraction. One problem that becomes particularly apparent at this point however is that we have limited tools available to us to work with those principles (Tait & Richardson, 2008; Stacey, 2010) and this has to some degree validated a charge that methods associated with complexity theory in the social sciences can deal well with the structural account but struggle to model the contribution of human agents. While agent based modelling pays explicit attention to agency in a more sophisticated way than rational action models, it nonetheless must model agency in terms of regularities, while 'free will' is expressed as noise in the system.

We need ways of describing complex relationships of structure and agency as based in non-equilibric systems that are emergent, continually reproducing but also changing. This encompasses a world external to actors, constituted by relatively permanent structures but at the same time constructed through the experiences and interpretations of actors within it. Bhaskar's critical realist ontology is concerned with these issues. He points out causes cannot be directly observed. Rather, what we observe empirically are regularities that result from generative mechanisms. Bhaskar's Transformational model of the society/person connection' identifies three levels through which social reproduction is achieved. At the base level society exists as a structural entity which socializes the individual. The agent however cannot be fully socialized and therefore acts, in different times and different contexts, to interpret the world and thereby to reproduce and to change it. Social roles perform the work of mediation between these structural and agentic moments, by instantiating those structures at the micro level. Although individuals are not in the simple position of creating society, they do reproduce or transform it.

So although agency is not always intentional and conscious, it is nevertheless effective in reproducing or changing the generative mechanisms that create the society at any point in time. At the same time, in order to act, the individual must work with pre-existing structures and mediating means of production. Using an Aristotelian frame Bhaskar argues that, for production to occur, both material and efficient causes must be operating. Drawing on both Weber and Durkheim he suggests that if society can be seen as the material cause of human action, praxis expresses both the conscious and unconscious dimensions of action through which the social world is produced and reproduced. These are each framed as dualities. So while society is at the same time the condition and the outcome of agency, praxis is both the conscious and unconscious production of the social world (Bhaskar, 1989). Agency is not, then, identified with specific individuals or groups but, rather, captures 'the capacity to transform' (*ibid.*: 35).

These propositions have significant consequences for how we go about doing research. They suggest, for example, that to render an explanation that accesses this dual character, rather than a focus of methodological individualism, the object of research is the field; the case; the system; and the capacity of agents to sustain a particular order or to bring about change.

The focus of this paper is on handling agency in a more nuanced way than either as regularities which suggests a structural quality or the randomness of 'noise'. It will look at how complexity based understanding of systems, describing systems at structural, organizational and personal levels, complements a discussion of the production of order through negotiation within and about system boundaries. While in developing negotiated order, Strauss (1963) focused on the study of organizations, building his theory from fieldwork in a hospital, he created a framework that can be extended to the wider social order and therefore deployed in any empirical study of a case. The remainder of this paper briefly outlines the theory but will focus primarily, through the example, on what negotiated order can bring to developing methods for handling complexity. An approach based in non-reductive and generative principles forms a valuable starting point for negotiated order but needs to be elaborated further to translate into a way of observing system regularities in their contingent manifestation shaped through meaningful action. The paper will suggest that negotiated order theory is useful as a practice theory (Maines, 1979) in keeping with Strauss' inductive approach. Its value to complexity study rests here, precisely because it is not simply a set of methods but rather a frame of reference, fundamentally organized in terms of the case study to describe a local, emergent and negotiated system.

Negotiated Order and Complexity Theory

In Negotiated Order (Strauss *et al.*, 1963) observable reality is constituted of the action of individuals in time and space, based in strategic (self-interest and collective interest) and habitual orientations in interaction with the (relatively permanent) institutions and organizations that are the carriers of structural relations. More significantly (consistent with the notion of agency as a capacity to transform) the approach points to both the need to observe individual action while at the same time regarding agency as not reducible to observable acts.

The concept of negotiated order was developed directly through empirical investigation (Strauss *et al.*,1963). Although founded in a symbolic interactionist perspective, Strauss suggested that negotiation processes need to be directly addressed to prevent the discussion of social order from being structureless 'as in game theory' or restricted to descriptive substantive accounts. Strauss' focus was on explaining order in organizations in which, he argued, order is an ongoing production of the actors involved. The relations and practices of organizations have a structural quality but are produced by actors in the continual process of making and remaking. For Strauss this signifi cant, structural component was observable. Hence, 'larger structural considerations need to be linked explicitly (original emphasis) with a more microscopic analysis of negotiations processes' (Strauss, 1978: xi). Responding to representations of the theory as constructivist (e.g., Day, 1977), Strauss later reiterated the signifi cance of both structural contexts and negotiation contexts to social explanation (Strauss, 1993). While structure is a central concept in establishing the positions from which individuals negotiate, giving them their regular and patterned nature, the outcomes of their interaction are shapedin history and are constantly revised. The order that results emerges from the combinations of daily decision-making and habitual actions of members. Although working in relation to structural constraint then, negotiation is the fundamental process of social order: 'a social order- even the most repressive- without some forms of negotiation would be inconceivable' (Strauss, 1978: ix). Negotiations are not simply conscious, purposeful and voluntaristic as the rational action frame implies. They take place within the context of pre-existing tacit agreements. Strauss distinguishes between structural context and negotiation context. 'The structural context bears directly on the negotiation context, but while the former is conceived of as a larger external structure, the latter refers more specifi cally to the structural properties entering very directly as conditions into the course of the negotiation itself (original emphasis)' (1978: 99). Negotiation contexts are patterned and therefore open to analysis. Strauss off ers a typology of properties of any negotiation context which focuses on the number and nature of negotiations, power relationships, interests and claims to legitimacy involved as well as the complexity of issues and opportunities for discontinuing negotiation (1978: 99-100). In keeping with his inductive approach he points out that rather than a logical construct the typology is built from observation and must respond to criteria of 'relevance not logic'.

Negotiated order and complexity theory therefore share a concern with interaction, history and contingency. Strauss seeks to develop an account based in order as a live and negotiated process rather than a formal or rule bound structure. This not only points to the signifi cance of local action and the relevance of interaction, but also off ers a lens for understanding why centrally planned rational policy implementation always gives rise to the unintended consequences evident in target-driven approaches in social policy, distorting the systems that they seek to control (Medd, 2001; Bevan & Hood, 2006). In Strauss' frame, external interventions disrupt the locally negotiated order.

The approach both maps well onto complexity concepts and adds to our ability to handle those concepts in the field (the need for such complexity consistent concepts is developed in Callaghan, 2008). Rather than modelling agency with a structural quality, or as the random noise in a system, negotiated order allows us to handle agency as comprising purposeful action in one dimension that can give rise to unintended consequences in another.

It allows us to take account of those interests and beliefs that inform action within the system. Thus the individual actor is seen not only as pursuing their personal and collective interests and beliefs, but also as working with discourses which may be more or less contested, highlighted earlier in, for example, Stacey's discussion of the contradiction of a dominant discourse of rational management set within an ostensibly evidence based environment (2010). Examples of these configurations of relationships are widely described in public policy research, but their analysis is severely constrained by being conducted through a rational action lens of observation.

Negotiated order and complexity theory share a set of ontological principles that can be drawn on to imply the methodological approach employed . First of these is the focus on the system and system boundaries. A fundamental ground of complexity theory is that the system and the interactions that occur within it is dealt with both as real and as the unit of analysis. The concept of the system implies boundaries albeit with permeable status, allowing some exchange with other systems. The value of Strauss' approach is in focusing on the order achieved within the system as the ongoing production of agents, working with the constraints imposed by structures and the existing distribution of power. In negotiated order this concept is operationalized in the focus on the case as the site of interaction in which both structure and agency are manifest. It is complexity consistent in recognizing the empirical significance of system boundaries and the implications of where boundaries are drawn, how permeable they are, for framing research questions and for the collection and analysis of data.

For both complexity and negotiated order theory context is central. Both identify the importance of geography and history (time) as interpenetrating in creating the conditions of the system as it manifests itself. The question of what constitutes a system and how it can be defined becomes a practical question implying particular methods and approaches to research. In dealing with the system we are describing a set of interactions of structure and agency in which actors' meanings form a central part of the reality that is being acted upon. As the unit of investigation, the case is a sampling universe with interaction at its core. Further, the determination of effective cause within the system arises from comparative observation so that theory is emergent.

The burden of Strauss work is to identify the ways in which people produce order within the constraints of structures but to suggest that this incorporates an understanding that the values and objectives of actors within the system vary. Systems working can be directly accessed through the negotiations of actors sometimes acting rationally in accordance with system goals, sometimes pursuing other rational objectives, sometimes as bearers of structural properties, but also acting in accordance with beliefs and values that are personally held. It is useful then, to consider how this conception of systems relates to the application of systems concepts in public policy.

Policy and systems working in the UK

Systems notions have become the currency of policy particularly in the last 20 years but their use in UK government policy has been rather naïve as evidenced by the Audit Commission definition of whole systems working (2002). Starting from a position that identified that 'whole system working' was necessary in health and social care to integrate services, the Audit Commission defined whole system working, among other things, as a position in which,' The following are all shared:

- Vision;
- Objectives;
- Action, including redesigning services;
- Resources, and;
- Risk.

(Audit Commission 2002, Steiner)

The insight immediately apparent from Strauss' work is that not only are vision and objectives not necessarily universally shared, but that interpretations of values and objectives vary depending upon embodied experience of power structures, available resources and perceptions of risk. While systems notions have been important in acknowledging the importance of the case (as expressing the particular interaction of geography and history), policy makers have sought to use them in combination with a rational and hierarchical discourse inimical to systems processes.

In the following section we draw from an example explicitly established in the UK as a means of achieving 'whole system' working for a vulnerable group of older people. The paper will describe the interaction of system at structural, organizational and individual levels, with agency shaped in terms of discourse, interests and beliefs. It will seek to show how the negotiated order perspective allowed observation of both structure and agency within the system.

THE SYSTEM AND ITS BOUNDARIES: ORGANIZATION, TIME AND PLACE

Using Esping Anderson's (1999) criterion for classifying welfare states, we can identify two different system models of health and social care services in the UK, founded in different relationships to decommodification of welfare. The NHS, being free at the point of delivery, has continued to largely fit with a social democratic model upon which much of the post war welfare state was founded. At the same time, government reforms since the 1980s have moved social care into the sphere of means testing and external and internal markets, representing more closely a neo-liberal model. As the two systems need to relate intimately in providing care/services to the person their demarcation lines are both problematic and very well recognized. Frequently, however, partly in response to the rational, hierarchical epistemology, the problems to be addressed have been diagnosed reductively as originating in professional intransigence or organizational boundary making.

When conceived as a nested system the two underpinning structural principles become the apparent, setting powerful grounds for system closure. Boundaries established by separate organizational structures and budgets manifest themselves in gaps in provision and duplication in service as organizations define their roles defensively to meet their own priorities. Such closure has been responded to in policy with attempts to overcome boundaries that focus on organizational and individual levels, without addressing the cause of fragmentation. Persistent system boundaries inevitably re-emerge, giving rise to system failures which have observable effect on the care of patients/users. A notorious example lay in the requirement for professionals to distinguish between the older person's need for assistance with bathing as being either a health or a social care need. While pointing to the specious nature of the distinction, professionals from each service felt organizational pressure to define need in terms of the other to protect limited budgets and valued core services.

As a consequence of the redefinition of state role in the 1980s toward individual responsibility, services shifted focus from early prevention to intervention only in response to higher level need. Accompanying this reduced role was an ethos of care rather than rehabilitation because eligibility criteria recognized only high dependency. It was in response to the failures occasioned by this context that the Intermediate Care system was established early in the new century. Its design was based in recognition that there were points at which older people with health care needs could be rehabilitated or supported to avoid entry into higher level support. This was both desirable for improving quality of life for the older person (in keeping with a growing emphasis on the discourse of dignity as independence) and as a cost saving measure reducing reliance on costly institutional care. An ageing population as well as recognition of the cost effectiveness of prevention reinforced the impetus to change approach.

The structural context for the policy then, was that the organizations that needed to work together around the user/patient to achieve this change were often unable to coordinate and/or had conflicting pressures on their budgets. Attempts to overcome structural divides through interpolation of ad hoc intermediate mechanisms had been tried, including pooled and joint budgetary arrangements and placing stress on the need for joined up and partnership working. Policy literature in the 1990s is largely concerned with problems of fragmentation and the solutions proposed above as well as in modifying professional demarcation of tasks and through blurring boundaries, co-location of professionals and multi-skilling. These solutions alone had not yielded the promise of a seamless service (Alford, 2009) and the 'whole system' idea became a popular prescription and a central idea in new labour policy from 1997 onwards.

The costly results of fragmentation in services for older people were delayed discharges from hospital and unnecessary admissions to residential care. Intermediate care was identified as a new whole system approach which would overcome the problem of lack of coordination by shifting the boundary from the line between health and social care to a line around a 'bridge' between hospital and home for older people. To remove the boundary, work that would either prevent admission or speed discharge would be undertaken within the Intermediate Care system and would be free of charge for a six week period. Effectively it shifted crisis social care from the neo-liberal to the social democratic system, although still operated in part by organizations working in a market frame. What emerged was a new system with a distinct set of boundaries which resolved some problems of integration but merely shifted others as this example will show.

In view of the limitations of space, the field research discussed below does not seek to describe the case studies but rather takes examples from empirical work undertaken by the author in one locality[1]. Its purpose is to illustrate the utility of the negotiated order perspective. It will therefore focus on the ways in which, while rationally planned as part of a national system, Intermediate Care developed in response to configurations of local factors including local agency.

Most of the examples described below focus on the 'Rapid Response' service as a nationally identified part of Intermediate Care intended to overcome failures in care that occurred through lack of timely coordination. It is highlighted here because, as a new service, it might be expected to be rationally, and therefore uniformly, structured to meet the new system goals. The Rapid Response service would embody the principles of Intermediate Care in creating a team with the necessary professional skills from health and social care sectors. It would have no standing case-load so that it could respond immediately to referrals. Rapid Response was therefore envisaged as a core service coordinating care in a timely way and moving people on to longer term health and social care services when appropriate. The structure and staff composition of the team however had to be determined locally to fit most cost effectively with other existing hospital and community services. This necessarily affected the way that the team could contribute to system goals.

Wider system factors had significant effects in shaping service delivery and therefore significantly affected the outcome for individuals receiving care. The organizational location and professional expertise in the team determined what help could be given in ways not always anticipated in planning the service. This is illustrated in one locality where the Rapid Response team, established

1. The fieldwork formed part of a national Evaluation of Intermediate Care carried out between 2002 and 2005 for the Department of Health. The research design was based on a comparative case study method across six localities. Fieldwork involved integration of qualitative data taken from non-participant observation, documentary analysis and qualitative interviews with staff and older people. The complexity frame for analysis was developed by the author. As the discussion here is primarily illustrative of methodological rather than substantive issues the examples are drawn from a single locality to focus on the usefulness of the negotiated order perspective.

to achieve speedy discharge from hospital, was sited in the hospital using the expertise of the hospital based physiotherapist. The unintended consequence of this organizational arrangement became apparent when a GP requested a walking aid for a frail older person in the community. The assessment that needed to be carried out required the expertise of staff whose role was restricted to the hospital setting. The consequence was that the older person was left at risk until a community physiotherapist could visit the following week; a circumstance the system design had been specifically intended to overcome. Such cases were often highlighted to strategic and managerial levels as a basis for reviewing organizational remit to negotiate a new way of coordinating structures around cases.

The negotiation context had significant impact on what was developed. At times, although system boundaries were recognized as inflexible, the power and interests of medical as compared with other professionals or managers and the consequences of adaptation for other parts of the system overwhelmed the apparent legitimacy/rationality of the proposed change. A further set of negotiations might then be undertaken to sustain the system but ameliorate its worst effects as will be discussed later. In either case, each of these rounds of negotiation and accommodation contributed to the local negotiated order and thus became the basis upon which a future order would be negotiated.

These issues underpinned different negotiations that were a daily part of system working and examples of their occurrence were observed throughout the research at individual organizational and wider system level. The significance of structural and negotiation contexts was clear, but directly observing them at the case level was essential to fully appreciate how agency shaped actual outcomes. This was illustrated in an observation of a specific negotiation process between two professionals in which competing discourses, power and organizational resources played a role. The example concerns an older lady who had fallen at home and had lain for several hours before being admitted to the Accident and Emergency department in a confused and distressed state. The Rapid Response worker was called in the late afternoon, when the doctor, having examined her cuts and bruises, had decided that she was medically stable and appropriate for discharge. Medical and social interpretations of the case pointed to conflicting outcomes and these positions were argued by doctor and Rapid Response worker. The doctor's intention, responding to pressure to avoid admissions that were unnecessary on medical grounds, was to discharge the lady to the care of a young adult grandson, as no out of hours domiciliary care was available. The Rapid Response worker considered this inappropriate because the woman was confused and would need help overnight with toileting, washing and dressing. While her physical needs could be met, having help from her grandson with intimate personal care would have implications for her dignity. In this particular case the Rapid Response worker was able to negotiate an overnight stay in hospital, with the undertaking that the care services would be in place first thing in the morning. The point here is that there was no necessary outcome from this exchange that could be read straightforwardly from system objectives. What was observed was the effect of opposing pressures on professionals giving rise to a debate about the patient's needs and appropriate care. The power of the medical profession as the gatekeeper for hospital beds is greater than that of a Rapid Response worker and, from observation, it seemed likely that a less determined stance from that worker would have resulted in a badly supported discharge home.

A further source of fragmentation in service that Intermediate Care was established to overcome was that created by different geagraphical boundaries. The impracticality of the notion of universal top down policy implementation was illustrated in the empirical research. One locality was operating two systems of intermediate care because an earlier reorganization of health services had merged two localities. Each operated with different configurations of health and social care services which could not be simply replaced by a unified structure. Two quite different planning and management systems were operating with different relationships and communication processes. The new service structures that emerged in each had to respond to pre-existing service configurations, geographical and demographic structures. In policy terms this is a familiar experience. Reorganization and redrawing of boundaries as a consequence of rational top down policy implementation gives rise to new negotiation contexts at the local level where the mismatches of organizational of geographies occur.

Further systemic sources of division were evident in differences in geographical boundaries between health and social care services within one rural locality. While ambulance services take people to the nearest hospital, social care services are defined by local authority boundary. The Rapid Response team, based in and working with anyone admitted to the hospital contained within in it a local authority care manager. Social services staff members worked to county council boundaries. So the team as a whole worked across an area that the care manager within it was not able to support. The resolution of this problem relied on judgement made at case level. For relatively simple and short term cases it was agreed that the Care Manager from within the team would provide the support. When the social care elements of follow-up support were likely to be complex however the case was referred on. As in other boundary issues these decisions were not made according to any simple criteria but were subject to the professional judgement of front-line workers and involved decisions made on the basis of predicting, sometimes unpredictable, care trajectories. The solutions therefore, here negotiated at practice level, had implications for resources, organizational responsibility and accountability which in reshaping the achieved order changed the system.

These examples demonstrate that structural divisions have consequences not only for organizational arrangements and practices but also for the ways in which work is undertaken at the individual level. They set some of the constraints and opportunities within which health and social care professionals, and private providers work and form the context for understanding the actual form of service provision. While a simple model of policy implementation would see front-line staff carrying out the instructions of policy designers, feeding back their experience to improve successive rounds of implementation, these examples demonstrate that front-line decisions significantly shape action and, in consequence, the nature of the system itself.

The impact of the persistence of system boundaries can be further exemplified in outline in the boundary created by time. While intermediate care removed some fragmenting boundaries, for others it simply moved the boundary that actors have to negotiate around. The time limit of six weeks intervention became a criterion by which people were assessed for appropriateness to enter. As it was embodied in the system design, the time boundary principally became counter-productive when rehabilitation was a reasonable objective but the intervention required to achieve it would take rather longer than the six weeks allotted. The six week period had been identified as sufficient for the majority of cases, but inevitably some people could reach similar rehabilitation objectives if support provided were in the medium rather than the short term. The disincentive to include those people in this dedicated service lay in the outcome targets by which the service was measured. Time further shaped the nature of service, for example in defining the workload of Rapid Response and the orientation of rehabilitative residential care to short term intervention.

Charting the changes around the six-week rule in itself exemplifies the process of local implementation of national policy. Initially the six-week rule was applied quite inflexibly to sustain the system as designed, because this was the basis of funding (and accounting in terms of outcome targets). Front-line staff felt this introduced rigidity into a system whose structure should be defined by its goal of rehabilitation. From very early in implementation it had been recognized as a problem, barring some people who were deemed appropriate from entry because they could not meet formal criteria of recovering speedily enough. As experience of the system developed, local policy makers and front line professionals began to negotiate around this boundary to achieve policy objectives rather than fulfil formal requirements. In one locality this became a point at which progress was evaluated, rather than an end point for service. While the problem was partly resolved at the local level by a joint policy decision to fudge the time boundary, this necessarily raised the question of; for how long? Not only was it a drain on the resources of the system, but, a system established and staffed around a short-term intervention could not always provide the best environment and care for older people who needed medium to long term support to achieve their goal.

Competing Discourses

It has been briefly mentioned above that actors' orientations to dominant discourses form a significant element of the negotiation context. The medical and social models of health exemplified contested discourses which were manifest in a number of ways in the Intermediate Care system. The principal focus of difference was action in keeping with the medical versus social models of health. This was exemplified in a debate about diversion from expensive medical procedures in which some argued that in being denied tests that younger people routinely had access to, older people are treated in a less advantageous way simply because they are older. An alternative view was that older people often have a number of health conditions which they may be unaware of and which may not affect their functioning. The latter argument suggests that the trauma or disorientation of admission to hospital, could be more detrimental than the failure to address ongoing, but low-level, medical problems. The two arguments suggested opposing actions for the intermediate care system. One (conceiving of health need in terms of the physical body) would only assess for Intermediate Care following full medical screening. The second (including in its definition the impact of intervention on the person) would avoid hospital admission entirely, maintaining treatment in the community with the support of GP and District Nurse. The division over this question could not be simply identified in terms of professional perspectives or more or less knowledgeable approaches to meeting older people's needs. Such issues form the broadest context for the whole system of health and social care and their particular resolution reflects not only the rational weight of argument but the priority of values and power of professionals within the system. Associated with broad medical/social debates were discussions that implied different objectives for older people's care.

The Intermediate Care system was most explicitly founded on a discourse of prevention and rehabilitation in order to change the mainstream system's high dependency orientation. This shift had observable effects on the operation of the system in terms of defining its 'success'. Staff in residential homes, and to a considerable degree those who had been employed in home care support, had been trained to work with high dependency because of the raised eligibility criteria that had operated since the 1980s. With the implementation of the new system the significance of these competing discourses became clear. At service level, different outcomes were reported for homes oriented toward rehabilitation than for those oriented to long-term care. Intervention to achieve rehabilitation of the older person was viewed by some workers as encouragement to maximise potential and by others as unfair pressure and even bullying. Private sector placements, in particular, frequently did not achieve rehabilitation objectives because staff had not been trained to encourage independence. This was also observed in private sector home care support. Rehabilitation is relatively costly in the short term because helping an older person to wash and dress themselves takes staff considerably longer than simply doing it for them. Some professionals pointed out that adopting a rehabilitation approach had resource implications that some private sector organizations might not support. In each of these settings the mix of views of staff and users, the ethos of the establishment and the training of staff contributed to the actual outcomes for older people. It highlighted the significance of observing the particular nature of the negotiated order which pertains in each establishment as part of the whole system.

The creation of a bounded system based in a rehabilitation discourse could, itself, have fragmenting effects on care. Professionals pointed out that defining a system in terms of a rehabilitation objective sometimes meant creating a boundary between rehabilitation and care, which in reality was not as clear as its rationale might suggest. The differences in need that could be matters of degree were being turned into matters of kind by system definitions. Professionals were required, in making an assessment for entry to intermediate care, to assess rehabilitation potential in unrealistically specific binary terms.

Thus far we have concentrated on professional action to illustrate that actors do not operate in terms of the system goals in the simple way that the 'whole system approach' identified at the beginning suggested. Numerous examples in the fieldwork attested to the significance, not only of these professional decisions, but also of the user defining their own role in the system. Users may not have substantial formal power, but their motivation toward rehabilitation is a crucial factor in its achievement. Further, the professional's expectations about user's confidence, motivation and ability to rehabilitate have a profound effect on what can be achieved. Boundaries are set by cultural expectations about what is appropriate as well as beliefs about what constitutes acceptable risk. It was an assumption of a system oriented to rehabilitation that users were motivated to return to their own home. This was implicit in the design of the system but was not universally true. Older people and their families' interests frequently focused on the balance between safety and enabling that was seen as risk taking.

CONCLUSION

The starting point for this paper was that we need to handle both structure and agency in developing explanations of complex social settings. The paper has used some illustrations from one empirical study to consider an approach to studying the relationship. The example was chosen both because it concerned a policy explicitly based in a systems approach and because the boundaries it was seeking to overcome were created by deep structures whose effects were manifest *within* the person as alternatively health or social care need. The case study demonstrated the interaction between new services, reorientation of professional roles and organizational relationships, partly responding to top down planning but at least as frequently, to the ways actors dealt with the issues presented at local policy and practice levels.

At the macro level system actors had twin imperatives. They worked with the constraints imposed by the need to contain costs and to reduce use of the expensive resources such as hospital care. At the same time they were required to provide services to meet the health and social care needs of the population. For managers and planners this had meant maximizing provision at the lowest possible cost and therefore limiting the definition of their organization's role as strictly as possible. This defensiveness resulted in the attempt to separate responsibilities for needs that are embodied within individuals. The inter-organizational frictions which resulted were frequently experienced by front-line staff as an obstructive attempt by the partner agency to deny responsibility. In part, this reflected professional context including the nature of accountability each worked with, expressed as autonomy versus performance management. Frequently however these divisions have been analyzed simply at the level of rationality and organizational structure, which fails to acknowledge the system dynamics that we have begun to explore here.

What negotiated order theory brings to this account is a way of handling the, often conflicting, rationalities upon which the system works and the ways in which actors manage those in their daily work. Observing not only the structural context, but the specific negotiation contexts in which these processes take place provides a basis for analysing the trajectories along which the system travels. The way in which order is produced becomes clear.

Previous rounds of negotiation established the system history and the particular outcomes or resolutions of produced tensions provide the basis for the next layer of negotiations. Comparison across cases highlights the constructed nature of each context. The comparative case study method proved essential to understanding the causes of these system variations, highlighting external structural causes and their effects in shaping the specific nature of the locality contexts. For the purpose of developing the argument this paper has focused on examples of the negotiations undertaken between agents to manage specific negotiation contexts, formed in the interaction between inter-organizational histories, legitimizing discourses and the relative power of actors in one locality. The value of recognizing negotiation context is that structures are observed as produced in that interaction. In recognizing the system as one established, not only in terms of the structural rules and boundaries, but also in terms of the negotiations of actors we can develop a clearer understanding of the ways in which each contributes to its changing form.

Negotiated order theory provides a framework to explain how structural pressures combine with actors' beliefs, power and interests to create the particular systems that emerge. The particular negotiated order is created from the interaction of explicit and recognized forces with the tacit agreements that form habitus. The latter reflects dominant ontological perspectives such as rational action theory and dominant substantive perspectives such as medical versus social models of health. As the results of negotiations are communicated between levels in the organizational hierarchy, they become part of the process of reproducing or revising the negotiated order explaining its dynamic character.

Further, implications emerge from this understanding of agency and social order for the production of empirical social science. In essence the resources of social theory have suggested a distinction between structure-based conceptions of agents as rule-following creatures versus a conception of agents as embodying a complex mixture of experience, orientation and motivation. This suggests that a notion of restricted complexity in which rules can be identified from which the range of emergence is defined may be problematic. In discussing social simulation models that have attempted to model emergence, Gilbert reinforces this point in highlighting the need to recognize both emergence at micro and macro levels to recognize the effects of feedback (1995). The problem of providing valuable knowledge when no simple prediction is possible is significant for an applied discipline. What this paper suggests is that despite its more modest ambition, valuable knowledge can be drawn as a basis for anticipating the consequences of policy decisions.

A final reflection this chapter makes is on the challenges of producing evidence to support the discussion of the relationship between methodological frameworks such as negotiated order and complex settings in which context is central. To have provided a full case illustration would have required a much lengthier discussion than a paper allows. This chapter has abstracted a series of examples from the case study to illustrate aspects of the use of the approach. The specific examples have been chosen in the belief that, across the span of public policy, analogous inter-organizational and inter-professional contexts will be recognizable, alongside different but equally dominant and contested discourses.

REFERENCES

Alford, J. (2009). Tackling inherently governmental obstacles to building inter-organizational trust. Paper presented to the 13th International conference IRSPM

Bevan, G. and Hood, C. (2006). What's Measured Is What Matters: Targets And Gaming In The English Public Health Care System. *Public Administration* 84(3): 517-538.

Bhaskar, R. (1997). *A Realist Theory of Science,* London Verso.

Bourdieu, P. (1977). *Outline of a Theory of Practice,* Cambridge, Cambridge University Press.

Bourdieu, P. (1990). *The Logic of Practice*, Cambridge, Polity.

Bourdieu, P. (1994). *Practical Reason: On the Theory of Action,* Cambridge, Polity.

Bourdieu, P. (1999). *The Weight of the World: Social suffering in contemporary society*, London, Polity.

Byrne, D. (1998). Complexity Theory and the Social Sciences, London Routledge.

Callaghan, G. (2008). Evaluation And Negotiated Order; Developing The Application Of Complexity Theory *Evaluation,* 14(4): 399-411.

Cilliers, P. (1998). *Complexity and Postmodernism: Understanding Complex Systems,* London Routledge.

Esping, A.G. (1990). Three worlds of welfare capitalism. Cambridge Polity.

Esping, A.G. (1999). Social Foundations of Post-industrial economies, Oxford University Press.

Fine, G.A. (1984). Negotiated orders and organizational cultures *Annual Review of Sociology* 10: 239-262.

Garfinkel, H. (1967). *Studies in Ethnomethodology*. Englewood Cliffs, NJ: Prentice-Hall.

Gerrans, P. (2005). Tacit Knowledge, Rule Following And Pierre Bourdieu's Philosophy Of Social Science, *Anthropological Theory*, 5(1): 53-74.

Giddens, A. (1979). *Central Problems in Social Theory*, London MacMillan.

Gilbert, N. (1995). Emergence in Social Simulation. In Gilbert, N.; Conte, R. 144–156. Conte, R. eds. 1995. Artificial Societies: The Computer Simulation of Social Life, London: University College Press.

Harvey, D.L. (2002). Agency and Community: A Critical Realist Paradigm. *Jnl. for the Theory of Social Behavior.* 32(2): 163-194.

Harvey, D.L. and Reed, M.H. (1994). The Evolution Of Dissipative Social-Systems *Jnl Soc Evol Syst* 17(4): 371-411.

Harvey, D. (2001). Chaos and Complexity: their bearing on social policy research. *Special Edition of Electronic Journal: Social Issues,* 1(2).

Haynes, P. (2001). Complexity, Quantification and the Management of Policy, *Special Edition of Electronic Journal: Social Issues,* 1(2).

Heylighen, F., Cilliers, P., and Gershenson, C. Complexity and Philosophy. Available at http://uk.arxiv.org/abs/cs.CC/0604072

Legrand, J. (2007). The Other Invisible Hand: delivering public services through choice and competition Princeton University Press. Princeton.

Lipsky, M. (1980). *Street Level Bureaucracy,* New York: Russel Sage Foundation.

Maines, D.R. (1979). Mesostructure and Social Process, *Contemporary Sociology,* 8(4): 524-527.

Medd, W. (ed). (2001). 'Complexity Science and Social Policy', *Special Edition of Electronic Journal: Social Issues,* 1(2).

Nicolis, G. and Prigogine I. (1989). *Exploring Complexity: An Introduction New York,* WH Freeman.

Parsons, T. (1937). Structure of Social Action, McGraw Hill.

Reed, M. and Harvey, D.L. (1992). *The new science and the old: complexity and realism in the social sciences* Journal for the theory of social behavior, 22: 56-79.

Skideskly, R. (2008). 'Where do we go from here?' *Prospect* 22nd October 2008.

Stacey, R.D (2010). Complexity and Organizational Reality, 2nd Ed, London Routledge.

Strauss, A., Schatzman, L., Ehrlich, D., Bucher, R. and Sabshin, M. (1963). 'The Hospital And Its Negotiated Order' in Freidson, E. (ed.), *The Hospital in Modern Society,* New York, Free Press.

Strauss, A. (1978). Negotiations: Varieties, Contexts, processes and Social Order, San Francisco, Jossey Bass

Strauss, A. (1982). "Interorganizational Negotiation." *Urban Life,* 11(3): 350-367.

Tait, A. and Richardson, K. (2008). "Confronting Complexity," *Emergence,* 10(2): 27-40.

WHERE IS AUTHOR BIO.

Chapter 4

ADDING A COMPLEX ADAPTIVE SYSTEMS APPROACH TO PUBLIC ADMINISTRATION RESEARCH

Erik W. Johnson (Arizona State University, USA)
Kevin C. Desouza (Arizona State University, USA)
Qian Hu (University of Central Florida, USA)
Rashmi Krishnamurthy (Arizona State University, USA)

The changing contour in public administration and public policy now only rarely reflects the traditional hierarchical form of government apparatus; instead a complex web of interactions among diverse stakeholders within the context of policy systems is increasingly common. Even with the emerging organizational forms and behavioral understandings, the field of public administration remains predominated by the "rationality" assumptions and positivist epistemology. This may have created structural blind research spots and allowed researchers to focus on management and policy solutions that do not fully take into consideration complexity of human behaviors and social systems. Thus, to advance systems thinking that is cognizant of these dimensions we can complement existing research with methodologies that explore and capture the underlying complexities in policy and human interactions. Methods traditionally associated with complex adaptive systems, like computational modeling, have great potential and provide tools to researchers to explore complex

web of interactions that are seldom linear. In this paper, we first briefly review existing research foci and epistemology in public administration and discuss challenges for policymaking and government management practice. Next, we introduce a complex adaptive systems approach to public administration research. And finally, we conclude that, rather than replacing one method with the other, contemporary management and policy challenges requires the use of multiple methods to understand the emergent complexities.

INTRODUCTION

Public administration has never been more challenging. Advances in communication lead to people and groups interacting in ways that put stress on traditional forms of administration within, across, between, and overlapping jurisdictional boundaries (Johnston, 2010). As individuals, organizations, and movements interact in with more sophistication, their expectations of anyone they interact with, including governments, similarly raise. With the sophistication in computational technologies, access to richer data on public agencies and their operations, and the emergence of participatory platforms that allow for innovative forms of citizen engagement, public administration scholars have little excuse to ignore the complexity of these new challenges. As active participants in this new reality, our research methods need to keep pace with this level of innovation if we are to take advantage of the new realities and opportunities emerging daily. To do so, we need to embrace a systems perspective of public administration, the notion of emergence (of which unintended consequences of actions is one element), and methodologies that allow us to model this complexity, these being the three main elements of the paper.

The research approaches to public administration and public policy remain predominantly inundated in quantitative methods that are rooted in positivist epistemology and a strong belief in the "rationality" assumptions about human behaviors (Heineman, Bluhm, Peterson, & Kearny, 1990; Stone, 2002; Adams, Brockington, & Dyson, 2003; Morcol & Ivanova, 2010). The predominance of quantitative methods in public administration and public policy largely results from an ingrained belief in value of efficiency and rationality. Despite harsh criticisms (Simon, 1955; Kahneman & Tversky, 1979), this kind of reasoning dominates our understanding of organizations (Simon, 1947; 1955) and the measurement systems for reforming government (Hood, 1991; Osborne

& Gaebler, 1992; Frederickson, 1996). The dominant economic analysis in public administration and public policy cannot fully capture the dynamics of human behavior and the complexity of social systems. This may have created structural blind spots for research and allowed researchers to focus on policy solutions which do not fully take into consideration the complexity of human behaviors and social interactions.

In this paper we explore some of the consequences of methodological dominance in any field and propose a complexity theory approach to complement existing research methods in public administration and public policy. The first section reviews existing research methods in public administration. The second section introduces complexity theory and its methods as a complementary approach to existing methods. And finally the paper concludes that rather than using one method over another, present policy challenges requires use of multiple methods to highlight latent dynamics in complex human interactions and the policy process.

EVOLUTION OF RESEARCH FOCI AND RESEARCH APPROACHES IN PUBLIC ADMINISTRATION

Debates over research methods and epistemology have been central to the evolution of the field of public administration due to the inherent diversity of subjects in this discipline. Newland (1994) argued that research in public administration has been a hard struggle because of its multidisciplinary foundation, foci on human behavior and social processes, and practice-orientation. Hence, research in public administration involves different foci at multiple levels. In this section we will review the evolution of research foci and approaches and epistemology in public administration and we reflect on contextual factors that contribute to the dominant status of economic analysis and quantitative approaches that are rooted in positivist epistemology and have explained away the dynamics of human behaviors and social systems.

Evolution of Research Foci and Research Methods

Earlier public administration researchers and practitioners emphasized businesslike approaches to administration and perceived efficiency as the core value for evaluating the success of public administration. Woodrow Wilson (1887), the founding father of public administration in his prominent essay, *The Study of Administration,* advocated efficiency as a central theme for administration and argued for businesslike administration in service delivery. Other scholars such as Taylor, Gulick, Urwick, White, and Fayol all endorsed developing management techniques rooted in the "science of administration" (Dhal, 1946; Waldo, 1948/2007; Denhardt 2002; Denhardt, Denhardt, & Aristigueta, 2002). These early public administration scholars argued that efficiency of the organization could be enhanced by institutionalizing the structures through unity of command, coordination, span of control, and division of labor. Early scholars with an overemphasis on developing a science of administration, largely neglected human interactions in organizations and treated humans as extension of machines (Dahl, 1946; Lasswell, 1955; Denhardt, Denhardt, & Aristigueta, 2002).

The 1960s and 1970s marked a widespread criticism of rationality and positivism (Amy, 1984). Waldo (1948/2007) argued that the principle of efficiency contradicts with the virtues of democracy, justice, freedom, and the role of citizens in the policy process. In contrast to the existing model of "economic man," Simon (1947,1955,1978) proposed the concept of "administrative man" who makes decisions based on "bounded rationality." The financial crisis and failure of welfare reforms in 1970s rekindled interest in research methods in public administration and public policy. Analyzing the research methods in public administration and public policy, White (1986) argued that although three prominent research methods in public administration and public policy are explanatory, interpretive and critical, explanatory research embedded in the philosophical tradition of positivism remains the predominant research methods used in public administration.

Despite the critiques over years, research approaches in public administration and public policy remains firmly rooted in positivism, empiricism, the fact-value dichotomy, and testable casual law (Amy, 1984; White, 1986; Morcol & Ivanova, 2010). Stone (2002) argued that one of the main reasons for the widespread use of quantitative methods by policy makers is due to the fact that these methods enable policy makers to use numbers to support their authority, illusions, and outcomes. Moreover, quantitative methods have been influential in creating shared beliefs that are supported by political and institutional practices (Johnston, Quigley, & Kim, 2010), which dominates the academic and practical knowledge-building in public administration and public policy. DeLeon (1994) argued that the shortcoming in present policy science can be attributed to an over-reliance on technical rationality and technocratic orientation. This over-reliance on technical rationality has led to a widening gap between the policy science and its original multidisciplinary character and goals of human dignity and democratic governance (Dahl, 1946; Asher, 1987; DeLeon, 1994). Criticizing positivist epistemology, DeLeon (1994) noted that "critical theory suggests that there are numerous valid avenues of policy inquiry, with the traditionally dominant natural sciences paradigm and the rational actor model being only two among many" (p. 82). Moreover, policy dialogues based on that traditions and conventional knowledge and preconceptions obscures policymakers from effectively addressing policy problems (Adams *et al.*, 2003). One-dimensionality created by quantitative methods rooted in positivist epistemology pose some serious threats to the policymaking process in the 21st century public administration.

The Slow Inclusion of Complex Adaptive Systems in PA Research Since the 1980s

Since the turmoil in 1980s, public administration included some alternatives from its traditional form of hierarchy. The financial crisis of 1970s and stagnating economy posed a great threat to the field of public administration. In response to a growing criticism against the government apparatus, public administration moved away from a "traditional hierarchical model" to a "networked form of governance" (McSwite, 1997; Harvey, 2006; Catlaw & Hu, 2009). The new form of governance favors inclusion of non-governmental organizations and private actors for efficient service delivery (Rhodes, 1996; Peters & Pierre, 1998; Kjaer, 2004). From the research perspective, the inclusion of non-state actors increases the complexity of the policymaking process. As interactions get more frequent and interests get more intertwined within, between, and across communities and jurisdictions for a wide range of policy scenarios, policymakers need to be able to anticipate, frame, and address diverse policy problems using systems and complexity frameworks and methodologies.

The changing contour of public administration has lead to increasing interdependencies at international, national, state, and local levels. As Koppell (2010) noted, the contemporary realities of public administration do not reflect the traditional boundaries of the field due to the emergence of quasi-governmental organizations in the policy arena, involvement of non-state actors in service delivery, and the growing importance of transnational organizations. The increasing interactions among policymakers with various interest groups with intertwined interests are resulting in creating a complex web of interaction. To arrive at a reasonable set of choices for policymaking, traditional methods neither help us to study the obvious (sometimes latent dynamics) nor encourage us to conduct comparative analyses, which might impact outcomes at local and global levels through time (Keller, Desouza, & Lin, 2010; Desouza & Lin, 2011; Kim, Johnston, & Kang, 2011).

The conventional research methods rooted in positivist epistemology mostly assume a linear relationship among factors and create discipline-based myopia. Organizational and professional training built on discipline knowledge in public administration predominantly remains rooted in positivism. While professional training helps increase managers' field knowledge, enhance management skills and build shared understanding and identity, it may institutionalize groupthink, discipline based myopathy or institutional isomorphism, as articulated below (Sandberg, 2011):

> *One form of institutional isomorphism is called normative isomorphism, and it is fueled primarily through the professionalization of the members of a particular organizational or occupational field. Professionalization leads to isomorphism by way of formal education and training and through networking or collaboration between members. More specifically, through formalized training programs, members acquire a shared body of knowledge that has been agreed upon. This knowledge is then implemented in organizations through the establishment of organizational norms. These norms spread throughout the organizational field when members commune with one another and share ideas oftentimes through meetings of professional associations.* (p. 9)

In other words, through mechanisms like accreditation processes that encourage a common curriculum, or attending conferences focusing on recurring themes, management professionals are trained to share similar, if not the same perspectives toward social phenomena. Although this is important for building an independent discipline and community of practice, this will neither attract individuals with different backgrounds nor encourage individuals to think diversely. However, diverse perspectives and thinking are crucial to address many existing management challenges (Page, 2006). Moreover, the present dominance of particular methods articulated through conferences and other academic interactions in an indication of how a field budgets their attention in favor of a particular methodology at the expense of other approaches.

The attention paid to positivist epistemology imposes certain types of questions and perspectives to be researched and obscures analyzing other alternatives, thus enabling structural blind spots to potential threats, challenges, and opportunities. An overdependence on economic models for predicting policy outcomes and the failure of market models has been blamed for not anticipating the recent financial collapse (Economist, 2010). When methodologies are limited or standardized, individual's behavior within an organization becomes stunted. Questions that can be asked are only the ones that can be defended with the same lens, and while behaviors become more predictable, there is also the threat of how the shared assumptions exclude contrarian viewpoints.

To address the changing dynamics in the policymaking process, we need additional perspectives to ameliorate the structural blind spots created by dependence on any particular research methods at the exclusion of others. For an analogy, the Netlogo model of fireflies shows that although a population can start with a set of fireflies flashing independently, if they mutually reinforce each other's behavior, then eventually, they will synchronize their behavior and all will flash at the same time. If the flashes of fireflies represented the curiosity of public administration scholars, we would observe that, for certain questions with agreed upon importance and established methodologies, an entire field would be ready to address them. However, what is of concern is the large absence of activity that is created between the flashes where no one is willing to go it alone. In the next section, we introduce complex systems thinking and its methods as an alternative methodology for exploring public administration research and ensuring the diversity of thought in public administration scholars.

SYSTEMS OF SYSTEMS THINKING FOR PUBLIC ADMINISTRATION

Public administration scholars examine *complex* institutions. Complexity is present due to the nature of goals, resource allocations, stakeholder interest, operational time frames, and the internal interdependencies among units. These complex institutions operate in *complex* environments. Environmental complexity is a function of the dynamic, unpredictable, and interdependent among institutional actors. Ignoring complexity is not an option for public administration scholars. Simply stated, it is in the study of complex interactions and turbulent environmental analyses wherein we are most likely to arrive at interventions that make a difference in practice. Public administration scholars are best served by thinking of institutions and even programs that they would like to study as *systems of systems*. The embedded element is critical to appreciate so as to arrive at knowledge and interventions that account for not only first-order effects, but second, third and even the *n*th order outcomes. Today, we have techniques and tools to help us study public administration problems (e.g., governance, resource allocations, etc.) within the systems of systems framework.

To truly appreciate the nature of the complex systems, five ingredients need to be carefully considered—*information, agents, objects, rules of interaction,* and the *environment* in which they live in. The collection of these five ingredients and the interactions between them constitutes the system. The interactions are non-linear, emergent, and cannot be specified a priori, thereby leading to a complex system. To study complex systems, one must begin with a thorough appreciation of how information emerges as it is created, transferred, and consumed between actors and objects within the environment (Desouza & Hensgen, 2005). Information remains a critical element that needs thorough examination in the public administration literature. Today, information cannot simply be relegated to reports that are prepared (e.g., after inquest) or evaluation outcomes (e.g., surveys). We must contend with information in multiple forms from videos to twitter feeds. Information is also generated in real-time and can be collected with precision at finer granularities. Information generation can take place through collaborations involving multiple agencies and even through public participation. Information can be analyzed by multiple parties in real-time, and can even be leveraged to solve problems in shorter time-cycles. Information needs to be a focus of study if we are to appreciate the complex nature of public agencies. After all, information is a critical input to decision-making. Towards this end, an examination of the *semiotics*, the study of anything that is related by signs or forms as represented in signals, words, images, sounds, gestures or objects, is valuable.

Semiotics

In the contemporary sense, signs contribute significantly and specifically to the construction of the *perceived* context of reality. And from such a perspective, semiotics determines the "logic" involved in a situation. Agencies may gather data as objects of interest to be synthesized or analyzed in an effort to calibrate appropriate actions. If, however, stages or steps otherwise associated with a semiotic ladder (Desouza & Hensgen, 2002, 2003, 2005) are prevaricated, the resulting context and maintenance of reality is logically incomplete. Such inattention to initial conditions often results in unanticipated results sometimes associated with chaos. How do signals from the external, or within the internal environment lead to actions? To full visualize the trajectory that signals take, one might consider tracing the evolution across the semiotic ladder, the stages of which are *morphological* (data), *empirical* (in-class information), *syntactical* (across-class information), *semantic* (knowledge or meaning), and *pragmatics* (actions).

Morphological

The initial morphological level is concerned with the aggregation of data. Data involves the gathering of "conditional facts" (or basic news bytes) from agents and objects of interest. This data is considered "conditional" as their source must be considered as reliable. It is important at this stage to gather rather than question the credibility of the source. Data will be accepted from multiple sources in multiple formats. As is common in the policy sector, the technical means of data gathering have been improved; however, as we have seen, nothing replaces human agents. This is because technical sources can gather precise data at a very high level while human agents can gather *richer* data at lower levels. The use of human agents to gather data is important at this level. We must have adequate *technology* resources to *aid* the human agents *rather than replace* them. The use of wireless communication devices, personal computers, and advanced devices such as at *pervasive* computing mechanisms, and wearable computers, will all be essential here. These devices will help a human agent gather the data and store it immediately for later processing.

While in the past public agencies had to contend with basic data gathering processes, today, the nature of data collection is undergoing fundamental changes. For one, data today is not only being collected in real-time, but most of it is in collected electronically thereby making it easier to store, process, and retrieve. Yet, the increase in data availability has not been matched with an equal level of innovation in how data can be used to guide decision-making in the public sector. Data is still examined through historical analytical frames, thereby limiting the sophistication of insights that one can generate.

The situation is complicated when we think about who has access to data. While in the past, governments had a significant amount of control over the data they generated. Today, with the pressure for greater transparency, opening up government data for citizen access has become fashionable. Open data initiatives not only provide a level of transparency into government operations, but also allow for citizens to use the data creatively to generate technology applications to address local, state, and global challenges. Various entities, both public and private, have utilized prizes, specifically prize-driven competitions, as incentives to promote the leveraging of the data and the design of applications. Consider the case of the Open Data initiative in Vancouver that led to the creation of the Van Trash application (http://vantrash.ca). This is an example of how the presence of data can be leveraged to better public services.

Finally, we also need to consider the public's role in data generation. While in the past, the public was passive in terms of data generation (i.e., they contributed data but seldom generated data themselves). Today, citizens are taking an active role in building data collection platforms that promote the sharing of data by the public. Consider, for example, IPAB (I Paid A Bribe)—a corruption tracking application/website funded by the Bangalore, India based non-profit Janaagraha (meaning "Public Revolution") that aims at enabling public participation in public governance issues. IPAB solely depends on the information provided by citizens of India who come to the website to report any bribe they have (or have not) paid. The website provides tools like bribe analytics, discussion forums, annual and quarterly reports, and tools that help citizens avoid corruption. There are other tools like Corruption Quiz, Citizen Speak, or Citizen Charters that allow citizens to discuss their problem and become more aware about their rights against corruption. The IPAB website has turned out to be an effective tool in tracking corruption and making citizens aware about their rights against corruption. According to IPAB, by using the website, the government has become aware about corruption worth 382 million rupees ($8.5 million). The website reports and analytic tools have also pointed out 21 states (409 cities) in India from where corruption has been reported on the website. As this example illustrates, public agencies are not the sole providers or aggregators of data that impact governance.

Empirical

Data processing occurs at the *empirical* level. Here we are concerned with processing "within" a class or context of data. Once data is aggregated, it must be triangulated against data from other sources. At this stage, questions involving the credibility of the sources are considered. Although suspect data sources cannot be ignored, their relevance to the context of investigation is related to the mission task. It is common to find empirical processing of data occurring within the local police departments. At this empirical layer, the role and need for technology increases and the use of secure, common repositories are advantageous. Chicago, for instance, has developed the Citizen Law Enforcement Analysis and Enforcement (CLEAR) system to tie various databases related to criminal records and activities into a user-friendly scheme to assist police in tracking and managing information regarding crimes in and around the city. This allows, for example, police to aggregate data related to gangs across various districts for the purpose of developing operational strategies based on logistics that would otherwise require an unrealistic level of detail and cooperation between interdepartmental agencies. The technology capacity of the FBI lags what is available in most public schools (CNN, 2002).

Challenges of data analysis at the empirical layer can be attributed to the fact that today we have to merge multiple forms of data to make sense of an issue. Previously, a department or agency could have focused on number crunching (e.g., census bureaus). However, today, in addition to numeric data, the department has to contend with data coming in from user feeds, viral videos, audio files, geographic information, etc. Technologies have also allowed us to analyze data with greater precision under shorter timeframes. Therefore, the public continues to pressure for more instant (nearly real-time) analysis of decisions and outcomes. The days of batch-processing information or delayed feedback are numbered. Mobile technologies have also added a layer of sophistication in terms of data provision.

Syntactical

At the syntactical level, we *share* and processed data "across" a class. An agent might exchange data in his/her possession with peers or with contacts at other intelligence agencies. Most of the breakdowns in intelligence communications and sharing during the events leading to 9/11 occurred at this level. Data gathered from the various agencies were closely *held* and not made available to peer-institutions, which severely lowered the chance for effective analysis. At this level, data is shared outside of one's local unit. In the state of Illinois, the next phase involving CLEAR is intended to enable the sharing of criminal data across the various cities and counties. The key determinant of the viability of such systems is the existence of standards and trust. *Standards* are important for ease of data sharing among the various local units. Failure to have standards for data-capture, sharing, and transfer will leave data to reside in *silos* within the organization. While standards are the technical dimension, trust is the personal dimension. *Trust* among the local agencies both horizontally and vertically must be ensured. *Horizontal trust* is among agencies at the same level (e.g., law enforcement agencies between counties and cities). *Vertical trust* is among federal and local agencies (e.g., between the FBI and local law enforcement agencies). Appropriate incentives must be in place to manage the flow of data in a timely manner. Although, unfortunately, this consideration often fails due to certain political considerations. Due to financial and resource pressures, public agencies have begun to outsource their data analysis work. This complicates the situation by ensuring that the goals of the outsourcers are aligned with those of the public agencies, privacy and security requirements are being met, and public service expectations are attended to. Non-traditional alliances are being formed such as the use of private military contractors/corporations (PMCs) for work that was previously under the purview of a nation's armed forces. Similarly, within the intelligence community, a significant portion of the analytical work is being conducted by private organizations such as Booz Allen Hamilton. While in the past it was assumed that the private organization would simply process data based on the rules provided by the public agency, today, the public agencies *depend* on the private sector to arrive at methods to process data and even generate data they need to inform decision-making. Taking the systems of systems approach will allow one to study how various factors such as power, trust, and distance between partners in an alliance can impact outcomes around information generation and provision.

Semantic

It is at the semantic level where *sense* is made from data and information. Questions such as "what is the meaning of this transmission?" or "what are these groups up to?" come into play. The sense made at this level may be considered in terms of "knowledge creation." Each piece of data gathered, and processed through the empirical and syntactical stages, is now placed in "context." This context will include the past behavior of the object or persons of interest, conditions in the environment, including opportunity, and various elements or conditions of historical importance.

The generation of meanings from data and events continue to get more complicated. The simple reason for this is the need to engage multiple diverse perspectives under very short time cycles. Moreover, since the perspective one brings to a situation is related to the goals and objectives they seek to advance, getting a group to interpret information is seldom easy (especially when each individual has a different agenda). Hence, today, the abundance of information has increased the likelihood of constant debates and deliberations with no end in sight. Information is often interpreted in a manner that suites the agenda of the individual, and can be manipulated to advance a cause. We need to employ platforms that promote richer discussions among stakeholders that might be spread across geographies, time zones, and organizations. The good news is that we do have technologies that have these capabilities. One good example is our recent project that examined how a dynamic computer simulation and information visualization environment can help people improve their understanding of uncertainty and complexity of water challenges in Phoenix. We asked participants to interact with the simulation that incorporated multidimensional data regarding water supply and demand in the Phoenix metropolitan area. We asked them to deliberate on the water problems facing Phoenix, factors influencing water sustainability, goals, and plans to address water problems. In our project, we not only used computer simulation to provide the deliberation context, but also explored how information display environments may facilitate people's deliberation and interactions, and help them gain a better sense of what various data and information mean for water sustainability for Phoenix.

Pragmatic

At the final stage of the semiotic ladder, action based on knowledge is expected. As an extension of the activities involved in this process, some activity intended to modify the behavior of the organization is expected. The presence of an environment where *evidence* that emerges from thoughtful consideration of previous stages of the semiotic ladder is valued becomes critical to arrive at sound actions. Failure to engage with the evidence in a meaningful manner to deliberate and execute actions will render the previous stage useless and diminish the value of the organization. Actions may involve something as subtle as a policy change or, when required, may be stated more dramatically; actions are calibrated to meet the perceived conditions of reality.

> *The law enforcement/national security divide is especially significant, carved deeply into the topography of American government. The national security paradigm fosters aggressive, active intelligence gathering. It anticipates the threat before it arises and plans preventive action against suspect targets. In contrast, the law enforcement paradigm fosters reactions to information provided voluntarily, uses ex post facto arrests and trial governed by rules of evidence, and protects the rights of citizens.* (Carter, Deuthch, & Zelikow, 1998: 82)

The above statement is important for two reasons. First, it makes the distinction between how knowledge is obtained and used for intelligence-security purposes versus law enforcement purposes. Second, there is the implication that the purposes of both national security and law enforcement might be better served if this distinction were removed. Pragmatic complexities arise from a number of factors including the number of actors that might be operating within a given space, the coordination and collaboration demands of projects, etc. As we have embraced the concept of networked governance (or networked structures for governance) the challenges only increase compared to the traditional vertically-integrated bureaucratic structures (Desouza *et al.*, 2008). While before we had one actor conducting pre-determined actions, today, we have multiple actors whose action sets undergo changes on a regular basis. In addition, it is impossible to clearly demarcate the boundaries of each agency and for each agency to control the trajectory of their actions due to the number of intermediaries between the source and destination (the customer) of actions in a networked setting.

In moving up the semiotic levels, the concept of emergence is succinct. Emergence can be defined as the construction of bottom-up order. Rather than imposing a priori schemas onto the generation of information, information emerges through interactions at the morphological, empirical, syntactical, and semantic levels. Desouza and Hensgen (2003b) apply the semiotic model towards the study of cyber terrorism, stating

> *It is not possible to prevent attacks of empirical or semantic natures, without first curtailing issues at the morphological layer. The majority of the acts at the lower level of the semiotic discourse are centered on efficiency from swift and highly cost efficient attacks. Conversely, at the higher levels, effectiveness or severity of an attack is more critical because it entails the broader and widespread destruction of networks. Events leading to terrorist activities can be viewed in terms of a natural progression. As we move through the semiotic layers, from the morphological to the pragmatic, the scope of terrorist activities increases because their sophistication and knowledge base, related to the system, must increase. We must therefore study not only activity involving cyber terrorism at the various semiotic levels, but also how these acts emerge and move across levels…* (p. 390)

Likewise, many challenges of public administration are a combination of the interaction across the semiotic levels and as such, require approaches that are well suited for these challenges instead of trying to explain them away.

A COMPLEX ADAPTIVE SYSTEMS APPROACH TO PUBLIC ADMINISTRATION AND PUBLIC POLICY

A complex adaptive systems approach goes beyond the traditional qualitative and quantitative methods of conducting research in the field of public administration. Approaches that embrace complex adaptive systems allow us to study the behavior of agents, objects, information, and rules of interaction and environments while appreciating emergence and dynamism. Systems (complexity) thinking and tools have their strengths in exploring the intertwined and dynamic relationships among different parties in the complex policy systems (Kim *et al.*, 2011). In the following section, we will address what make a complexity theory approach unique and what a complexity theory approach can contribute to theory development and theory testing in the field of public administration and public policy. Studying complex adaptive social systems using techniques from the computational modeling space (e.g., agent-based modeling) can be particularly useful to advance public policy and administration research. Specifically, the use of dynamic, emergent, methodologies allows us to ask different kind of questions, some of which were not possible with traditional methods, and also seek out answers that are significantly more sophisticated and robust.

Raising New Questions

The use of agent-based models (ABM) forces us to articulate the *assumptions* we make about our research problems. Many studies applied regression-based techniques to analyze large sets of data. However, not all scholars carefully scrutinize the data to ensure that the assumptions for these techniques are upheld, e.g., the distribution of the data (Gill & Meier, 2000). AMB allows the researchers to test out how outcomes might change if certain assumptions are altered, re-considered, or even disregarded certain assumptions. As an example, consider the case of dismantling terrorist networks (Keller *et al.*, 2010). A popular assumption made by policymakers after 9/11 was that eliminating the heads of terrorist networks would lead to an eradication of the menace. This is an overly simplistic and linear thought. Through the use of agent-based modeling approach, Keller *et al.* (2010) were not only able to show the unintended and potentially more harmful, consequences of this strategy, but also were able to uncover latent insights through *visualizing emergent interactions*. For instance, while resources in a terrorist network are dispersed among various agents, they gradually converge to a few leaders as the network evolves. Therefore, over time as a terrorist group matures, it does not operate similar to a traditional hierarchy. Visualizing the emergent interactions allows the researchers to open up the proverbial black box and understand the mechanics that drive interactions among the components of their model. Continuing with the terrorist network study, Keller *et al.* (2010) were able to uncover that a leader-focused elimination strategy reaches its peak performance when the network is still growing. At this stage of the evolution of the terrorist group, a relatively large number of candidates compete for leadership roles, and, hence, are more likely to be caught; also, their eradication at this level is likely to deter others from ascending into these roles. However, once the terrorist network is mature, there are few leaders and they are well protected, making the leader-focused strategy less efficient and effective. These types of insights are an outcome of temporal slicing of data at fine granularities that are not possible with traditional research approaches. Agent-based methods allow researchers to follow the *interactions of agents and objects* from a detailed level through emergent behavior and to group, organizational, or environmental outcomes. As an example, agents can be endowed with a given set of resources and permitted to conduct a set of actions. It is through the interactions of agents, in conjunction with the global constraints of the environment in which they operate, that the overall global order of the system emerges. The end-state of the system can then be dissected to trace how the evolution took place, what were the critical inter-dependencies or tipping points that significantly altered the outcomes. These can then be further explored through experimentation.

Theory Development and Theory Testing

ABM can also be used for theory development and theory testing (Davis, Bingham, & Eisenhardt, 2007). For theory development, ABM offers the researchers a means to identify how interactions among the various constructs evolve in a dynamic context. Ideally, if ABM is used for theory construction, the theory being constructed should encompass some level of dynamism in terms of how the variables interact, and hopefully the value-added from the ABM employment points to the fact that these interactions are emergent and non-linear. In the context of theory testing, the use of simulation models provides us with a framework whereby we can study and test the dynamic evolution of interactions among a set of constructs. As noted by Desouza and Lin (2011), "computational modeling and experimentation is also powerful in exploring the impacts of different scenarios on the targeted social system. Since explicit specification of conditional parameters is always required for simulation experiments, policy makers can systematically study certain scenarios by setting them as model parameters" (p. 14).

Validation and Ethics in Modeling

Like ABM and other methods, complex adaptive systems methods need to be valid and reliable. When applying the simulation to studying complex social interactions, relationships and processes, scholars need to follow the basic rules and standard practice so that simulation models can be evaluated, replicated and extended further (Garson, 2009; Gilbert & Tioitzsch, 2005).

Yet, unlike other methods, validation is a more important step in developing simulations. Researchers make efforts not only to verify whether the models were programmed correctly, but also to go through different types and stages of verification (Garson, 2009; Gilbert & Tioitzsch, 2005). Researchers check whether the data, patterns, processes, and all other findings generated through the simulation models can correspond to the real-world scenario, and whether the simulation can be replicated in another similar context or a broader context. Validation is not a binary process but rather a continuous effort to increase the confidence of the user in the results of the model for a specific purpose (Johnston *et al.*, 2008). A model validated with regard to one specific dimension is not validated along all dimensions and it is the researchers' obligation to be clear to communicate which processes they have undertaken to validate which dimensions of the model.

Another important aspect in developing and using simulation models is ethical considerations. Ethics is a crucial topic in public administration research and praxis (Svara, 2007); however, ethics in simulation modeling receives little attention. Simulation is not only a useful tool for social science research, but also a novel medium for education and communicating with policymakers. The recent decades have seen the increasing use of simulation in many classrooms, libraries, and museums (Binstadt *et al.*, 2007). Various computer simulations have been used to inspire students' interests, encourage participation, and enhance student learning due to the real-time interactions and intuitive manner of information presentation in simulations. Compared with the large number of studies examining the effectiveness of simulation in improving learning outcomes, the study on unintentional outcomes of using simulation receive little attention. One caveat to this discussion is that dynamic simulation models may influence people's perceptions in ways that was unintended and unnoticed by the model builders.

Facilitating Conversations

Many practitioners and analysts describe the 21st century policymaking environment as increasingly evidence-averse. If true, then researchers face an increasing challenge to be aware of where discovery of a new perspective takes place. The long dominant model is that a researcher discovers truth and then shares it with those with power. However, the more efficient delivery system is to engineer environments where those in power make the "ah-ha" discoveries for themselves.

Participatory model building with practitioners uniquely served to not only advance research on the topic of interest, but also allow stakeholders to take part in the modeling process, which enables collective interaction and helps "create self-referential conditions whereby each user's decision modifies the other users' environment, and influences his or her own expected outcomes" (Barreteau *et al.*, 2007: 187). Additionally, many computational models also communicate with and inform practitioners through interaction and visualization. The ability to visualize the system contexts and the life cycle of phenemenon with ABM lead to various insights that would not have been possible through traditional analytical approaches (Johnston, Kim, & Ayyangar, 2007). First, policymakers and researchers could witness how early differences can propagate over time to lead to long-term outcome differences. Second, similar outcomes or phenomena that occur for a variety of reasons could be differentiated and understood. Third, once identified, strategies to mitigate such threats could be tested and compared.

For practitioners working in complex and networked environments, the CAS approach opens up the research process and develops a shared language through interactions with the model to better communicate its findings. Consequently, by interacting with the model, practitioners develop a forum to discuss how to address the challenges inherent in public administration. Barreteau, Le Page, and Perez (2007) argue that interactive simulations have immense potential to improve decision-making skills and helps participants to jointly explore the possibility space, building up a shared understanding of key concepts and searching creatively for solutions (Haung & Hutema, 2009). Role-playing through computer simulation can reduce the distance among players and are uniquely suited for raising stakeholder's awareness of complex issues and process and facilitate dialogue about difficult to experience conditions (Barreteau, Le Page, & Perez, 2007; Camargo, Jacobi, & Ducrot, 2007).

Perhaps more important than the how is why a computational approach consistently leads to changes in the conversation. Questioning and critical thinking within groups is necessary to develop the capacity for groups to be open to new ways of viewing the world. Allowing group participants to identify the relevant questions fosters an inquisitive environment. Rather than presenting the evidence, individuals create the evidence themselves. This is crucial for adopting new perspectives for the following reason:

Learning answers without learning questions produces a kind of ideology in which everything is already settled, in which contingencies appear as necessities, in which social constructs appear as natural inevitabilities, in which everything "is what it is" and nothing more or less. (Bowker, 2010)

Simulation modeling can provide an evidence-rich research context for questions raised during interactive dialogue. In terms of what leads to actual changes in policy, administration, and management of governance programs in programs we have used this approach with, we were able to trace more changes back to the conversations about the model than from the practitioners reviewing our peer-reviewed findings (Kim *et al.*, 2011). In the former, the conversation was coproduced by all participants as they interacted and created the model; in the latter the practitioners were reading about what someone else discovered. It is our experience then that a policy informatics approach can be used to change the conversation by engineering the conditions for ah-ha moments to be discovered by those in power.

CONCLUSION

To effectively navigate the issues and problems of contemporary public administration and public policy, we need to broaden the conversation beyond conventional research approaches rooted in positivist methodology. In this paper, we introduced a complex adaptive systems approach to public administration and public policy research. This approach allows for a relaxation of the "rationality" assumptions of human behaviors, and critiqued the assumed linear relationships between factors and static status of social systems. This approach allows studying emergent phenomena that result from the interactions of human behaviors and dynamic social systems. To help understand the nature of complex systems, we also introduce the study of semiotics to explain how information emerges from the raw data and from the interactions among data, human agents, and the environment. Finally, we highlight how participatory model building creates an environment where the actionable discoveries result as a byproduct of participating in the model building process itself. Overall, we are not arguing to replace the conventional research methods in our field. Instead, we think the complexity theory approach can complement the existing research frameworks and help researchers to exploit new connections between variables, to raise new questions, and to explore innovative approaches to governance and management while including practitioners in the production of knowledge.

REFERENCES

Ackoff, R.L. (1967). Management misinformation systems. *Management Science,* 14(4): 147-156.

Adams, W., Brockington, D., Dyson, J., and Vira, B. (2003). Managing tragedies: Understanding conflict over common pool resources. *Science,* 302(5652): 1915-1916.

Amy, D.J. (1984). Toward a post-positivist policy analysis. *Policy Studies Journal,* 13(1): 210-211.

Ascher, W. (1987). Editorial: Policy sciences and the economic approach in a post-positivist era. *Policy Sciences,* 20(1): 3-9.

Barreteau, O., Le Page, C., and Perez, P. (2007). Contribution of simulation and gaming to natural resource management issues: An introduction. *Simulation Gaming*, 38(2): 185-194.

Binstadt, E. S., Walls, R. M., White, *et al.* (2007). A comprehensive medical simulation education curriculum for emergency medicine residents. *Annals of Emergency Medicine,* 49(4): 495-504.

Bowker, M. H. (2010). Teaching students to ask questions instead of answering them. *The NEA Higher Education Journal,* Fall 2010**,** 127-134.

Camargo, M. E., Jacobi, P. R., & Ducrot, R. (2007). Role-playing games for capacity building in water and land management: Some Brazilian experiences. *Simulation & Gaming,* 38(4): 472-493.

Carter, A. B., Deutch, J. M., & Zelikow, P. D. (1998). Catastrophic terrorism: Tackling the new danger. *Foreign Affairs, 77*(6): 80-94.

Catlaw, T., & Hu, Q. (2009). Legitimacy and public administration: Constructing the American bureaucratic fields. *American Behavioral Scientist,* 53(3):, 458-481.

Dahl, R. (1946). The science of public administration: Three problems. *Public Administration Review,* 7(1): 1-11.

Davis, J. P., Bingham, C. B., & Eisenhardt, K. M. (2007). Developing theory through simulation methods. *Academy of Management Review,* 32(2)*:* 480-499.

DeLeon, P. (1994). Democracy and the policy sciences: Aspirations and operations. *Policy Studies Journal*, 22(2):, 200-212.

Denhardt, R. (2002). *Theories of public organization* (5th ed.). Boston, MA: Wadsworth.

Denhardt, R. B., Denhardt, J. V., & Aristigueta, M. P. (2002). *Managing human behavior in public and non-profit organizations*, Thousand Oaks, CA: Sage Publications.

Desouza, K. C., and Hensgen, T. (2002). On "information" in organizations: An emergent information theory and semiotic framework. *Emergence,* 4(3): 95-114.

Desouza, K. C., and Hensgen, T. (2003a). Every citizen a missile: The need for an emergent systems approach for law enforcement. *Government Information Quarterly,* 20(3): 259-280.

Desouza, K. C. and Hensgen, T. (2003b). A semiotic emergent framework for addressing the reality of cyber terrorism. *Technology Forecasting and Social Change*, 70(4): 385-396.

Desouza, K. C., and Hensgen, T. (2005). *Managing information in complex organizations: Semiotics and signals, complexity and chaos*. Armonk, NY: M.E. Sharpe, Inc.,

Desouza, K. C., and Lin, Y. A. (2011). Towards evidence-driven policy design: Complex adaptive systems and computational modeling. *The Innovation Journal: The Public Sector Innovation Journal*, 16(1): Article 7.

The Economist. (2010, July 22.) Agents of change. *The Economist*. http://www.economist.com/node/16636121.

Frederickson, H. (1996). Comparing the reinventing government movement with the new public administration. *Public Administration Review*, 56(3): 263-270.

Garson, G. D. (2009). Computerized simulation in the social sciences. *Simulation & Gaming*, 40(2): 267-279.

Gilbert, N., & Tioitzsch, K. G. (2005). *Simulation for the social scientist* (2nd ed.). Columbus, OH: McGraw-Hill Education.

Gill, J and Meier, Kenneth. (2000). Public administration research and practice: A methodological manifesto. *Journal of Public Administration Research and Theory*, 10(1): 157-200.

Haung, C., & Hutema, D. (2009). Learning through games? Evaluating the learning effect of a policy exercise on European climate policy. *Journal of Air Transport Management*, 17(3): 199-205.

Harvey, D. (2007). *A brief history of neoliberalism*, New York, NY: Oxford University Press.

Heineman, R. A., Bluhm, W. T., Peterson, S. A., & Kearny, E. N. (1990). *The world of the policy analyst*. Chatham, NJ: Chatham House.

Hood, C. (1991). A public management for all seasons? *Public Administration*, 69(1): 3-19.

Johnston, E. (2010). Governance Infrastructures in 2020. *Public Administration Review*, 70(1): s122-s128.

Johnston, E., Nan, N., Hicks, D., & Zhong, W. (2008). Between implementation and outcomes, growth matters: Validating an agent-based modeling approach for understanding collaboration process management. *The Innovation Journal: The Public Sector Innovation Journal (A special issue on Complexity and Innovation)*, 13(3), Article 7.

Johnston, E., Quigley, P., & Kim, Y. (2010) Crossing the border: Exploring complexity methods in establishing international borders and other evidence-based policy decisions. In K. Martinás, D. Matika, A Srbljinović (Eds), Complex societal dynamics: Security challenges and opportunities. Amsterdam, Netherlands: IOS Press.

Johnston, E., Kim, Y., and Ayyangar, M. (2007). Intending the Unintended: The Act of Building Agent-based Models as a Regular Source of Knowledge Generation. *Interdisciplinary Description of Complex Systems, Special issue on Poverty and Human Development*. 5(2): 81-91.

Kahneman, D. and Tversky, A. (1979). Prospect theory: An analysis of decision under risk. *Econometrica,* 47(2): 263-292.

Keller, J. P., Desouza, K. C., and Lin, Y. A. (2010). Dismantling terrorist networks: Evaluating strategic options using agent-based modeling. *Technology Forecasting and Social Change,* 77(7): 1014-1036.

Kim, Y., Johnston, E., & Kang, H. S. (2011). A computational approach to managing performance dynamics of networked governance systems. *Public Performance & Management Review,* 34(4): 580-597.

Kjaer, A. (2004). *Governance.* London: Polity Press.

Koppell, J. (2010). Administration without borders. *Public Administration Review,* 70(1): 46-55.

Lasswell, H.D. (1955). Current studies of the decision process: Automation versus creativity. *The Western Political Quarterly* 8, 381-399.

McSwite, O. C. (1997). *Legitimacy in public administration: A discourse analysis.* Thousand Oaks, CA: Sage Publications.

Morçöl, G. and Ivanova, N. P. (2010). Methods taught in public policy programs: Are quantitative methods still prevalent? *Journal of Public Administration Education,* 16(2): 255-277.

Newland, C.A. (1994). Foreword. In *Research in public administration:Reflections on theory and practice,* edited by Jay D. White and Guy B. Adams (pp. ix-xii). Thousands Oaks, CA: Sage Publications.

Osborne, D. and Gaebler, T. (1992). *Reinventing government: How the entrepreneurial spirit is transforming the public sector.* Reading, MA: Addison-Wesley Publishing Company, Inc.

Page, S.E. (2006). *The difference: How the power of diversity creates better groups, firms, schools, and societies.* Princeton, NJ: Princeton University Press.

Peters, B.G. and Pierre, J. (1998). Governance without government? Rethinking public administration. *Journal of Public Administration Research and Theory,* 8(2): 223-243.

Ramaprasad, A., & Rai, A. (1996). Envisioning management of information, omega. *The International Journal of Management Science,* 24(2): 179-193.

Rhodes, R. A. W. (1996). The new governance: Governing without government. *Political Studies,* 44(4): 656-667.

Sandberg, B. (2011). *Problematizing service in the nonprofit sector: From methodless enthusiasm to professionalization* (dissertation). Arizona State University, Phoenix, Arizona.

Svara, J. H. (ed.). (2007). *The ethics primer for public administrators in government and nonprofit organizations*. Sudbury, MA: Jones & Bartlett Learning.

Shannon, C. E., & Weaver, W. (1949). *A mathematical model of communication*. Urbana: University of Illinois Press.

Simon, H. A. (1946). The proverbs of administration. *Public Administration Review,* 6(1)*:* 53-67.

Simon, H. A. (1955). A behavioral model of rational choice. *The Quarterly Journal of Economics,* 69(1)*:* 99-118.

Simon, H. A. (1978). Rationality as process and as product of thought. *The American Economic Review,* 68(2)*:* 1-16.

Stone, D. (2002). *Policy paradox: The art of political decision making* (Rev. ed.). New York: W.W. Norton.

Waldo, D. (1948/2007). *The administrative state: A study of the political theory of American public administration.* City?, NJ: Transaction Publishers.

Wilson, W. (1887). The study of administration. *Political Science Quarterly,* 2(2): 197-222.

White, J. D. (1986). On the growth of knowledge in public administration. *Public Administration Review,* 46(1)*:* 15-24.

Erik W. Johnston is an associate professor in the School of Public Affairs at Arizona State University and the director of the Center for Policy Informatics. His research looks at how technology enables more sophisticated policy analysis and governance infrastructures, and promotes collective decision making. Dr. Johnston earned a PhD in Information and a Certificate in Complex Systems from the University of Michigan. He is a two-time NSF IGERT fellow, in the Socio-Technical Infrastructure for Electronic Transactions (STIET) and Institutions, Diversity, Emergence, Adaptation, and Structures (IDEAS) programs.

Kevin C. Desouza serves as the Associate Dean for Research at the College of Public Programs (COPP) and is an associate professor in the School of Public Affairs at Arizona State University. Immediately prior to joining ASU, he directed the Metropolitan Institute in the College of Architecture an Urban Studies and served as an associate professor at the Center for Public Administration and Policy within the School of Public and International Affairs at Virginia Tech. From 2005-2011, he was on the faculty of the University of Washington (UW) Information School and held adjunct appointments in the UW's College of Engineering and at the Daniel J. Evans School of Public Affairs.

At UW, he co-founded and directed the Institute for Innovation in Information Management (I3M); founded the Institute for National Security Education and Research, an inter-disciplinary, university-wide initiative, in August 2006 and served as its director until February 2008; and was an affiliate faculty member of the Center for American Politics and Public Policy. He holds a visiting professorship at the Faculty of Economics, University of Ljubljana. He has held visiting positions at the Center for International Studies at the London School of Economics and Political Science, the University of the Witwatersrand in South Africa, the Groupe Sup de Co Montpellier (GSCM) Business School in France, and the Accenture Institute for High Business Performance in Cambridge, Massachusetts (USA). Desouza has authored, co-authored, and/or edited nine books and has published more than 125 articles in prestigious practitioner and academic journals. His work has also been featured by a number of publications such as Sloan Management Review, Harvard Business Review, Businessweek, Washington Internet Daily, Computerworld, KM Review, Government Health IT, and Human Resource Management International Digest. Desouza has received over $1.4 million in research funding from both private and government organizations. For more information, please visit: http://www.kevindesouza.net.

Qian Hu is an assistant professor in School of Public Administration at University of Central Florida. Qian received her PhD from Arizona State University in 2011. Her research interests include collaborative public management, e-governance, policy informatics, and strategic and performance management. Qian's work has been published in academic journals such as American Behavioral Scientist, Journal of Public Affairs Education, and Journal of Nanoparticle Research.

Rashmi Krishnamurthy is a doctoral student in the School of Public Affairs at Arizona State University. Her research is focused on the role of information technology in collaborative decision making. Her current project includes application of virtual reality technology in understanding the determinants of collaborative decision making under conditions of uncertainty as they relate to water scarcity issues.

Chapter 5

ASSESSING THE APPLICABILITY OF QUALITATIVE COMPARATIVE ANALYSIS FOR THE EVALUATION OF COMPLEX PROJECTS

Stefan Verweij (Erasmus University Rotterdam, NLD)
Lasse M. Gerrits (Erasmus University Rotterdam, NLD)

This paper introduces and assesses Qualitative Comparative Analysis (QCA) as a complexity-informed framework for evaluating complex infrastructure projects coupled with area development. This is done in four steps. First, the properties of infrastructure and area development are discussed. When combined with a complexity perspective, it follows that the context of a project is explanatory for its outcomes. Secondly, prerequisites for a evaluation framework following this point of departure are developed. Thirdly, common infrastructure evaluation methods are assessed against these requisites. It follows that a comparative case-based approach is most suitable to study the relationships between context and outcomes in projects. Fourthly, fuzzy set Qualitative Comparative Analysis (fsQCA) is introduced and assessed related to the developed requisites. The paper concludes with a discussion of the further development of fsQCA.

ISSUES IN ASSESSING TRANSPORTATION INFRASTRUCTURE PROJECTS

Transportation infrastructure projects are characterized by a lingering history of budget overruns, time delays and public resistance. The Dutch Elverding Committee concluded that decision making on infrastructure development in the Netherlands takes 11 years on average (Advisory Committee VBI, 2008), and Flyvbjerg *et al.* (2003a) found that cost overruns above 40 percent are common and that cost overruns above 80 percent are not uncommon. A European study more than a decade earlier showed similar results (WRR, 1994). Already in the 1970s and 1980s the effectiveness, efficiency and legitimacy of infrastructure development was heavily debated (De Hoo, 1982). This indicates that the complexity underlying infrastructure development is persistent and that there are no quick fixes to the issues associated with such development. This instigated a continuous analytical and professional search for a better understanding of infrastructure development, both in terms of causes and solutions. For instance, the Netherlands Institute for Transport Policy Analysis (KiM) strongly advocates the need for more and better infrastructure development evaluations (especially ex-post) emphasizing the need for a better understanding of why projects perform as they do (KiM, 2009). However, as also noted by Flyvbjerg *et al.* (2003a), it is more difficult to assess 'what' causes than 'that' causes. Hence, evaluation tends to be focused on comparing before and after, often inadequately incorporating the influence of contextual local conditions to infrastructure projects (Sanderson, 2000). According to Sanderson (2000), dominant evaluation methodologies impede policy learning since these account insufficiently for the complex nature of the policy system. Similar issues are raised by KiM, which states that ex-post evaluations are not just performed marginally, but that current evaluations are subject to methodological deficiencies (KiM, 2009; PBL & KiM, 2010), which can be related to the misfit between the way infrastructure development projects are understood and the methodologies used to evaluate them. Consequently, there is a need for evaluation methodologies that do justice to the complex nature of infrastructure projects.

In-depth case studies can pay ample attention to the contextual nature of such projects. However, evaluation aims to improve present infrastructure development practice and this requires transferability of lessons learned, and focusing on one particular context of one particular case does not allow analyzing patterns that may reoccur over different cases. At this point, the requirements of contextuality and generalization seem to be at loggerheads with each other, as will be shown in the next section. On the one hand, current dominant methods often fall short in terms of discounting local conditions in cross-case comparative studies. On the other hand, generalization is difficult from in-depth case studies. This paper introduces and assesses Qualitative Comparative Analysis (QCA) as a research approach for evaluating complex infrastructure projects that aims to reconcile the above dilemma. It combines a focus on generic patterns that are researched in variable-oriented studies with a focus on idiosyncratic events researched in case-based studies. However, before presenting this method, it is first necessary to elaborate our understanding of complex infrastructure projects, since establishing a fit between this and the method presented is pivotal. Hence, this article aims to answer the following question: what are the ontological, epistemological and methodological components of a method for assessing how both idiosyncratic events and recurring patterns influence transportation infrastructure development projects, and what should a coherent evaluation method based on these components consist of?

GENERALIZATION VERSUS CONTEXTUALITY IN TRANSPORT INFRASTRUCTURE STUDIES

Flyvbjerg *et al.* (2003a; see also Flyvbjerg *et al.,* 2002, 2003b, 2004, 2005; Flyvbjerg, 2007b, 2009; Næss *et al.*, 2006) performed a large quantitative comparative study of 258 large infrastructure projects in 20 different countries. The study revealed a number of patterns. For instance, actual costs are on average 45 percent higher than estimated, cost overrun is a global phenomenon and cost performance has not improved over the past 70 years. Although this study and its findings are of indisputable importance in accounting for the 'that' of infrastructure development performance, it cannot, as stated by Flyvbjerg *et al.* (2003a), explain the 'what' of cost overrun. Albeit they discern some more specific patterns by distinguishing, among others, between rail, road and fixed link projects (and between different types within these categories), this large-N study, by its very nature, cannot account in great depth for the influence of local conditions on cost performance. For example, Flyvbjerg *et al.* (2003a: 19) state that:

> *For the Channel Tunnel, changed safety requirements were a main cause of overrun. For the Great Belt link, environmental concerns and accidents with*

flooding and a devastating fire made the budget balloon. For the Øresund link, it proved more costly than estimated to carve major new transport infrastructure into densely populated Copenhagen, and so on.

In general, the study points to many similarities amongst projects in terms of their performance. For instance, the Øresund link and Channel Tunnel are quite similar regarding the percentage of cost overrun (i.e., respectively 70 and 80 percent) (Flyvbjerg *et al.*, 2003a). However, the causal paths leading to those results are different. An ex-post evaluation performed by Anguera (2006) into the cost performance of the Channel Tunnel reveals that, in addition to the findings of Flyvbjerg *et al.*, it is a combination of factors leading to cost overrun, for instance: a clear project owner was absent from the outset, the unforeseen advent of low cost airlines led to reduced ridership, political events between the French and British governments, difficult ground conditions, transport related incidents such as the PanAm crash at Lockerbie, and so forth. Obviously, although similarities undoubtedly exist, such specific conditions often cannot account for the cost overrun for the Øresund link project.

This brief example illustrates the tension between generic patterns and contextuality. Large-n quantitative studies such as those performed by Flyvbjerg *et al.* provide lasting insights into generic patterns such as cost development. The downside is that these studies do not allow analyzing the idiosyncratic nature of such projects in detail, while specific events may significantly contribute to project outcomes. Studies such as those performed by Anguera are case-oriented since these provide important insights in the unique and idiosyncratic nature of individual projects. However, these case-oriented researches have little to contribute in terms of patterns recurring in different cases. By their very nature, both types of studies thus have their merits and demerits. Table 1 provides an illustrative list, excluding the above mentioned references by Flyvbjerg *et al.*, of transportation infrastructure development studies that fit the above distinction.

Understanding Complex Infrastructure Projects

If projects are called complex it usually means that they are perceived as difficult. However, the word 'complex' is in fact a multi-layered concept that needs further elaboration. In abstract terms, developing infrastructure means performing modifications to an existing system. Any built area is a system because it consists of interacting three-dimensional units (e.g., rooms,

Variable-oriented	Focus on generic patterns	Hecht & Niemeier (2002); Hsieh *et al.* (2004); Irfan *et al.* (2011); Kaliba *et al.* (2009); Lee (2008); Magnussen & Olsson (2006); Manavazhi & Adhikari (2002); Odeck (2004); Polydoropoulou & Roumboutsos (2009); Welde & Odeck (2011); Yang (2007).
Case-oriented	More attention to idiosyncratic events	Anguera (2006); Cantarelli *et al.* (2010); Han *et al.* (2009); Peters (2010); Priemus (2007); Rigden (1983); Van Marrewijk *et al.* (2008); Walter & Scholz (2007).

Table 1 *Variable- And Case-Oriented Studies In Transport Infrastructure Development*

buildings, assemblages of buildings), two dimensional units (i.e., the layout or distribution of the three-dimensional units), and linear units or transport networks linking the three-dimensional units thereby largely determining the area's layout. Together, these form an urban syntax that is specific to any given area (Marshall, 2009). While there are some things that are often in common in many areas (e.g., where the main railway station is located in relation to the town center), there are also many differences that are particular to any local situation (De Roo & Schwartz, 2001). An infrastructure developer wishing to change something to a specific situation, therefore, deals with an urban syntax that is a mixture of generic elements (i.e., elements that also appear elsewhere) and specific elements (i.e., elements that only appear in this local situation). Over time, the interaction of generic elements with specific elements leads to a local situation that is unique in its particularity, even though it still has recognizable elements (Byrne, 1998, 2001, 2003, 2005; Marshall, 2009). The complexity of a built area is further increased if one takes into account that there exists social tissue, i.e., individuals and social groups living, working, traveling and recreating in any given area. The urban syntax and its social components are nested in the way that properties of the subunit (e.g., a street) reflect the properties of the whole (e.g., a district) but not to the extent that both levels are exact copies (Marshall, 2009). Thus, the built order that exists locally is emerging from the specific local conditions. As a consequence, infrastructure developers have to deal with a unique local area that nonetheless exhibits similarities with other areas. Developing infrastructure and built areas is, therefore, not a matter of applying generic planning or managerial rules to that area, because such application would miss out the specific local conditions. It is pivotal for the development of infrastructures in built areas that this specific pattern of local conditions and generic developments is researched and understood. Indeed, these local conditions often play a significant role in the development of infrastructures. For instance, certain ground conditions, residential areas or nature areas affect the development and outcomes of projects.

In sum, this perspective on the development of built areas focuses on a number of dimensions. First, transportation infrastructure development takes place within a specific interacting mix of local conditions (e.g., difficult ground conditions) and generic patterns (e.g., economic crisis) that occurs in any given location. Second, this points to the fact that the causal relationships between site-specific conditions and generic developments and the elements within are poorly known and, if known, only for that specific time and place. Thus, known causal relationships specific to a certain area are by definition case-specific. Indeed, the unique nature of built systems implies that other systems are constituted differently, although the emerged order can be quite similar. Third, the emergent nature of any built area implies that it is the result of longitudinal development. That is, it is the result of past changes and events that are to some extent path-dependent. Taking these points together, this paper understands built areas as complex systems (Batty, 2010). The next question then is: how should this complexity as such be named, understood and researched? This triple question is answered in the following section, leading to the formulation of requisites for complexity-informed evaluation.

Foundations for Understanding and Researching Situated Complexity

There are two basic ways of how complexity can be characterized. Byrne (2005) and Cilliers (2001), among others, make a distinction between simplistic or generic versus complex or situated complexity. Generic complexity focuses on the emergence of complex processes and structures from a limited set of variables. It assumes a general set of rules from which emergent complexity flows (Buijs *et al.*, 2009). Although elegant, this approach does not do justice to the types of project evaluation for which this methodology is being developed. Built areas as systems are open by definition (De Roo, 2010), meaning that their composition and behavior is constituted through interaction with their environment, resulting, as stated above, in their specific local (i.e., situated) mixture of generic and specific elements. So, any approach to understanding them requires that they are treated as such, which implies a situated complexity approach (Gerrits, 2011). The premise of built areas as open systems assumes that an explanation for the development (or lack thereof) of a project can be found in its contextuality, i.e., that the local conditions hold explanatory variables (cf. Mjøset, 2009). Therefore, Buijs *et al.* (2009) use the denominator 'situated complexity' to focus on the explanatory value of the contextuality of a phenomenon. Although some argue that a research methodology should start from either the generic or situated approach (e.g., Bar-Yam, 1997), Buijs *et al.* (2009) argue that a case can and should be made for a systematic in-depth comparison across systems. The idea behind this is that (open) systems "do not operate according to general rules applied in all contexts" (2009: 37), but that nevertheless a systematic comparison can reveal both differences and similarities between the operations of different systems. Situated complexity focuses on both recurring patterns over multiple systems and idiosyncratic events that are local to a particular system, that both determine how systems develop over time. The research methodology presented in this paper starts from that premise.

The second part of the triple question concerns the way complexity is understood, which is basically a question of how reality can be understood. The classical divide is between positivism and postpositivism. To some extent, this divide coincides with the difference between generic and situated complexity. Positivism is primarily concerned with determining general rules in reality by taking it apart in discrete components, which coincides with the aims of the research program into general complexity. However, postpositivism has many different sub-strands that range from the extreme relativism of social constructivism to the more realist thesis of negotiated subjectivism (Byrne, 2003; Haynes, 2001; Uprichard & Byrne, 2006) or critical realism (Guba & Lincoln, 1989). The common theme within those strands is that the contextuality is explanatory for what is being observed, which coincides with situated complexity.

The fact-value dichotomy that underlies the positivist stance has been thoroughly undermined (Bateson, 1984; Byrne, 2002; Fischer & Forester, 1987; 1993). Complex causality is always subject to interpretation and consequently debatable as every interpretation carries with it normative judgments (Williams, 2009). If it is accepted that systems are open, than it follows that systems' boundaries do not exist a priori but that any individual will develop a particular demarcation or set of boundary judgments about the system which includes and excludes variables (i.e., reducing real complexity) that may be connected but not perceived as such by the observer (Cilliers, 2001). In other words, there is no unambiguous separation between systems and their context. Situated complexity is therefore not confined to presupposed systems' demarcations but intersects all system representations by respondents. In other words, the observer is as much part of the complexity as the system or agents that are observed. This reduction of the real complexity is both compromising the research and inevitable in keeping it manageable (Cilliers, 2001, 2005). It requires that multiple perspectives on a particular system are taken into account. If a multitude of observers can develop a multitude of boundary judgments about what is taken into account or not with a multitude of perspectives about how something is being perceived, chances are that a larger part of the complexity of the system is captured (Cilliers, 1998, 2005). This type of thinking implies a convergence of the fact-value dichotomy. However, it does not imply that the postmodernist stance, where subjective storytelling is all that remains, is carried over in this position. It means that it is understood that explanation is possible, as long as it is accepted that it is local in time and place (Byrne, 2001, 2003, 2005; Morçöl, 2001). Although specific to a given locality, cause and effect relations do exist and can be known through respondents' perception (i.e., it is agent-bound). Causality can still be determined in terms of change and response that render certain effects that can be observed (cf. Bryman, 2004; Hammersley, 2008, 2009). The ontological point of departure in this paper is therefore complex realism (cf. Byrne, 2002; Harvey, 2009; Reed & Harvey, 1992).

So, how should this complexity be researched? As situated complexity from the perspective of complex realism is the point of departure in this paper, the adherent methodology should focus on both recurring patterns over multiple systems and on system's peculiarities. Therefore, the first criterion for a complexity-informed evaluation framework is [1] the extent to which it manages to balance between in-depth understanding and reductionist generalization. Second, since studying situated complexity requires first an in-depth understanding of cases, [2] the method has to be case-based. Moreover, since single case studies cannot be employed for explanatory purposes in other cases and, therefore, do not allow statements about patterns across systems, a comparative multiple-case study approach is required. Third, as stated above, situated complexity focuses on the explanatory value of contextuality. Indeed, complex systems are open systems; generic patterns and specific events interact and coevolve. Therefore, it can be inferred that, in methodological terms, [3] the method should allow observation and analysis of complex interaction between variables. This is elaborated in the following sections. Finally, as argued above, since complex systems are anything but static (i.e., a built area is the result of longitudinal development), [4] the method has to be able to account for the influence of time on the constitution of situated complexity (i.e., complex dynamics).

Towards a Complexity Informed Case-Comparative Framework

The requisites mentioned in the previous section are covered only partially by the two general evaluation approaches presented before. Variable-oriented studies such as those performed by Flyvbjerg *et al.* (2003a) examine relationships between more general features of infrastructure projects. These features are conceived as variables (e.g., project type, topography and cost overrun) and the correlations between them are tested. Hereby, it is possible to deduce empirical generalizations about structural processes, relevant to a larger number of cases (Ragin, 1987). For instance, cost underestimation and overrun are higher in rail projects and within the set of rail projects, overrun in developing nations is more pronounced than in North America and Europe (Flyvbjerg, 2007a). "However, the simplifying assumptions that make this approach possible often violate commonsense notions of causation and sometimes pose serious obstacles to making interpretative statements about specific cases or even about categories of cases" (Ragin, 1987: xiii). That is, the study cannot account for the idiosyncratic nature of specific cases, as the study of Anguera (2006) can. Thus, variable-oriented studies are by their very nature not case-oriented. Furthermore, correlational methods are not equipped to deal with contextuality; correlational methods do not allow for complex causality. For example, in the research performed by Flyvbjerg *et al.*, the importance of the context is recognized by pointing to the fact that cost overruns are due to different circumstances (cf. citation in second section). However, this is not reflected in the generic patterns that appear in the research. Finally, variable-oriented studies can account for time. For instance, Flyvbjerg *et al.* (2003a) conclude that infrastructure development performance has not improved in the past 70 years. However, such studies have a hard time including complex dynamics, such as the influence of particular events (e.g., the crash at Lockerbie) on the course of developments in a specific case.

Case-based methods, however, are by their nature sensitive to the complexity, diversity and (historical) uniqueness of cases (Ragin, 1987). Projects are treated holistically and not as collections of parts; case studies are sensitive to contextuality and temporality. For instance, Anguera (2006) is able to discuss in detail the Channel Tunnel project, its performance, and the influential factors thereon just because of the focus on a single case study (see Section 2). However, in the words of Aus, "most case studies (...) could maliciously be qualified as theoretical 'data dumps'. One of the methodological reasons for this rather unfortunate state-of-the-art is that single case studies can hardly be employed for explanatory purposes" (2009: 175). Hence, the methodology needs to be case-comparative to allow for causal inference (Aus, 2009), for studying patterns across cases. However, when case study material is analyzed and compared in case-comparative studies, this often happens rather loosely and non-formalized (Rihoux, 2006). Often, comparative studies, necessarily limited in the number of cases, result in an overview of the most important similarities and differences. This comparative process is often not formalized in the sense that little insight is and can be provided into the way it was performed, including decisions that were made influencing the outcome of the comparative process. The rich data represents many possible causal conditions that are often hard to disentangle and there is a danger of introducing possible biases. Consequently, the scientific value of these studies is often questioned (Ragin, 1987). As Table 2 below summarizes, both approaches have their strengths and weaknesses.

Requirements	Both in-depth and generalization	Both comparative and case-based	Contextuality	Temporal
Variable-oriented	Generic patterns	Comparative, not case-based	At most limited	At most limited
Case-oriented	In-depth case peculiarities	Not comparative, case-based	Yes	Yes

Table 2 *Variable-Oriented Versus Case-Oriented Approach*

In the next section we discuss fsQCA as a hybrid third research approach that aims to integrate the strengths of both the above approaches, thereby mitigating their weaknesses.

QUALITATIVE COMPARATIVE ANALYSIS AND FUZZY SETS

Qualitative Comparative Analysis (QCA)—an umbrella term that captures three different types of comparative methods (Rihoux & Ragin, 2009), namely crisp set QCA (csQCA) (Rihoux & De Meur, 2009), multi-value QCA (mvQCA) (Cronqvist & Berg-Schlosser, 2009) and fuzzy set

QCA (Ragin, 2009)—aims to integrate the case-oriented and variable-oriented approaches and argues that scientists do not have to choose between the "understanding of complexity and knowledge of generality" (Ragin *et al.*, 2003: 324). QCA can be used to "achieve a systematic comparison across a smaller number of individual cases (e.g., a sample of between 10 and 30 cases) in order to preserve complexity, and yet being as parsimonious as possible and illuminating otherwise often hidden causal paths on a micro level" (Rihoux & Lobe, 2009: 228). It is a comparative case-based approach that allows for the examination of multiple causal configurations (Byrne, 2009). This configurational approach implies first that most often *combinations* of causal conditions (i.e., variables) produce a certain outcome. For instance, Anguera (2006) presents several factors that together affected cost overrun in the Channel Tunnel project. Second, it implies that several *different combinations* may produce the outcome. For instance, the Øresund link and the Channel Tunnel have similar outcomes, but their paths towards that outcome are different. Third, it implies that certain factors can have different effects in *different contexts*. For instance, the effects of the advent of low cost airlines differ between the Channel Tunnel and the Øresund link due to their different contexts and natures. Grofman & Schneider (2009) and Schneider & Wagemann (2010) refer to these characteristics of complex causality as respectively conjunctural causation, equifinality and multifinality. They add to this the notion of asymmetric causality, i.e., that the presence and absence of outcomes require different explanations. In sum, the approach ticks all of the boxes in Table 2, excluding, as discussed later, the last one concerning temporality.

It is important to clarify that QCA is not just a method; it is first of all a research approach (Rihoux, 2003). Central to this approach is the dialogue between theoretical ideas and empirical evidence (Berg-Schlosser & De Meur, 2009; Berg-Schlosser *et al.*, 2009; Ragin, 1987, 2000, 2008a; Yamasaki & Rihoux, 2009). In our view this means that this research approach is neither purely inductive nor deductive, but essentially a two-way-street whereby (inductive) case-based empirical data is ordered with the help of theory for the selection and construction of cases and variables. In QCA, variables are reframed as causal conditions or sets. Set theory is a "mathematical calculus for dealing with collections of objects and certain relationships among these objects. At its most basic, a set is simply a list of objects" (Smithson & Verkuilen, 2006: 4). For instance, the cases Channel Tunnel, Øresund link and Great Belt link, are all members of the set 'transportation infrastructure projects'. What makes set theory interesting for case comparative studies is that sets can be intersected (i.e., the set operator 'logical and'—referring to conjunctural causation) and unified (i.e., the set operator 'logical or'—referring to equifinality and multifinality). QCA is able to systematically compare and analyze these conjunctions of sets (i.e., variables in different contexts, also known as causal recipes) (Ragin, 1987, 2000, 2008a; Smithson & Verkuilen, 2006). Thus, contextuality is deemed as explanatory for case outcomes. However, social phenomena such as infrastructure development are often difficult to grasp in terms of sets. For instance, consider defining set boundaries of concepts such as 'democracy' and 'legitimacy' as opposed to defining the set of 'transport modes'. Producing a list of projects that fit a certain type of transport mode is probably easier than making a list of legitimate public participation processes in infrastructure development projects (cf. Smithson & Verkuilen, 2006). Therefore, it is important to use theoretical and substantive knowledge to substantiate the construction of sets.

Using set theory implies a focus on set relations instead of correlations. That is, instead of studying the net-additive effects of variables, QCA studies the necessity and sufficiency of relations between (combinations of) sets and the outcome condition (Ragin, 1987, 1999, 2000, 2008a, 2008b; Schneider & Wagemann, 2010). A condition is necessary if it has to be present for the outcome to occur, indicated by the outcome being a subset of the causal condition. Suppose condition B is the set 'cost overrun' and condition A is the set 'construction delay'. Then, Figure 1 shows that every case that exhibits cost overrun (i.e., is in set B) also exhibits construction delay. If a case does not exhibit A, then it cannot be in set B. Thus, according to this figure, construction delay is a necessary condition for the outcome to occur. A condition is sufficient if it can produce the outcome by itself, indicated by the condition being a subset of the outcome. Now suppose that B is the set 'construction delay' and A is 'cost overrun'. Then, Figure 1 shows that every case that exhibits construction delay (i.e., is in set B) also exhibits cost overrun. If construction delay is the only condition at play, then this figure indicates that it alone is sufficient to produce cost overrun. However, as explained by Ragin (2000), necessary and sufficient conditions can and most often are combined within a causal recipe since causation is complex. This means that often there are no pure necessary or sufficient conditions for an outcome to occur. Nevertheless, conditions can be necessary in certain contexts (i.e., configurations). Such conditions

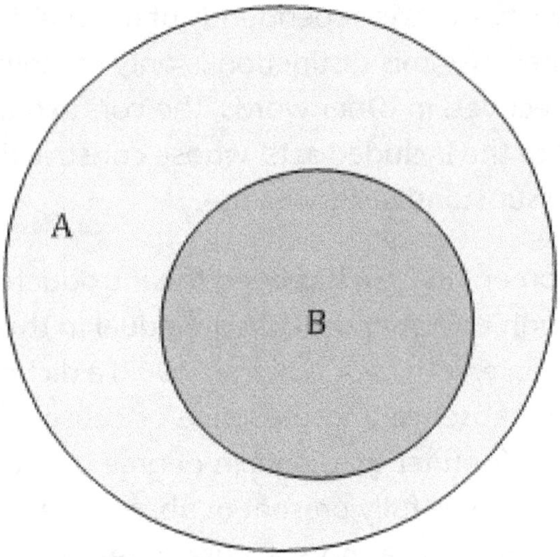

Figure 1 *Set Relations*

are called INUS conditions, which can be defined as an "*insufficient* but *non-redundant* part of an *unnecessary* but *sufficient* condition (Mackie, 1980: 62, original italics). For example, imagine that the Channel Tunnel, Great Belt link and Øresund link have different 'scores' on three causal conditions A, B and C. Suppose that the Channel Tunnel exhibits conditions A and B, the Great Belt link conditions B and C and the Øresund link conditions A and C. Thus, there are three intersected sets: A*B, B*C and A*C. The * sign indicates 'logical and'. These three different 'paths produce the outcome cost overrun. This implies 'logical or', indicated by a + sign. This can be written as a Boolean expression, namely:

$$A*B + B*C + A*C \rightarrow \text{cost overrun}$$

Thus, A and B, or B and C, or A and C result in cost overrun. This means first that none of the three conditions is individually sufficient since in none of the three cases the outcome is produced by a single condition. Second, it means that none of the three conditions is necessary. For instance, condition A is not necessary for cost overrun to appear, since it can also appear in the combination B*C. It does mean, however, that A is an INUS condition: it is a not sufficient (i.e., it cannot produce the outcome by itself) but a non-redundant (i.e., it is a necessary condition in both the combinations A*B and A*C) part of an unnecessary (i.e., A*B and A*C are not necessary since cost overrun also appears in B*C) but sufficient (i.e., A*B and A*C are sufficient for cost overrun to occur) condition. Finally, it is important to note that "neither necessity nor sufficiency exists independently of theories that propose causes" (Ragin, 2008b: 42), because this distinction is only meaningful in the context of theoretical perspectives. In other words, the contextuality of the research method is confined to the included sets, whose construction is substantiated with theoretical and substantive knowledge.

A recent development in QCA has been the introduction of fsQCA (Ragin, 2000; 2008a). Although csQCA and fsQCA are equal in their set-theoretic and configurational rationale, with csQCA (Ragin, 1987) a dichotomous distinction is made between the absence and presence of causal conditions in a case whereas fsQCA allows for finer gradients in degree of set membership (i.e., a variable does not need to be fully present or absent in a case) (cf. Ragin, 2000; 2008a; 2009) and therefore is a more accurate descriptor of the complexity exhibited by infrastructure projects. For instance, with csQCA, an infrastructure project can be fully in or fully out the set 'cost overrun' (i.e., it can score 0 or 1). However, one might argue that a cost overrun of 70 percent in the Øresund link project is substantially different from a cost overrun of 110 percent in the Great Belt project (see Flyvbjerg *et al.*, 2003a). Therefore, the latter may be more in the set 'cost overrun' then the former. This can be formalized by assigning different set memberships to these cases for cost overrun (e.g., 1.0 for the Great Belt project and 0.75 for the Øresund project). Consequently, the set 'cost overrun' constitutes a difference in kind.

When each case is assigned fuzzy set membership scores on each of the conditions, the researcher can move beyond mere theoretical (i.e., set construction) and empirical (i.e., case scoring) description to comparative analysis. The first step is the construction of the truth table. The fundamental unit of analysis is the truth table row (Ragin, 1999). Each row presents a unique theoretical combination of variables (i.e., a configuration of intersected sets). A truth table, then, lists al the logically possible combinations of causal conditions—expressed by the exponential formula 2^k ('k' being the number of conditions) since a condition can be both present and absent—and sorts the cases according to these conditions (i.e., assessing the empirical presence of the combinations). Next, for each causal recipe the outcome value is defined. For illustrative purposes, a hypothetical truth table with three conditions is depicted below. Note that this truth table shows hypothetical dichotomized crisp set data. Due to limited space available, this chapter does not allow for elaborating on the procedures for transforming fuzzy set data to crisp set data.

Condition A	Condition B	Condition C	Outcome	Distribution of cases
1	1	1	1	
1	1	0	1	
1	0	1	1	
1	0	0	1	
0	1	1	1	
0	1	0	0	
0	0	1	0	
0	0	0	0	

Table 3 *Hypothetical Truth Table*

These procedures and their rationale and advantages compared to a crisp set approach are discussed in Ragin (2008a: 124-144).

Note that the process described so far takes the researcher from (inductive) qualitative thick-case descriptions via the construction of sets and the assignment of set membership scores towards the construction of the truth table, thereby completing the inductive-deductive circle. One might argue that this discounts the unique and complex nature of cases. However, it actually formalizes the comparative procedures thereby providing insights therein. This increases the repeatability of case-oriented comparative research. Evaluators or researchers still need to make simplifying choices but these are made visible through the different steps in the QCA research approach. Thus, importantly, the method takes the researcher from in-depth understanding of complex systems to the identification of recurring patterns over multiple systems, without undermining the essence of complex systems: their contextuality and complex causation.

Next, the truth table can be minimized to produce a so-called 'solution' (i.e., a statement about generic patterns across the cases). That is, maximum complexity was assumed a priori, and this complexity is now brought back to its core. This minimization process is structured by using Boolean algebra. Its basic procedure can be summarized as follows: "if two Boolean expressions differ in only one causal condition yet produce the same outcome, then the causal condition that distinguishes the two expressions can be considered irrelevant and can be removed to create a simpler, combined expression" (Ragin, 1987:93). For example, the previous hypothetical truth table gives the following Boolean expressions:

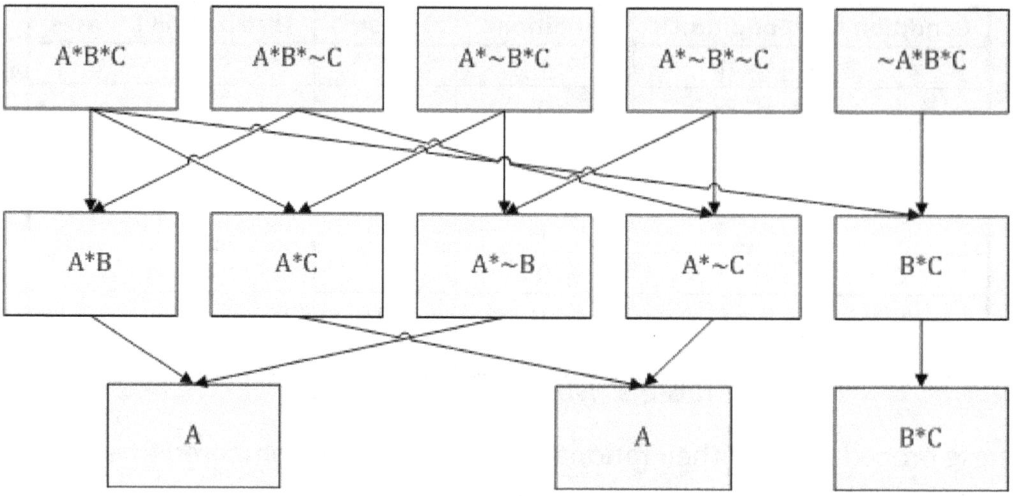

Figure 2 *Boolean Minimization*

A*B*C + A*B*~C + A*~B*C + A*~B*~C + ~A*B*C → cost overrun

The tilde sign (~) indicates the absence of a condition. This expression exhibits five empirically observed paths towards cost overrun. As a next step it can be evaluated what the sufficient, necessary and/or INUS conditions are in this expression. The first two expressions can be minimized to A*B: whether condition C is present or not, cost overrun appears nonetheless. This minimization procedure is displayed in Figure 2 and results in the following solution formula:

A + B*C → cost overrun

It is pivotal that the resulting formula is not applied mechanically in concluding the comparative analysis. As stated above, QCA is first of all a case-based qualitative approach. Hence, the solutions formula should be interpreted in light of the individual cases: does it make sense? In conclusion, this section aimed to provide some insights in the rationale and procedures of fsQCA. Obviously, it is worth discussing many more—such as limited diversity/ logical remainders, counterfactual analysis and the measures consistency and coverage—but unfortunately this is not possible due to the limited space available.

DISCUSSION

This paper set out to identify an evaluative framework that would address the complexity of developing transportation infrastructure projects, thereby contributing to the discussion of complexity in evaluation (cf. Callaghan, 2008) and of QCA in evaluation (cf. Befani *et al.*, 2007; Marx, 2005). It was found that the contextuality of such projects is explanatory for the outcomes and that a suitable framework would need to address this contextuality. Some prerequisites were developed and it followed that an ideal type evaluation framework integrates the depth of case-based approaches with the possibilities to generalize using variable-based approaches. This ideal type was found in QCA, specifically fsQCA. Some may argue that it is too much of a compromise, but as a framework it maintains the integrity of the conceptualization of infrastructure development presented in this paper.

While fsQCA fulfills three of the four requisites, it should be acknowledged that it does less well on the fourth, i.e., the time dimension. In essence, the method is static (Rihoux, 2003) and therefore does not capture the dynamics of complex systems to their full extent (cf. Gerrits, 2011). Some provisional workarounds are put forward by, among others, Rihoux (2003) and De Meur *et al.* (2009), for instance by capturing the time dimension by repetitive deployment of the method, by interpreting the time dimension, by conceptualizing time as (part of) a set, or by complementing QCA with other methods. Also, attempts are being made to develop a distinct type of QCA (Caren & Panofsky, 2005). However, all options are compromises and it should be concluded that, while fsQCA is strong in mapping the systemic complexity of infrastructure development, it performs weak on time dynamics.

The approach helps bringing together the generalizations that could be derived from systemic comparison and the in-depth idiosyncratic nature of single systemic cases (cf. Buijs *et al.*, 2009). It follows that this approach is limited in the number of conditions that can be taken into account, since the logically possible number of combinations increases exponentially. It also follows that the addition of a new case to the (restricted) set of cases being compared could lead to different outcomes. Although both issues are not unique to QCA—one could argue that such limitations are inherent to research into social reality—they should not be ignored. Regarding the limited number of conditions being considered, Rihoux (2003) and De Meur *et al.* (2009) suggest that a possible remedy is to carry out multiple routines, i.e., building an increasingly clearer set of conditions by going through the process multiple times. In this way, the researcher is able to find out, in a very transparent way (Rihoux *et al.*, 2009), what conditions do not matter or to find conditions that yield the same or similar results. These conditions can then be excluded from the analysis respectively grouped together in macro variables. Another promising approach is provided by Schneider & Wagemann (2006) who propose to analyze remote (i.e., contextual) and proximate factors separately in a first stage. In a second stage, influential remote and proximate factors are brought together in a single analysis. The second issue, the possibility of arriving at different conclusions after adding new cases, is actually part of the philosophy behind QCA and its roots in systemic thinking. Rihoux (2003) states that with small-n studies, the researcher does not strive to identify a single central tendency that could get closer to reality the more cases are added. Rather, it helps focusing on different causal pathways to an outcome and how such pathways are linked to individual cases. Adding a new case may lead to the discovery of a new pathway. More common comparisons are aimed at finding the variable that controls for differences and similarities in multiple cases. Following the discussion on generic and situated complexity, such a search is beside the point. Every case has its unique pathway and comparison should be used to highlight the particularities of the pathways. Note that QCA moves back and forth between generic and situated complexity. It is also through this perspective that thinking in terms of dependent or independent variables is replaced by thinking of variables as conjunctions (Aus, 2009), which bears a closer resemblance to social reality.

Summarizing the argument in this paper, it can be concluded that fsQCA is a promising approach that largely meets the evaluation requirements set in this paper. Some issues will have to be dealt with, most prominently the time dimension. A next step would be to use fsQCA in a concrete evaluation research to test how it performs when analyzing complex infrastructure development projects.

ACKNOWLEDGEMENTS

This paper was made possible with funding from the Next Generation Infrastructures knowledge program. The authors would like to thank Christopher Koliba, Asim Zia, Frank Boons, Nanny Bressers, Jasper Eshuis, Mike Duijn and Göktuğ Morçöl for their valuable feedback on an earlier version of this work.

REFERENCES

Advisory Committee VBI (2008). *Sneller en beter: Advies commissie versnelling besluitvorming infrastructurele projecten*, The Hague, NL: Ministery of Transport.

Anguera, R. (2006). "The Channel Tunnel: An ex-post economic evaluation," *Transportation Research Part A: Policy and Practice*, ISSN 0965-8564, 40(4): 291-315.

Aus, J.P. (2009). "Conjunctural causation in comparative case-oriented research," *Quality & Quantity*, ISSN 1573-7845, 43(2): 173-183.

Bar-Yam, Y. (1997). *Dynamics of complex systems*, ISBN 9780201557480.

Bateson, N. (1984). *Data construction in social surveys*, ISBN 9780043120224.

Batty, M. (2010). "Complexity in city systems: Understanding, evolution and design," in G. de Roo and E.A. Silva (eds.), *A planner's encounter with complexity*, ISBN 9781409402657, pp. 99-122.

Befani, B., Ledermann, S. and Sager, F. (2007). "Realistic evaluation and QCA: Conceptual parallels and an empirical application," *Evaluation*, ISSN 1356-3890, 13(2): 171-192.

Berg-Schlosser, D. and De Meur, G. (2009). "Comparative research design: Case and variable selection," in B. Rihoux and C.C. Ragin (eds.), *Configurational comparative methods: Qualitative comparative analysis (QCA) and related techniques*, ISBN 9781412942355, pp. 19-32.

Berg-Schlosser, D., De Meur, G., Rihoux, B. and Ragin, C.C. (2009). "Qualitative comparative analysis (QCA) as an approach," in B. Rihoux and C.C. Ragin (eds.), *Configurational comparative methods: Qualitative comparative analysis (QCA) and related techniques*, ISBN 9781412942355, pp. 1-18.

Bryman, A. (2004). *Social research methods*, ISBN 9780199264469.

Buijs, M.J., Eshuis, J. and Byrne, D. (2009). "Approaches to researching complexity in public management," in G.R. Teisman, M.W. van Buuren and L.M. Gerrits (eds.), *Managing complex governance systems: Dynamics, self-organization and coevolution in public investments*, ISBN 9780415459730, pp. 37-55.

Byrne, D. (1998). *Complexity theory and the social sciences: An introduction*, ISBN 9780415162968.

Byrne, D. (2001). *Understanding the urban*, ISBN 9780333724293.

Byrne, D. (2002). *Interpreting quantitative data*, ISBN 9780761962625.

Byrne, D. (2003). "Complexity theory and planning theory: A necessary encounter," *Planning Theory*, ISSN 1741-3052, 2(3): 171-178.

Byrne, D. (2005). "Complexity, configurations and cases," *Theory, Culture & Society*, ISSN 1460-3616, 22(5): 95-111.

Byrne, D. (2009). "Complex realist and configurational approaches to cases: A radical synthesis," in D. Byrne and C.C. Ragin (eds.), *The SAGE handbook of case-based methods*, ISBN 9781412930512, pp. 101-119).

Callaghan, G. (2008). "Evaluation and negotiated order: Developing the application of complexity theory," *Evaluation*, ISSN 1356-3890, 14(4): 399-411.

Cantarelli, C.C., Flyvbjerg, B., Van Wee, B. and Molin, E.J.E. (2010). "Lock-in and its influence on the project performance of large-scale transportation infrastructure projects: Investigating the way in which lock-in can emerge and affect cost overruns," *Environment and Planning B: Planning and Design*, ISSN 1472-3417, 37(5): 792-807.

Caren, N. and Panofsky, A. (2005). "TQCA: A technique for adding temporality to qualitative comparative analysis," *Sociological Methods and Research*, ISSN 1552-8294, 34(2): 147-172.

Cilliers, P. (1998). *Complexity and postmodernism: Understanding complex systems*, ISBN 9780415152877.

Cilliers, P. (2001). "Boundaries, hierarchies and networks in complex systems," *International Journal of Innovation Management*, ISSN 1363-9196, 5(2): 135-147.

Cilliers, P. (2005). "Complexity, deconstruction and relativism," *Theory, Culture & Society*, ISSN 1460-3616, 22(5): 255-267.

Cronqvist, L. and Berg-Schlosser, D. (2009). "Multi-value QCA (mvQCA)," in B. Rihoux and C.C. Ragin (eds.), *Configurational comparative methods: Qualitative comparative analysis (QCA) and related techniques*, ISBN 9781412942355, pp. 69-86.

De Hoo, S. (1982). *Besluitvorming en rijkswegenaanleg: Analyse van het besluitvormingsproces over rijkswegenprojecten*. Leiden, NL: Sociologisch Instituut Rijksuniversiteit Leiden.

De Meur, G., Rihoux, B. and Yamasaki, S. (2009). "Addressing the critiques of QCA," in B. Rihoux and C.C. Ragin (eds.), *Configurational comparative methods: Qualitative comparative analysis (QCA) and related techniques*, ISBN 9781412942355, pp. 147-166).

De Roo, G. (2010). "Being or becoming? That is the question! Confronting complexity with contemporary planning theory," in G. de Roo and E.A. Silva (eds.), *A planner's encounter with complexity*, ISBN 9781409402657, pp. 19-40.

De Roo, G. and Schwartz, M. (eds.) (2001). *Omgevingsplanning, een innovatief proces: Over integratie, participatie, omgevingsplannen en de gebiedsgerichte aanpak*, ISBN 9789012089067.

Fischer, F. and Forester, J. (eds.) (1987). *Confronting values in policy analysis: The politics of criteria*, ISBN 9780803926172.

Fischer, F. and Forester, J. (eds.). (1993). *The argumentative turn in policy analysis and planning*, ISBN 9780822313724.

Flyvbjerg, B. (2007a). "Cost overruns and demand shortfalls in urban rail and other infrastructure," *Transportation Planning and Technology*, ISSN 1029-0354, 30(1): 9-30.

Flyvbjerg, B. (2007b). "Policy and planning for large-infrastructure projects: Problems, causes and cures," *Environment and Planning B: Planning and Design*, ISSN 1472-3417, 34(4): 578-597.

Flyvbjerg, B. (2009). "Survival of the unfittest: Why the worst infrastructure gets built - and what we can do about it," *Oxford Review of Economic Policy*, ISSN 1460-2121, 25(3): 344-367.

Flyvbjerg, B., Bruzelius, N. and Rothengatter, W. (2003a). *Megaprojects and risk: An anatomy of ambition*, ISBN 9780521009461.

Flyvbjerg, B., Skamris Holm, M.K. and Buhl, S.L. (2002). "Underestimating costs in public works projects: Error or lie?," *Journal of the American Planning Association*, ISSN 1939-0130, 68(3): 279-295.

Flyvbjerg, B., Skamris Holm, M.K. and Buhl, S.L. (2003b). "How common and how large are cost overruns in transport infrastructure projects?," *Transport Reviews*, ISSN 1464-5327, 23(1): 71-88.

Flyvbjerg, B., Skamris Holm, M.K. and Buhl, S.L. (2004). "What causes cost overrun in transport infrastructure projects?," *Transport Reviews*, ISSN 1464-5327, 24(1): 3-18.

Flyvbjerg, B., Skamris Holm, M.K. and Buhl, S.L. (2005). "How (in)accurate are demand forecasts in public work projects? The case of transportation," *Journal of the American Planning Association*, ISSN 1939-0130, 71(2): 131-146.

Gerrits, L.M. (2011). "A coevolutionary revision of decision making processes: An analysis of port extensions in Germany, Belgium and The Netherlands," *Public Administration Quarterly*, ISSN 0734-9149, 35(3): 309-329.

Grofman, B. and Schneider, C.Q. (2009). "An introduction to crisp set QCA with a comparison to binary logistic regression," *Political Research Quarterly*, ISSN 1938-274X, 62(4): 662-672.

Guba, E.G. and Lincoln, Y. S. (1989). *Fourth generation evaluation*, ISBN 9780803932357.

Hammersley, M. (2008). "Causality as conundrum: The case of qualitative inquiry," *Methodological Innovations Online*, ISSN 1748-0612, 2(3).

Hammersley, M. (2009). "Challenging relativism: The problem of assessment criteria," *Qualitative Inquiry*, ISSN 1552-7565, 15(1): 3-29.

Han, S.H., Yun, S., Kim, H., Kwak, Y.H., Park, H.K. and Lee, S.H. (2009). "Analyzing schedule delay of mega project: Lessons learned from Korea Train Express," *IEEE Transactions on Engineering Management*, ISSN 0018-9391, 56(2): 243-256.

Harvey, D.L. (2009). "Complexity and case," in D. Byrne and C.C. Ragin (eds.), *The SAGE handbook of case-based methods*, ISBN 9781412930512, pp. 15-38.

Haynes, P. (2001). "Complexity, quantification and the management of policy," *Social Issues*, ISSN 1474-2918, 1(2).

Hecht, H. and Niemeier, D.A. (2002). "Comparing transportation project development efficiencies: The California department of transportation and the California county sales tax agencies," *Transport Policy*, ISSN 0967-070X, 9(4): 347-356.

Hsieh, T., Lu, S. and Wu, C. (2004). "Statistical analysis of causes for change orders in metropolitan public works," *International Journal of Project Management*, ISSN 0263-7863, 22(8): 679-686.

Irfan, M., Khursid, M.B., Anastasopoulos, P., Labi, S. and Moavenzadeh, F. (2011). "Planning-stage estimation of highway project duration on the basis of anticipated costs, project type, and contract type," *International Journal of Project Management*, ISSN 0263-7863, 29(1): 78-92.

Kaliba, C., Muya, M. and Mumba, K. (2009). "Cost escalation and schedule delays in road construction projects in Zambia," *International Journal of Project Management*, ISSN 0263-7863, 27(5): 522-531.

KiM Netherlands Institute for Transport Policy Analysis (2009). *Na het knippen van het lint: Het ex-post evalueren van infrastructuur*, The Hague, NL: Ministery of Transport.

Lee, J. (2008). "Cost overrun and cause in Korean social overhead capital projects: Roads, rails, airports, and ports," *Journal of Urban Planning and Development*, ISSN 1943-5444, 134(2): 59-62.

Mackie, J.L. (1980). *The cement of the universe: A study of causation*, ISBN 9780198246428.

Magnussen, O.E. and Olsson, N.O.E. (2006). "Comparative analysis of cost estimates of major public investment projects," *International Journal of Project Management*, ISSN 0263-7863, 24(4): 281-288.

Manavazhi, M.R. and Adhikari, D.K. (2002). "Material and equipment procurement delays in highway projects in Nepal," *International Journal of Project Management*, ISSN 0263-7863, 20(8): 627-632.

Marshall, S. (2009). *Cities, design & evolution*, ISBN 9780415423298.

Marx, A. (2005). "Systematisch comparatief case onderzoek van evaluatieonderzoek," *Tijdschrift Voor Sociologie*, ISSN 0777-883X, 26(1): 95-113.

Mjøset, L. (2009). "The contextualist approach to social science methodology," in D. Byrne and C.C. Ragin (eds.), *The SAGE handbook of case-based methods*, ISBN 9781412930512, pp. 39-68.

Morçöl, G. (2001). "What is complexity science? Postmodernist or postpositivist?," *Emergence: Complexity & Organization*, ISSN 1532-7000, 3(1): 104-119.

Næss, P., Flyvbjerg, B. and Buhl, S.L. (2006). Do road planners produce more 'honest numbers' than rail planners? An analysis of accuracy in road-traffic forecasts in cities versus peripheral regions," *Transport Reviews*, ISSN 1464-5327, 26(5): 537-555.

Odeck, J. (2004). "Cost overruns in road construction - what are their sizes and determinants?," *Transport Policy*, ISSN 0967-070X , 11(1): 43-53.

Peters, D. (2010). "Digging through the heart of reunified Berlin: Unbundling the decision-making process for the Tiergarten-Tunnel mega-project," *European Journal of Transport and Infrastructure Research*, ISSN 1567-7141, 10(1): 89-102.

PBL Netherlands Environmental Assessment Agency and Netherlands Institute for Transport Policy Analysis (2010). *Met de kennis van nu: Leren van evalueren: Een casestudy: A5 verlengde westrandweg*, The Hague/Bilthoven: PBL & KiM.

Polydoropoulou, A. and Roumboutsos, A. (2009). "Evaluating the impact of decision-making during construction on transport project outcome," *Evaluation and Program Planning*, ISSN 0149-7189, 32(4): 369-380.

Priemus, H. (2007). "Development and design of large infrastructure projects: Disregarded alternatives and issues of spatial planning," *Environment and Planning B: Planning and Design*, ISSN 1472-3417, 34(4): 626-644.

Ragin, C.C. (1987). *The comparative method: Moving beyond qualitative and quantitative strategies*, ISBN 9780520066182.

Ragin, C.C. (1999). "Using qualitative comparative analysis to study causal complexity," *Health Services Research*, ISSN 1475-6773, 34(5): 1225-1239.

Ragin, C.C. (2000). *Fuzzy-set social science*, ISBN 9780226702773.

Ragin, C.C. (2008a). *Redesigning social inquiry: Fuzzy sets and beyond*, ISBN 9780226702759.

Ragin, C.C. (2008b). *User's guide to fuzzy-Set/Qualitative comparative analysis*, Tucson, Department of Sociology, University of Arizona. Http://www.u.arizona.edu/~cragin/fsQCA/software.shtml.

Ragin, C.C. (2009). "Qualitative comparative analysis using fuzzy sets (fsQCA)," in B. Rihoux and C.C. Ragin (eds.), *Configurational comparative methods: Qualitative comparative analysis (QCA) and related techniques*, ISBN 9781412942355, pp. 87-122.

Ragin, C.C., Shulman, D., Weinberg, A. and Gran, B. (2003). "Complexity, generality, and qualitative comparative analysis," *Field Methods*, ISSN 1552-3969, 15(4): 323-340.

Reed, M. and Harvey, D.L. (1992). "The new science and the old: Complexity and realism in the social sciences," *Journal for the Theory of Social Behaviour*, ISSN 1468-5914, 22(4): 353-380.

Rigden, S.R. (1983). "Project management and the construction of the Hong Kong mass transit railway," *International Journal of Project Management*, ISSN 0263-7863, 1(3): 136-141.

Rihoux, B. (2003). "Bridging the gap between the qualitative and quantitative worlds? A retrospective and prospective view on qualitative comparative analysis," *Field Methods*, ISSN 1552-3969, 15(4): 351-365.

Rihoux, B. (2006). "Qualitative comparative analysis (QCA) and related systematic comparative methods: Recent advances and remaining challenges for social science research," *International Sociology*, ISSN 1461-7242, 21(5): 679-706.

Rihoux, B. and Lobe, B. (2009). "The case for qualitative comparative analysis (QCA): Adding leverage for thick cross-case comparison," in D. Byrne and C.C. Ragin (eds.), *The SAGE handbook of case-based methods*, ISBN 9781412930512, pp. 222-242.

Rihoux, B. and Ragin, C.C. (2009). "Introduction," in B. Rihoux and C.C. Ragin (eds.), *Configurational comparative methods: Qualitative comparative analysis (QCA) and related techniques*, ISBN 9781412942355, pp. xvii-xxv.

Rihoux, B., Ragin, C.C., Yamasaki, S. and Bol, D. (2009). "Conclusions - the way(s) ahead," in B. Rihoux and C.C. Ragin (eds.), *Configurational comparative methods: Qualitative comparative analysis (QCA) and related techniques*, ISBN 9781412942355, pp. 167-178).

Sanderson, I. (2000). "Evaluation in complex systems," *Evaluation*, ISSN 1356-3890, 6(4): 433-454.

Schneider, C.Q. and Wagemann, C. (2006). "Reducing complexity in qualitative comparative analysis (QCA): Remote and proximate factors and the consolidation of democracy," *European Journal of Political Research*, ISSN 1475-6765, 45(5): 751-786.

Schneider, C.Q. and Wagemann, C. (2010). "Standards of good practice in qualitative comparative analysis (QCA) and fuzzy sets," *Comparative Sociology*, ISSN 1569-1330, 9(3): 397-418.

Smithson, M. and Verkuilen, J. (2006). *Fuzzy set theory: Applications in the social sciences*, ISBN 9780761929864.

Uprichard, E. and Byrne, D. (2006). "Representing complex places: A narrative approach," *Environment and Planning A*, ISSN 1472-3409, 38(4): 665-676.

Van Marrewijk, A., Clegg, S.R., Pitsis, T.S. and Veenswijk, M. (2008). "Managing public-private megaprojects: Paradoxes, complexity, and project design," *International Journal of Project Management*, ISSN 0263-7863, 26(6): 591-600.

Walter, A.I. and Scholz, R.W. (2007). "Critical success conditions of collaborative methods: A comparative evaluation of transport planning projects," *Transportation*, ISSN 1572-9435, 34(2): 195-212.

Welde, M. and Odeck, J. (2011). "Do planners get it right? The accuracy of travel demand forecasting in Norway," *European Journal of Transport and Infrastructure Research*, ISSN 1567-7141, 11(1): 80-95.

Williams, M. (2009). "Social objects, causality and contingent realism," *Journal for the Theory of Social Behaviour*, ISSN 1468-5914, 39(1): 1-18.

WRR Scientific Council for Government Policy (1994). *Besluiten over grote projecten*, The Hague: Sdu.

Yamasaki, S. and Rihoux, B. (2009). "A commented review of applications," in B. Rihoux and C.C. Ragin (eds.), *Configurational comparative methods: Qualitative comparative analysis (QCA) and related techniques*, ISBN 9781412942355, pp. 123-146.

Yang, C. (2007). "Primary influential factors in the management of public transport projects in Taiwan," *Canadian Journal of Civil Engineering*, ISSN 1208-6029, 34(1): 1-11.

Stefan Verweij M.Sc. studied Public Administration at the Erasmus University Rotterdam. He now works as a PhD Student at the research group *Governance of Complex Systems* (GOCS) at the Department of Public Administration, Erasmus University Rotterdam. His research interests concern: complexity, management of infrastructure and area development projects and qualitative research methods.

Lasse M. Gerrits, Ph.D. studied Public Administration and Urban Planning at the Erasmus University Rotterdam and now works as associate professor at the research group *Governance of Complex Systems* (GOCS) at the Department of Public Administration, Erasmus University Rotterdam. His research interests concern the complexity of public decision-making and the development of the urban. He also teaches courses in complexity and systems.

Chapter 6

COMPLEXITY FRIENDLY" MESO-LEVEL FRAMEWORKS FOR MODELING COMPLEX GOVERNANCE SYSTEMS

Christopher Koliba (University of Vermont, USA)
Asim Zia (University of Vermont, USA)

Five theoretical frameworks of policy and governance systems including the multiple policy streams, punctuated equilibrium, institutional analysis and development, advocacy coalition, and governance network frameworks are presented as some of the meso-level, "complexity friendly" frameworks that incorporate whole systems properties, tend to avoid reductionism, allow for emergence and path dependencies, and accommodate dynamic systems characteristics. The extent to which these frameworks are accommodating to complex adaptive systems modeling and how they may be used to generate lines of inquiry that may be tested, within limits, using complex adaptive systems modeling is assessed. Further, the potential utilization of complex systems modeling tools to inform a meta-theoretical research program to compare and refine alternate theoretical frameworks is discussed. Implications for a meta-theoretical research and modeling program are explored.

INTRODUCTION

Complexity has always been a part of every day public administration and policy practices (Klijn & Snellen, 2009: 17). Just as the field of physics "has discovered complexity by complicating its own language of description," (Tsoukas, 2005: 236), we argue here that public administration and policy has come to complexity science in much the same way. Policy and governance systems organized around the framing of public problems, the deliberation of policy alternatives, and/or the implementation of public policies are complex social systems of a particular type. They are guided by dynamics that are governed by certain political and administrative practices undertaken by social and institutional agents representing a variety of public, private and nonprofit sector organizations and interests. The meta-theoretical underpinnings of complexity theory, coupled with the computational tools and modeling capacity being utilized in complexity science, provides the field of public administration and policy with a growing capacity to employ this "new" language to study observable governance phenomena. In this article we discuss how the language and science of complexity may be combined with some of our existing "complexity friendly" policy and governance frameworks to provide the basis for a meta-theoretical program focused on modeling meso-level dynamics. We assess whether and how these frameworks are accommodating to complex adaptive systems modeling and how they may be used to generate lines of inquiry that can be tested, within limits, using complex adaptive systems modeling. We then explore a meta-theoretical research program for comparing and refining alternate theoretical frameworks with respect to their adequacy in accounting for non-linearity, lags, inertia, cross-scale interactions and complex feedback loops observed in policy and governance systems.

The application of computer simulation modeling to address the kinds of questions of concern to the public administration field has begun to emerge. Agent-based models have been constructed of collaborative governance groups and combined with game theory to yield studies that examine some of the fundamental tenants guiding the establishment of voluntary ties (Axelrod & Cohen, 1999), with specific inferences drawn to administrative practice (Knott, Miller and Verkuilen, 2003; Johnston *et al.*, 2008; Nan *et al.*, 2008; Johnston *et al.*, 2010). There is also a growing body of research that is focused on "pattern-oriented" (Grimm *et al.*, 2005) computer simulation models that employ agent-based models to study complex governance networks carrying on specific policy functions, such as transportation planning (Zia *et al.*, 2011) and watershed governance (Zia *et al.*, 2010). The more recent application of agent-based modeling techniques have complimented the extensive tradition of system dynamic modeling found within the field for the past twenty or so years (Richardson, 1991).

Computer simulation models found within the literature are built on observations of social and political phenomena occurring at the micro scale (as within the "gaming" dynamics of personal interactions occurring at the interpersonal scale), the meso scale (as within and across multi-group, organizational or institutional arrangements), or the macro scale (occurring as resource flows across different system components). Miller and Page (2007) describe the process of modeling as an, "attempt to reduce the world to a fundamental set of elements (equivalent classes) and laws (transition functions), and on this basis ... understand and predict key aspects of the world... Modeling proceeds by deciding what simplifications to impose on the underlying entities and then, based on those abstractions, uncovering their implications" (Miller & Page, 2007: 40, 65). The effort to translate a conceptual model drawn up in abstract into an adequate simulation of real world phenomenon requires that model development be anchored in and calibrated to empirically observed phenomena. To give shape to these conceptual models, researchers turn to theories and frameworks. Models of micro-scale personal and household interactions may be guided by assumptions found within other disciplines and theoretical traditions. Although these theories have been applied to develop models of administrative behavior (Johnston *et al.*, 2010), we will not focus on this suite of theories in this article. Nor will we focus on the larger, macro- scale frameworks common to political science and macro-economics that are amendable to complexity systems modeling. Instead, we focus here on those meso-scale frameworks that may be used to study and simulate complex governance systems.

COMPLEXITY FRIENDLY, MESO-SCALE POLICY AND GOVERNANCE FRAMEWORKS

A body of complementary, yet incommensurate, meso-scale theoretical frameworks have evolved to describe the relationship of coalition formation and policy development; the role of institutional rules in structuring human interaction; the persistence of network ties established to achieve particular policy goals; the role of feedback and equilibrium in the formation and implementation of public policies; and the coupling of policy streams. The most widely known and respected framework is the institutional analysis and development (IAD) framework first developed by Nobel Laureate Elinor Ostrom (2005). The IAD framework draws on institutionalism and neo-institutionalism theories, game theory, transaction cost theory, and common resource pool theory to craft a description of multi-institutional systems that explain the crafting of public policy as ultimately an institutional design problem in complex "action arenas." Ostrom (2005) emphasizes the roles that rules play in structuring governance arrangements. Drawing on her empirical analysis of natural resource management networks she makes a compelling argument in favor of more decentralized concentrations of power and authority to enhance performance in some institutional contexts and conditions. Other comprehensive frameworks may be found in John Kingdon's multiple streams framework (1984), Baumgartner and Jones' policy subsystem and punctuated equilibrium framework (1993), and Paul Sabatier and associates' advancement of the advocacy coalition framework (Sabatier & Jenkins-Smith, 1993).

Some of these frameworks impose homogenous assumptions about human decision making behaviors, such as expected utility maximizing behaviors in IAD, while others assume more unpredictable, chaotic decision making behavior, such as those found in the multiple streams framework. Another difference that arises across these frameworks concerns the balance between individual behavior and institutional norms and rules. ACF focuses attention on the role that common belief networks play in powerful advocacy coalitions. While IAD focuses more attention on the role that operational, collective choice and constitutional rules play in shaping multi-institutional arrangements.

A second grouping of theoretical frameworks common to public administration and policy are built on the foundation of basic network architecture of nodes and ties (Rhodes, 1997; Kickert *et al.*, 1997; Milward & Provan, 1998; Agranoff & McGuire, 2003; Sorensen & Torfing, 2005; Koliba *et al.*, 2010). Several of these frameworks account for the multi-scalar dimensions of social networks—meaning, they account for the fact that these networks are populated by individual people, groups, organizations and networks of organizations. The nature of the ties between these agents are premised on the types of resources that flow between them (Rhodes, 1997), the kinds of managerial strategies employed (Agranoff & McGuire, 2003) and the mixed structures of administrative authority (Koliba *et al.*, 2010) that persist within and across them.

In the next section we argue that these frameworks are compatible with the core features of complex adaptive systems. We refer the reader to many excellent books that have focused on the properties of complex adaptive *social* systems (Katz & Kahn, 1978; Kiel, 1994; Marion, 1999; Miller & Page, 2007; Dennard *et al.*, 2008). We distill these properties down to six distinct, but interrelated features: holism; emergence; feedback loops; self organization; nonlinearity; and path dependency[1]. In order for a theoretical framework to be "complexity friendly," it must be able to:

- Avoid simple reductionism, addressing the holistic properties of complex systems;
- Accommodate the emergence of new structures and functions;
- Accommodate the existence of feedback, stocks & flows, inputs and outputs;
- Allow for the self organization of the system as a whole or parts of the subsystem;
- Allow for dynamic, nonlinear interactions that lack clear cause and effect relationships, and;
- Accommodate time and path dependencies.

We argue that all of the theoretical frameworks discussed in this article are, to one degree or another, amendable to system dynamics and network architecture, and are therefore "complexity friendly." All of these theoretical frameworks are grounded in a system dynamics logic, allowing for the existence of feedback loops, stocks and flows, and certain assumptions about input and output flows. All of these frameworks account for the roles that individual social agents, groups of agents and organizations play in the whole system.[2] We now provide a brief overview of each of these frameworks, highlighting the features that make them particularly applicable to complex systems modeling.

1. We do not describe these features in detail here and refer the reader to the books cited above and the many excellent articles that have appeared outlining the basic characteristics of complex adaptive systems.

2. We are quick to note, however, that the frameworks selected here do not represent all of the possible complexity friendly public administration and policy theories that exist. In the concluding section we argue that the epistemological foundation of complexity science need not be viewed as mutually exclusive of the wide ranging, domain-specific theories of public administration and policy studies.

Multiple Policy Streams

John Kingdon's policy stream model, Figure 1, does not assume linearity, nor rational behavior on the part of policy actors. Problems, policies and politics streams may couple, and in fact, need to couple for agendas to be set and policy windows to open. Kingdon recognizes that policy streams are created and directed through social networks and indirectly asserted that social networks form within individual streams, and provide a basis for the coupling of streams (1984). Kingdon recognizes that a number of policy actors, including interest groups, academia, media, and political parties coordinate actions within and across the policy stream. He grounds the policy stream model in the coordinated actions that arise during the pre-enactment phases of policy selection and design, although we may recognize the coupling of streams across all facets of the policy process.

Zahariadis observes that, "Much like systems theory, [policy streams] views choice as the collective output formulated by the push and pull of several factors... It shares common ground with chaos theories in being attentive to complexity, in assuming a considerable amount of residual randomness, and in viewing systems as constantly evolving and not necessarily settling into equilibrium (Kingdon 1984: 219)" (2007: 66). The extent to which problem, policy and politics streams are coupled is something that complexity science can shed light on, as coupling comes about through a generative process of interlocking feedback loops occurring across each stream. A wide range of

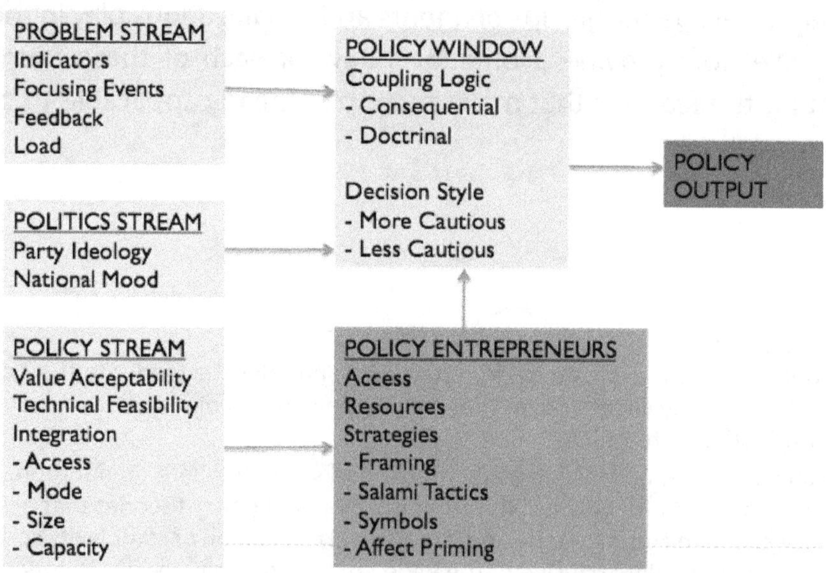

Figure 1 *Structural features of the MS Framework* **(***adapted from Kingdon, 1984***)**

actors are mobilized within each stream. Some of these actors span more than one stream—for example, a legislator may become convinced of the importance of a particular problem or policy solution and work to align the politics stream with a particular problem definition or policy tool. These dynamics can be modeled using systems dynamics tools.

Punctuated Equilibrium

Baumgartner and Jones' punctuated equilibrium framework is predicated on the assumption that, "... policymaking both make leaps and undergo periods of near stasis as issues emerge on and recede from public agenda" (True, Jones & Baumgartner, 2007: 157). Their view of policy subsystems draws extensively from the description of nonlinear systems dynamics. Their theory relies most directly on system dynamics models because of its reliance on feedback loops as a critical feature of subsystem dynamics.

The key features of interest for complex systems models concerns the relationship between policy subsystems in punctuated equilibrium theory. This theory assumes that,

> *Political systems, like humans, cannot simultaneously consider all the issues that face them, so policy subsystems can be viewed as mechanisms that allow the political system to engage in parallel processing (Jones, 1994). Thousands of issues may be considered simultaneously in parallel within their respective communities of experts. This equilibrium of interests does not completely lock out change. Issue processing within subsystems allows for a politics of adjustment, with incremental change resulting from bargaining among interests and marginal moves in response to changing circumstances* (True, Jones & Baumgartner, 2005: 158-159).

Much like the Advocacy Coalition Framework (ACF), punctuated equilibrium theory recognizes the role that certain actors or combinations of actors play in establishing system-wide equilibrium. These entanglements of subsystems are more than likely comprised of stable sets of institutional actors and rules. However, these same actors will likely produce "a plethora of small accommodations and a significant number of radical departures from the past" (True, Jones & Baumgartner, 2005: 156). The ranges of small, short term accommodations and long term radical departures from the stable state must be placed within the context of the system as a whole. Those using punctuated equilibrium theory often rely on changes within the outputs or inputs of the whole systems over time to demonstrate phase transitions. Substantial deviations from the kind of variations attributable to small accommodations are noted. When radical changes to relatively stable patterns are noted, explanations are sought using system dynamics logic.

Institutional Analysis and Development (IAD)

Although there are many facets to Ostrom's IAD framework, as shown in Figure 2, we highlight two of the major contributions it makes to the study of complex adaptive social systems here. These two facets concern the role that "rules-in-use" and rule making play in the structuring and functioning of these systems; and the "action arenas" through which these rules combine to structure action. Ostrom distinguishes between three types of rules: (a) *operational rules* that govern day-to-day activities of appropriators; (b) *collective choice rules* concerning overall policies for governing common pool resources and how those policies are made, and (c) *constitutional choice rules* that establish who is eligible to determine collective choice rules. The operational functions of any social system are governed by a complex array of operational rules, norms, habits and customs. Collective choice theory has long been viewed as a central feature of resource exchange frameworks. Collective choice is shaped by individual and collective interests all needing to be balanced in order to create an optimal level of autonomy and dependence.

Ostrom has focused much of her attention on how these rules shape social interaction and cautions that, "The capacity of humans to use complex cognitive systems to order their own behavior at a relatively subconscious level makes it difficult at times for empirical researchers to ascertain what the working rules for an ongoing action arena may actually be in practice" (Ostrom, 2005: 19). The combining, commingling and competition of rules operating at various levels can be modeled using system dynamics modeling. These dynamics are represented as "rule-in-use" in her model shown in Figure 2.

The extent to which these rules guide the behaviors of those social agents in the IAD framework is predicated on how authoritative they are. We have noted how the self-organizing capacity of autonomous agents are shaped by decision rules and relational scripts. According to Ostrom's approach, self-

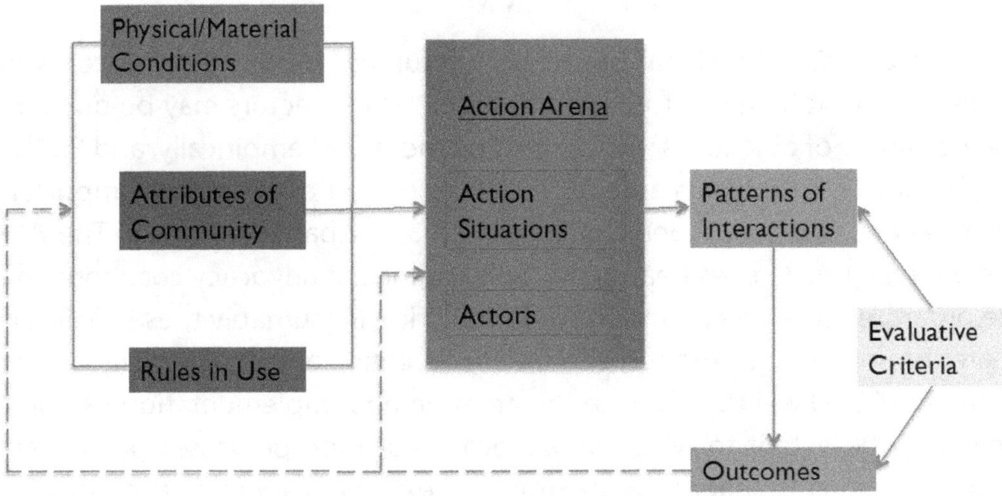

***Figure 2** An Overview Of IAD Framework (Ostrom, 2007)*

organized governance systems "need to match rules that impose costs in a rough proportion to the likely positive payoffs that appropriators are likely to obtain over time..." (2005: 234). Ostrom's emphasis on rational collective action is subject to the kind of critiques that have been raised regarding rational action more generally. Pierson argues that, "... we should generally exercise considerable skepticism about assertions that institutional arrangements will reflect the skilled design choices of rational actors. Instead, we should anticipate that there will often be sizable gaps between the ex ante goals of powerful political actors and the actual functioning of prominent institutions" (Pierson, 2004: 15).

A second major dimension of the IAD model concerns the role of action arenas as spaces where social agents commingle with institutional rules of many forms to generate certain activities or events. Complex policy and governance systems will likely be comprised of many action arenas, each of which plays somewhere between a minor to major role in determining the outputs of a whole system. Variation in the structures of these action arenas becomes a critical consideration in the IAD framework. Ostrom (2005) has argued, quite effectively, how the composition of these action arenas has a considerable impact on a system's performance. The dynamics that unfold in strategic action arenas may be modeled using an agent based modeling framework (Johnston *et al.*, 2011; Zia *et al.*, 2011).

Advocacy Coalition Framework

The Advocacy Coalition Framework (ACF), assumes, "(1) that belief systems are more important than institutional affiliation, (2) that actors may be pursuing a wide variety of objectives, which must be measured empirically, and (3) that one must add researchers and journalists to the set of potentially important policy actors (Sabatier & Jenkins-Smith, 1993)" (Sabatier, 2007: 5). The ACF, shown in Figure 3, relies heavily on the existence of advocacy coalitions that are organized around common belief networks. Presumably these coalitions share common mental models of problem definition and policy solutions, and share a political will to influence the creation and implementation of public policies. The extent to which an advocacy coalition possesses power over other coalitions is shaped by parameters, external events, and constraints and resources available to a policy subsystem. The framework operates on the basic premise of system dynamics: inputs shaping outputs with potential feedback loops shaping the nonlinear, recursive nature of the system. Those who have worked to advance the ACF have tended to downplay the role that institutional rules play in shaping the actions of the policy subsystem. Emphasis is placed on the influence that the advocacy coalitions operating within a subsystem play. The dominant drivers of coalition behavior are the "core beliefs" of coalition members.

The ACF exhibits some network qualities because it allows for the possibility to understand coalitions as collections of individual agents—each of whom contributes to the stability or instability of the coalition's core beliefs. These agents have the capacity to influence one another. These influences are nonlinear. it is also possible to view individual advocacy coalitions as agents unto themselves. Coalitions are characterized by the emergence of bottom-up influences. According to Sabatier and his associates, these bottom-up properties take precedence over top down and externally driven institutional rules and norms (Sabatier & Weible, 2007).

Governance (and Policy) Networks

The first application of network theory to policy arenas rested on institutional arrangements and the resources exchanged between them (Rhodes, 1997). The subsequent "network turn" in public administration and policy studies has been marked by descriptions of policy networks (Rhodes, 1997; Kickert *et al.*, 1997; Koopenjan & Klijn, 2004), and certain forms of

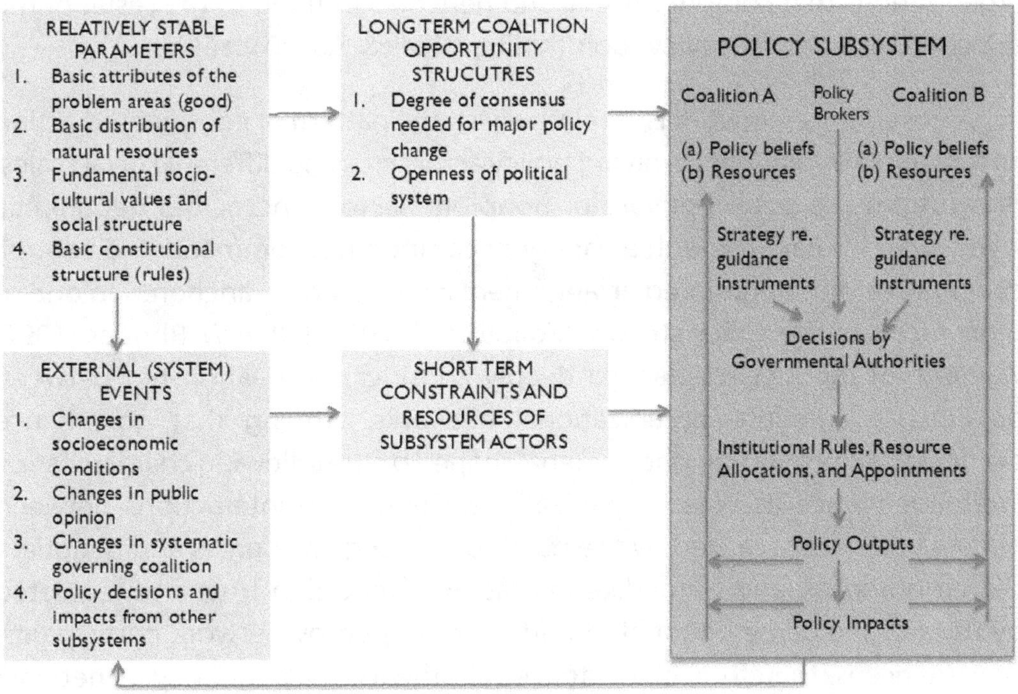

Figure 3 *Advocacy Coalition Framework (Sabatier & Weibe, 2007)*

intergovernmental relations (Wright, 2000). Inter-organizational networks have also been described as public sector networks (Agranoff, 2005), governance networks (Sorensen & Torfing, 2005; Koliba, *et al.*, 2010), and public management networks (Milward & Provan, 2006; Agranoff, 2007).

Inter-organizational networks have also been described in terms of the functions that they perform, including service contracts, supply chains, ad hoc, channel partnerships, information dissemination, civic switchboards (Goldsmith & Eggers, 2004), problem-solving, information sharing, capacity building, and service delivery (Milward & Provan, 2006), learning and knowledge transfer (McNabb, 2007), or civic engagement (Yang & Bergrud, 2008). Within this body of network literature we find the following similarities: networks facilitating the coordination of actions and/or exchange of resources between actors within the network; network membership being drawn from some combination of public, private and non-profit sector actors[3]; networks carrying out one or more policy function; networks exist across virtually all policy domains; networks are mostly defined at the inter-organizational level, they are also described in the context of the individuals, groups *and*

3. With the obvious exception of inter-governmental networks, which may be described as networks of governments of different geographical scope.

organizations that comprise them; and networks forming as the result of the selection of particular policy tools (Koliba, Meek & Zia, 2010: 60).

A governance network is "a relatively stable pattern of coordinated action and resource exchanges involving policy actors crossing different social scales, drawn from the public, private or nonprofit sectors and across geographic levels; who interact through a variety of competitive, command and control, cooperative, and negotiated arrangements; for purposes anchored in one or more facets of the policy stream" (Koliba *et al.*, 2010: 60). A.W. Rhodes (1997) was one of the first scholars to deeply consider the relationship between governance and inter-organizational networks, arguing that governance occurs as "self-organizing phenomena" shaped by the following characteristics: Interdependence between organizations; Continuing interactions between network members, caused by the need to exchange resources and negotiate shared purposes; and Game-like interactions, rooted in trust and regulated by rules of the game negotiated and agreed upon by network participants. Governance is, therefore, characterized by the interdependency of network actors, the resources they exchange, and the joint purposes, norms, and agreements that are negotiated between them. Considerations of network governance leads to an inevitable consideration of the bargaining and cooperative systems of more "horizontally arranged" ties, in addition to the traditional "vertically oriented" command and control systems of mono-centric government systems (Kettl, 2006: 491). Mixed-form governance networks may incorporate all forms of administrative authority and rely on the basic architecture of networks: nodes, ties and whole network characteristics. These structures may be combined with the specific characteristics of network actors and the nature of their ties to determine how network structures lead to certain network outputs (Koliba *et al.*, 2010).

TOWARD A META-THEORETICAL RESEARCH PROGRAM: CHALLENGES AND OPPORTUNITIES

Despite the paradigmatic and theoretical incommensurability of these frameworks, theoretical advancement in policy and administrative sciences will require us to develop meta-theoretical procedures to test these theories in field settings. For comparing *theories*, Schlager (2007) proposed to compare boundaries and scope of inquiry through a model of the individual, collective action, institutions and treatment of the policy change. For comparing *frameworks*, Schlager (2007) proposed to focus on type of

TYPE OF VAR.	VARIABLE	DESCRIPTORS
Agents (Nodes)	Social scale	Individual; Group; Organizational/Institutional; Inter-organizational
	Social sector (organizational level)	Public; Private; Nonprofit
	Geographic scale	Local; Regional; State; National; International
	Role centrality	Central – peripheral; Trajectory
	Capital resources actor provides (as an input)	Financial; Physical; Natural; Human; Social; Cultural; Political; Knowledge
	Providing accountabilities to....	Elected representatives; Citizens and interest groups; Courts; Owners/Shareholders; Consumers; Bureaucrats/Supervisors/Principals; Professional Associations; Collaborators/Partners/Peers
	Receiving accountabilities from...	See above
	Performance/Output and Outcomes Criteria	Tied to policy function and domain
Ties	Resources Exchanged/ Pooled	Financial; Physical; Natural; Human; Social; Cultural; Political; Knowledge
	Strength of tie	Strong to weak
	Formality of tie	Formal to informal
	Administrative authority	Vertical (command and control); Diagonal (negotiation and bargaining); Horizontal (collaborative and cooperative); Competitive
	Accountability relationship	See above
Whole Network	Policy tools	Regulations; Grants; Contracts; Vouchers; Taxes; Loans/loan guarantees, etc.
	Operational functions	Resource exchange/pooling; Coordinated action; Information sharing; Capacity building; Learning and knowledge transfer
	Policy functions	Define/frame problem; Design policy solution; Coordinate policy solution; Implement policy (regulation); Implement policy (service delivery); Evaluate & monitor policy; Political alignment
	Policy domain functions	Health, environment, education
	Macro-level governance structures	Lead organization; Shared governance; Network administrative organization
	Network configuration	Inter-governmental relations; Interest group coalitions; Regulatory subsystems; Grant and contract agreements; Public-private partnerships
	Properties of network boundaries	Open – closed; Permeability
	Systems dynamics	Systems-level inputs; processes; outputs and outcomes

Table 1 *Governance Network Framework (Koliba et al., 2010)*

actors, variable development, unit of analysis, and levels of analysis. While these criteria provide important mechanisms to ascertain the commensurability of policy theoretical frameworks in terms of their boundaries and scope or models of individuals and institutions, these descriptive criteria do not provide adequate methodology to "test" whether one theory better explains public policy and administrative systems.

We have argued in this article that the challenges with the linear causality are accentuated by the social construction of belief systems and the active, emergent, adaptive behaviors of individuals. Some have argued that individuals in policy systems have such heterogeneous behaviors that we cannot, ever, develop "calibrated" and "valid" models of them. We may view this challenge from a meta-theoretical standpoint by juxtaposing a belief-based model of the individual (ACF) against a game-theoretical model of the individual (IAD). These models of individual behavior may be coherent within the context of their specific theories, however, this "coherence" does not tell us whether belief-based models are better than game-theoretical models in describing and explaining the behaviors of actors in policy systems. We propose a long term meta-theoretical research program to understand public policy and governance systems that employs complex systems modeling tools to "test" the explanatory power of these theories in a large variety and contexts of policy domains.

The treatment of space and time plays a key role in the structuring of such meta-theoretical research program. Building our capacity to design computer simulation models that captures the spatial and temporal dimensions of complex systems and tests policy and governance frameworks will ultimately help us to address questions relating to policy and governance system stability, innovation and collapse. Developing our capacity to understand how complex social systems adapt is of tremendous importance.

The long term development of a meta-theoretical program for modeling the complexity of policy and governance systems will likely hinge on three critical questions:

1. How incommensurable are policy and governance theoretical frameworks? To what extent are they compatible or comparable?

2. What are the spatial and temporal boundary conditions set for empirical studies of policy and governance systems? Where does the system begin and end? What gets included and left out of the model?
3. To what extent do assumptions concerning meta-patterning predetermine outcomes?

It has long been noted that policy and governance systems exist within (and sometimes across) virtually all policy domains (Baumgartner & Jones, 1993). The relative resiliency of any complex policy or governance system needs to be gauged against certain outputs that are produced by these whole systems.

The capacity of computer simulation models of complex policy and governance systems to lead to accurate forecasting and predicting particular policy outcomes is predicated on a "deep uncertainty" that characterizes our current state of understanding of complex social systems. Bankes (2002) characterizes this deep uncertainty arising as, "the result of pragmatic limitations in our ability to use the presentational formalisms of statistical decision theory to express all that we know about complex adaptive systems and their associated policy problems" (p. 7263). Rendering the kind of validity needed to confirm or disprove empirical hypotheses requires the development of large samples and a set of empirically-normed instruments and measures.

The extent to which simulations can be used to undertake theory testing in these contexts remains to be seen. We need to ask, just what kind of inferential reasoning is possible given the inherent non-linearity and uncertainty characteristics of complex social systems? The deep uncertainty that characterizes simulated experimentation needs to be accounted for. The level of systemic error that is possible in computer simulation models can potentially be quite large. Modelers refer to this as "noise" in the model. Although efforts can be made to reduce the noise of a model, the propensity for large systemic error virtually assures us that the error rates of simulation models of social systems far exceed levels of statistical significance found in more linear regression models. We are mindful of why these error rates may be higher in social systems, than they are in the more predicable (but still uncertain) areas of natural and biological systems. Social agents maintain a certain level of autonomy in most social systems. The capacity of individual social agents to exert their own free will inevitably lead to a certain level of unpredictability. Agent based modelers account for this unpredictability in ascribing probability functions to agent behavior that are, ideally, calibrated to empirical observations. Modelers must still make a wide range of choices in building their models, as they are boundedly rational as well. They make choices around what elements to incorporate into the model and should be prepared to defend those choices (Batty, 2005).

We argue that those looking to apply complexity science and theory to the study of policy and governance phenomena will be pressed to avoid the kinds of positivist and postpositivist legacies found in most computer simulation models of complex systems (Buijs, Eshuis & Byrne, 2009). The ontological tensions that arise between the "naïve realist" or "empiricist epistemologies" found in the positivist traditions of science and the social constructionism common to more qualitative, interpretive sciences may be found in virtually all attempts to apply complexity science to the study of social phenomena. Rather than call for the purging these models of rationalist assumptions, we need to recognize how these tensions play out within computer simulation models of policy and governance systems (Richardson, 2008).

Although space precludes a deeper discussion here, we recognize that the development of models that are calibrated to observed patterns hinges on our capacity to render a "thick description" of these phenomena. To adequately develop models that may be able to test the validity of certain meso-level theoretical constructs requires finer and finer grain analysis of social phenomena extending across all levels of social scale (individual, group, or organizations). In order to study and test the efficacy of theories relating to the social construction of policy streams, the path dependency of stable and destabilized systems, the decision making structures within action arenas, the belief networks of advocacy coalitions, or the emergent features of network composition, we will need to develop mixed method studies that combine elements of quantitative and qualitative social science approaches. As our capacity to undertake data mining of textual and narrative data expands, the opportunities to understand the phenomenological traces of nuanced social interactions intensifies. These advancements will deepen our capacity to develop finer and finer grained analysis of social systems and, in the long run, allow for the integration of theories and frameworks drawn from not only the kinds of policy and governance theories highlighted here, but also extending into our theories of management as well.

CONCLUSIONS

Those who have written about the promise and limitations of developing computer simulations of policy and governance systems note that the purpose for undertaking this may not lie in predicting, "the future state of a given system, but to understand the system's properties and dissect its generative mechanisms and processes, so that policy decisions can be better informed and embedded within the system's behavior, thus becoming part of it" (Squazzoni & Boero, 2010: 3). Under the rubric of computer generated decision support systems, the very process of providing feedback about a systems' dynamics and network structure back to critical agents in the system itself becomes an important component of decision making and action. These models can be used "when policy makers need to *learn from science* about the complexity of systems where their decision is needed," as well as "when policy makers need to *find and negotiate certain concrete ad hoc solutions*, so that policy becomes part of a complex process of management that is internal to the system itself" (Squazzoni & Boero, 2010: 6). There is, indeed, a long history of employing computer simulation modeling to stimulate systems thinking (Mitroff *et al.*, 1974).

The complexity of most policy and governance systems will not easily reveal itself to modelers. In many instances the complexity of a governance system may remain a mystery to those who are most intimately involved in its daily operations. We believe that by partnering with the actors who contribute to the management of a policy or governance system a thicker description of the system may be rendered. These thick descriptions may, in turn, be used to design calibrated models of operating policy and governance systems.

We have asserted here that Newtonian assumptions regarding rational action and the linearity of causes and effects are not sufficient enough to describe, evaluate and design effective policy and governance systems. The persistence of wicked problems, from funding for health care, to managing the stability of our financial system or climate, to the preservation of our ecosystems and social welfare, suggests that we may cling exclusively to the Newtonian approach to science at our peril. A complex systems approach, on the other hand, may open up new vistas to confront these wicked public problems.

ACKNOWLEDGEMENT

The authors wish to recognize the support of the James M. Jeffords Policy Research Center at the University of Vermont for the writing of this chapter.

REFERENCES

Agranoff, R. (2007). *Managing within networks: Adding value to public organizations*. Washington, DC: Georgetown University Press.

Agranoff, R., & McGuire, M. (2003). *Collaborative public management: New strategies for local governments*. Washington, D.C.: Georgetown University Press.

Axelrod, R., & Cohen, M. (1999). *Harnessing complexity: Organizational implications of a scientific frontier*. New York: Free Press.

Bankes, S. (2002). Tools and techniques for developing policies for complex and uncertain systems. *PNAS*. 99(3): 7263-7266.

Baumgartner, F. R., & Jones, B. D. (1993). *Agendas and instability in American politics*. Chicago and London: The University of Chicago Press.

Baumgartner, F. R., & Jones, B. D. (eds.). (2002). *Policy dynamics*. Chicago: University of Chicago Press.

Batty, M. (2005). *Cities and complexity: Understanding cities with cellular automata, agent-based models and fractals*. Cambridge, MA: MIT Press.

Buijs, J.M., Eshuis, J., & Byrne, D. (2009). Approaches to researching complexity in public management. In Teisman, G., van Burren, A. and Gerrits, L. (eds.) (2009). *Managing complex governance systems: Dynamics, self-organization and coevolution in public investments*. New York: Routledge Press, pp. 37-55.

Dennard, L., Richardson, K.A. & Morcol, G. (eds.) (2008). Complexity and policy analysis: Tools and methods for designing robust policies in a complex world. Goodyear, AZ: ISCE Press.

Goldsmith, S., & Eggers, W. (2004). *Governing by network*. Washington D.C: Brookings.

Grimm, V. Revilla, E., Berger, U., Jetsch, F., Mooij, W., Railsback S., Thulke, H., Weiner, J., Wiegand, T., and DeAnelis, D. (2005). Pattern-oriented modeling of agent-based complex systems: Lessons from Ecology. *Science*. 310(5750): 987-991.

Jones, B.D. (1994). *Reconceiving decision-making in democratic politics: Attention, choice and public policy*. Chicago: University of Chicago Press.

Johnston, E., Hicks, D., Nan, N., & Auer, J. C. (2010) Managing the Inclusion Process in Collaborative Governance. *Journal of Public Administration Research and Theory 2010*. doi: 10.1093/jopart/muq045

Johnston, E., Nan, N., Hicks, D., & Zhong, W. (2008) Between implementation and outcomes, growth matters: Validating an agent-based modeling approach for understanding collaboration process management. Special Journal of

Complexity and Innovation of *The Innovation Journal.* 13 (3).

Katz, D., & Khan, R. (1978). *The social psychology of organizations.* New York: Wiley.

Kettl, D. (2006). Managing boundaries in American administration: The collaborative imperative. *Public Administration Review.* 66(6): 10-19.

Kickert, W. J. M., Klijn, E.-H., & Koppenjan, J. F. M. (eds.). (1997). *Managing complex networks: strategies for the public sector.* London: Sage Publications.

Kiel, L. D. (1994). Managing chaos and complexity in government: A new paradigm for managing change, innovation and organizational renewal. San Francisco: Jossey-Bass.

Kingdon, J. (1984). *Agendas, alternatives, and public policies.* New York: Harper Collins.

Klijn, E.-H., & Snellen, I. (2009). Complexity theory and public administration: A critical appraisal. In Teisman, G., van Burren, A. and Gerrits, L. (eds.) (2009). Managing complex governance systems: Dynamics, self-organization and coevolution in public investments. New York: Routledge Press.

Knott, J.H., Miller, G.J. & Verkuilen, J. (2003). Adaptive incrementalism and complexity: Experiments with two-person cooperative signaling games. *Journal of Public Administration Research and Theory.* 13(3): 341-366.

Koliba, C., Meek, J. & Zia, A. (2010). *Governance networks in public administration and public policy.* Boca Raton, FL: CRC Press/Taylor & Francis.

Koliba, C., Zia, A., & Lee, B.H.Y., (2011a). Governance Informatics: Managing the Performance of Inter-Organizational Governance Networks. *The Innovation Journal: The Public Sector Innovation Journal.* 16(1).

Koppenjan, J., & Klijn, E. (2004). *Managing uncertainties in networks.* London: Routledge.

Marion, R. (1999). *The edge of organization: Chaos and complexity theories of formal social systems.* Thousand Oaks, CA: Sage.

McNabb, D. E. (Ed.). (2007). *Knowledge management in the public sector: A blueprint for innovation in government.* Armonk, NY: M.E. Sharpe.

Miller, J. H., & Page, S. E. (2007). *Complex adaptive systems: An introduction to computational models of social life.* Princeton: Princeton University Press.

Milward, H., & K. Provan (2006). A manager's guide to choosing and using collaborative networks. *IBM Center for the Business of Government.*

Milward, H., & Provan, K. (1998). Principles for controlling agents: The political economy of network structure. *Journal of Public Administration Research and Theory,* 8(2), 203-222.

Mitroff, I. I., Betz, F., Pondy, L. R., & Sagasti, F. (1974). On managing science in the systems age: Two schemes for the study of science as a whole systems phenomenon. *Interfaces,* 4(3): 46-58.

Nan, N., Johnston, E., & Olson, J. (2008) Unintended consequences in central-remote office arrangement: A study coupling laboratory experiment with multi-agent simulation. *Computational & Mathematical Organization Theory*. 14 (2): 57-83.

Ostrom, E. (2005). *Understanding institutional diversity*. Princeton, NJ: Princeton University Press.

Pierson, P. (2004). *Politics in time: History, institutions, and social analysis*. Princeton, NJ: Princeton University Press.

Pollitt, C. (2009). Complexity theory and evolutionary public management. In Teisman, G., van Burren, A. & Gerrits, L. (eds.) (2009). *Managing complex governance systems: Dynamics, self-organization and coevolution in public investments*. New York: Routledge Press, pp. 213-230.

Rhodes, R. (1997). *Understanding governance: Policy networks, governance, reflexivity and accountability*. Buckingham: Open University Press.

Richardson, G. (1991). *Feedback thought in social science and systems theory*. Philadelphia: University of Pennsylvania Press.

Richardson, K.A. (2008). On the limits of bottom-up computer simulation. In Dennard, L., Richardson, K.A. & Morcol, G. (eds.) (2008). *Complexity and policy analysis: Tools and methods for designing robust policies in a complex world*. Goodyear, AZ: ISCE Press, pp. 37-54.

Sabatier, P. A., & Jenkins-Smith, H.C. (1993). *The advocacy coalition framework: An assessment*. Boulder, CO: Westview Press.

Sabatier, P.A. (2007b). Fostering development of policy theory. In Sabatier, P. A. (ed.). (2007). *Theories of the policy process* (Second ed.). Boulder: Westview Press, pp. 321-336.

Sabatier, P. A., & Weible, C. M. (2007). The advocacy coalition framework: Innovations and clarifications. In P. A. Sabatier (ed.), *Theories of the Policy Process*. Boulder, CO: Westview Press, pp. 189-220.

Schlager, E. (2007). A comparison of frameworks, theories and models of policy processes. In Sabatier, P. A. (ed.). (2007). *Theories of the policy process* (Second ed.). Boulder: Westview Press, pp. 293-319.

Sorensen, E., & Torfing, J. (2005). The democratic anchorage of governance networks. *Scandinavian Political Studies*. 28(3): 195-218.

Squazzoni, F. & Boero, R. (2010). Complexity-friendly policy modeling. In Ahrweiler, P. (ed.) *Innovation in complex systems*. London: Routledge.

True, J.L., Jones, B.D., & Baumgartner, F.R. (2007). Punctuated-equilibrium theory: Explaining stability and change in public policymaking. In Sabatier, P. A. (ed.). (2007). *Theories of the Policy Process* (Second ed.). Boulder: Westview Press, pp. 155-188.

Tsoukas, H. (2005). Complex knowledge: Studies of organizational epistemology. New York: Oxford University Press.

Wright, D. S. (2000). *Models of national, state, and local relationships*. Washington, D.C.: CQ Press.

Yang, K., & Bergrud, E. (Eds.). (2008). *Civic engagement in a network society*. Charlotte, NC: Information Age Publishing.

Zahariadis, N. (2007). The multiple streams framework: Structure, limitations, prospects. In Sabatier, P. A. (ed.). (2007). *Theories of the policy process* (Second ed.). Boulder: Westview Press, pp. 65-92.

Zia, A., Metcalf, S., Koliba, C. & Widener, M. (2010). Management of complex governance networks: Opportunities and challenges for agent-based models of policy analysis. American Society of Public Administration Annual Conference. San Jose, CA.

Zia, A., Koliba, C., Metcalf, S., & Widener, M. (2011) Agent-based models of Cross-Jurisdictional Governance Networks: Simulating the Emergence of Project Prioritization Patterns Under Alternate Policy Theoretical Frameworks and Network Structures. *"Challenges of Making Public Administration and Complexity Theory Work" (COMPACT) Conference*. Rotterdam, The Netherlands.

Christopher Koliba is an Associate Professor in the Community Development and Applied Economics Department at the University of Vermont (UVM) and the Director of the Master of Public Administration (MPA) Program. He possesses a Ph.D. and an MPA from Syracuse University's Maxwell School of Citizenship and Public Affairs. His research interests include governance networks and complex adaptive systems, organizational learning and development, action research methods, civic education, and educational policy. His current research program focuses on the development of complex adaptive systems models of regional planning, watershed governance, food systems, and transportation planning, and smart grid energy networks. Chris is the lead author of Governance Network in Public Administration and Public Policy published in the fall of 2010 by Taylor & Francis and has published over 30 articles in such prominent peer reviewed journals as Public Administration Review, the Journal of Comparative Policy Analysis, International Journal of Public Administration, Administration & Society, American Journal of Evaluation, Ecological Economics, Educational Policy, and the American Behavioral Scientist. He, along co-authors Russell Mills and Asim Zia, is the 2011 Marshall E. Dimock Award winner for best lead article in the Public Administration Review. He has served as principal investigator or co-principal investigator in research and program development grants from the National Science Foundation, the United States Department of Transportation, the United States Department of Agriculture, the United States Department of Education, the Corporation for National and Community Service, and the Spencer Foundation. He is chair of the Complexity and Network Studies section of the American Society of Public Administration.

Asim Zia is currently serving as Assistant Professor in the Department of Community Development and Applied Economics at the University of Vermont. He has a Ph.D. in Public Policy from the Georgia Institute of Technology. His research is focused on the development of computational and complex-systems based approaches for Policy Analysis, Governance Informatics and Adaptive Management. His recent peer-reviewed publications have appeared in journals such as Ecology and Society, Sustainability, Public Administration Review, Conservation Biology, Journal of Comparative Policy Analysis, Public Understanding of Science, The Innovation Journal, Transportation Research Part D, and Public Performance and Management Review. Most recently, he has co-authored with Chris Koliba and Jack Meek a book entitled "Governance Networks in Public Administration and Public Policy", published by CRC Press in 2010.

Chapter 7

GOVERNANCE NETWORK ANALYSIS: EXPERIMENTAL SIMULATIONS OF ALTERNATE INSTITUTIONAL DESIGNS FOR INTERGOVERNMENTAL PROJECT PRIORITIZATION PROCESSES

Asim Zia (University of Vermont, USA)
Christopher Koliba (University of Vermont, USA)
Yuan Tian (University of Saskatchewan, CAN)

Increasingly, governance networks are comprised of actors from a diverse array of public agencies at multiple scales and private and non-profit sectors to facilitate decision making on complex public policy problems, such as building sustainable transportation infrastructure, reducing green house gas emissions and improving water quality in freshwater lakes. This paper introduces "governance network analysis" as a policy analytical tool to compare alternate institutional designs of intergovernmental decision-making processes that are aimed at addressing complex public policy problems. A prototypical Agent Based Model (ABM) of transportation governance network in the state of Vermont is presented to demonstrate an application of governance network analysis in real world public policy settings. This ABM simulates the dynamics of transportation project prioritization processes under alternate inter-governmental institutional rule structures and assesses their impacts on financial investment flows

from federal to state, regional and local scale governments. The current version of the ABM is limited to simulating roadway projects that are primarily funded through US Surface Transportation Program (STP) and Interstate Maintenance Program (IMP). Multiple focus groups, individual interviews, and analysis of federal, state and MPO transportation project and program data informed the development of this prototypical ABM. This paper presents the results from experimental simulations to test system-wide effects of alternate institutional designs on the differential emergence of roadway project prioritization patterns and funding allocations across regions and local towns. Implications are drawn for the application of governance network analysis to study alternate institutional designs of inter-organizational governance networks for addressing complex public policy problems.

INTRODUCTION: GOVERNANCE NETWORK ANALYSIS

Increasing complexity of many governance and public policy problems has led to the evolution of "novel" public-public, private-private and public-private partnerships at multiple geographical scales. In the broader literature on public policy, public administration, political science, and public management, these partnerships are broadly characterized as "governance networks" (Bogason & Musso, 2006; Coen & Thatcher, 2008; Haas, 2004; Ingram *et al.*, 2005; Jones *et al.*, 1997; Kickert *et al.*, 1997; Klijn, 1996; Klijn & Skelcher, 2007; Koliba & Zia, 2009; Koliba *et al.*, 2010; Lowndes & Skelcher, 1998; Meyer & Baltes, 2004; O'Toole, 1997; Park, 1996; Provan & Kenis, 2007; Skelcher, 2005; Sorensen, 2002; Sorensen & Torfing, 2005; Torfing, 2005; Zia & Koliba, 2011). Koliba *et al.* (2010) define governance networks as relatively stable patterns of coordinated action and resource exchanges; involving policy actors crossing different social scales, drawn from the public, private or non-profit sectors and across geographic levels; who interact through a variety of competitive, command and control, cooperative, and negotiated arrangements; for purposes anchored in one or more facets of the policy stream. Empirical examples of governance networks include, but are not limited to, watershed partnerships, airshed partnerships, regional planning networks, climate action partnerships, foodshed partnerships, health networks, smart grid governance networks and so forth.

The shift from "government to governance" across multiple levels of government, and among public, private and non-profit actors, through the emergence of dynamic governance networks in different public policy streams has led to calls for "governance of these governance networks" "meta-governance" of these networks, "network management" or simply "governance network analysis". Four broad clusters of questions have been identified for governance network analysis:

1. Formation: How are governance networks formed? Who is included and who is excluded from these governance networks? What are the goals of these governance networks?
2. Operation: How do governance networks operate? What type of activities are performed by governance networks? What do governance network actors decide and how? How different institutional arrangements and socio-economic structures affect the operations of governance networks?
3. Performance and accountability: How do meta-governors manage the performance of governance networks? How could accountability flows be democratically anchored in governance networks?
4. Sustainability: How could governance networks be sustained across spatial and temporal scales? What type of institutional arrangements could be facilitated by meta-governors to enable sustainability of effective and democratically-anchored governance networks?

In the past literature, classical research methods, such as case studies, interviews, and surveys, have been used for governance network analyses. While these classical methods are useful in providing some insights about analyzing governance networks, we argue in this study that computational approaches, especially agent based modeling, provide a powerful way to understand, model, steer and meta-govern the governance networks. Experimental simulations generated by validated and calibrated agent based models of governance networks could be useful in the next generation management sciences of policy informatics (Johnston & Kim, 2011) and governance informatics (Koliba *et al.,* 2011). Furthermore, computational modeling of observed governance networks provides a powerful tool to test alternate public policy and management theories (Koliba & Zia, 2011). In this study, we demonstrate the governance informatics and policy informatics capabilities of computational models of intergovernmental governance networks by simulating roadway project prioritization processes in the state of Vermont; thus primarily focusing on question 2 described above about the operations of governance networks. In the next section, we briefly describe the federal and state transport policy context of this intergovernmental transportation planning network that governs its dynamic operations in the specific context of intergovernmental project prioritization. Next, we present research methods that were used to generate project prioritization processes and decision heuristics of multi-level agents in the simulated model. In "The Structure of the Agent Based Simulation Model", we present the fundamental structure of the stochastic, multi-level agent based model. We then present findings from experimental simulations of alternate institutional designs followed by a discussion on the limitations of the current simulation model and possible next steps to further improve and generalize it. Lastly, we discuss the implications of these findings for transportation policy implementation, in particular, and public policy analysis and management, in general; and section 8 presents our conclusions.

FEDERAL AND STATE TRANSPORT POLICY CONTEXT FOR VERMONT'S INTERGOVERNMENTAL TRANSPORTATION PLANNING NETWORK

Since 1991, federal transportation laws emphasize two concepts resonating from the US DOT down to the local municipality. First, Congress is interested in developing an intermodal transportation system, where citizens can safely use multiple forms of private and public transit, "It is the policy of the United States to develop a National Intermodal Transportation System that is economically efficient and environmentally sound, provides the foundation for the Nation to compete in the global economy, and will move people and goods in an energy efficient manner." Intermodal Surface Transportation Efficiency Act of 1991 (ISTEA) PL 102–240, December 18, 1991, 105 Stat 1914 §2, codified at 1 U.S.C. § 134. Second, to qualify for federal transportation funds projects need to be carried out with the cooperation of state and local governments: the planning process needs to be continuing, comprehensive and cooperative. This principle of "3C's has been in place since the Federal Aid Highway Act of 1962. Pub. L. No. 87-866, § 1, 76 Stat. 1145, 1148 (1962). The detailed planning system intertwines municipalities, state transportation agencies, and the US DOT. The circular framework requires a hierarchal planning system supplemented by federal funding for infrastructure development. In Vermont, VTrans is delegated the tasks assigned to the State. Also, the Chittenden County MPO (CCMPO), the only Vermont MPO, and nine other regional planning commissions collaborate with the State to incorporate the regional transportation planning needs.

The planning process and financial structure developed in ISTEA, TEA-21, and SAFETEA-LU, creates both a complementary and competitive partnership between the regional commissions/MPO and VTrans. Federal legislation directs the state to act as an intermediary between the MPOs/regional commissions and the US DOT. ISTEA required each state transportation agency to accumulate each TIP from MPOs and assemble a Statewide Transportation Improvement Program (STIP). Similar to the MPO, the state must demonstrate the need for a project, the financial requirements, and the proposed funding location. Also, each state is required to create a statewide Long Range Plan (SLRP). Congress's choice to require both the state and individual MPOs to create a LRP and TIP demonstrates its intent to commence planning at the local level while requiring involvement with all agencies and individuals with expertise in the transportation sector. Still, the collaborative network for decision-making reveals a constant competitive struggle between the MPOs/RPCs and state agencies.

First, inherent in the federal legislation, the state has the discretion to either sub-allocate specific money to an individual municipality for implementing projects, or directly finance, organize, and implement each project. The state typically chooses to complete each project, with the MPO/RPC acting as a consulting partner. This financing and implementation structure creates conflicts between the state and MPO over desired funded projects. While the MPO desires specific projects, the state may choose to go down a separate path. For example, the current VTrans financing structure promotes a "road to affordability." VTrans desires to give priority to projects that upgrade the existing system, and strays away from any new development or capacity improvements. This structure could preempt the MPO's/RPCs' efforts to try to alleviate local congestion with innovative projects, such as walkable communities with pedestrian and bicycle-friendly environments. Because the state legislature retains final say in the STIP project prioritization, the state can prevent local projects from gaining federal financing.

Second, the state's ultimate decisions remain a political process that lacks transparency. The Governor and state legislature's final prioritization decisions remain difficult for planning commissions and the general public to comprehend. In Vermont, VTrans makes an effort for the CCMPO, RPCs and individual citizens to understand the project prioritization process by using technical data and spreadsheets, illustrating a quantified reason for implementing a project. VTrans quantifies each project with a scoring equation, puts the data into a spreadsheet, and provides the table to the legislature. Vermont statute, 19 V.S.A. 10(b), requires the legislature to evaluate the projects with this technical data. Yet, the final decision remains a political process because the legislature can disregard the quantified information when choosing the projects to complete.

While the transportation laws generate competition between the CCMPO/RPCs and VTrans, the LRP and TIP processes require the two entities to collaborate. The CCMPO/RPCs develop their LRP and TIP in partnership with VTrans, and VTrans develops the SLRP and STIP in cooperation with the CCMPO/RPCs. Further, the STIP incorporates all projects included in the MPO's/RPCs TIP. If the state desires to fund projects inside an MPO/RPC jurisdiction, then the project must be placed onto both the MPO's/RPC's TIP and VTran's STIP. Otherwise, the US DOT will not allocate funding to the state.

Under ISTEA, the STIP must address statewide transportation and tourism needs. 23 U.S.C. 135(d) requires the state to incorporate MPO/RPC TIPs; establish efficient planning for non-urban communities; and plan systems to improve bridges, roadways, and general transportation safety. In creating the STIP, the state accepts all TIP requested projects so long as "full funding can reasonably be anticipated to be available for such project within the time period contemplated for completion of the project." ISTEA. 23 U.S.C. 135(f) (2). Therefore, each state presents the US DOT with their proposition to improve the statewide transportation needs in both urban communities and rural areas, and maintain the interstate highways.

While TEA-21 maintained ISTEA's programs and policies emphasizing local involvement and intermodal transportation, the legislation focused transportation development on promoting competition within the global economic market. With TEA-21, Congress concentrated on maintenance and improvement for roadway surface projects "that will serve the mobility needs of people and freight and foster economic growth within and through urbanized areas." TEA-21. 23 USC 135(a) (1). Further, TEA-21 recognized the state's need to communicate with both MPOs and local officials of nonmetropolitan areas, and accorded the nonmetropolitan regions a voice. The legislation required the state to coordinate with both municipal government officials and nonmetropolitan Regional Planning Commissions.

SAFETEA-LU requires the state to develop a SLRP and STIP that is "continuing, cooperative, and comprehensive to the degree appropriate, based on the complexity of the transportation problems addressed." SAFETEA-LU. 23 U.S.C. 135(a) (3). The state must continue to coordinate with each MPO/RPC in creating a STIP. Also, SAFETEA-LU inherited the state's general considerations for the planning process from TEA-21, except making security and safety separate concerns and promoting consistency between municipal zoning, statewide land use regulations, and the STIP. SAFETEA-LU also requires each state to consult with "local agencies responsible for land use management, natural resources, environmental protection, conservation, and historic preservation." SAFETEA-LU. 23 U.S.C. 135(1). When available, the SLRP takes into account municipal zoning maps, statewide planning maps, and "inventories of natural and historic resources." SAFETEA-LU. 23 U.S.C. 135(2). Further, the state – after consulting with the Federal and state wildlife, land management, and regulatory agencies – must include potential environmental mitigation activities into the SLRP. These two additions in SAFETEA-LU, coupled with the expansive environmental language in 23 U.S.C. 135 (d)(1)(E), demonstrate that the state must carefully consider environmental effects caused by the transportation sector.

The US DOT apportions money to each state in accordance with the approved STIP. The Federal Highway Administration (FHWA) authorizes the apportionment and certifies to the state that funding is available to them for each program. Once the US DOT apportions the money to the state, the agency can only reacquire the funds through congressional action or lapsing. ISTEA required that the apportionment to a state be based on the ratio between the population in that specific state's urbanized areas in comparison to the total population in all urbanized areas throughout the country. Yet, the formula established a floor, with each state required to receive, at a minimum, half of 1% of the total amount apportioned. The amount apportioned to each state primarily depended on the population size within that specific state, with each state guaranteed some federal money.

The ISTEA established equations to determine the amount of funding to apportion to each state, and how each state should use the funds. The state continues to collect all capital distributed by the US DOT, and apportions the money to the several regions of that state. After the state provides funding to the specific projects financed through legislative earmarks, the state is instructed to authorize 62.5% of the remaining money to large urban areas and small rural communities. The state uses the remaining 37.5% at their discretion, with the discretionary funding based on the STIP. The project-financing framework requires the state, within the STIP, to designate the amount of funding desired for each project from a specific US DOT managed fund. The US DOT requires the state to use the apportioned money within four years, or the remaining amount lapses and the US DOT reserves the authority to redirect those funds. 23 U.S.C. 144(e). The Highway Bridge Program (HBP) illustrates one exception to the required percentage apportionments. The HBP stipulates that at least 15% of the state's apportionment must be used for public bridge projects not part of a Federal-aid Highway.

The state acquires money for each individual project through one of several Federal-funding categories. The US DOT informs the states of the amount of capital available in each fund every fiscal year, how the money is distributed, the available length of time to use the allocation, and the eligible activities for using the funds. Further, Congress can change each program through modification, addition, or elimination. For example, Congress broadened the ISTEA funded HBP in SAFETEA-LU to include systematic preventative maintenance. Also, Congress added the Highway Safety Improvement Program into SAFETEA-LU, to provide for strategic highway safety planning; replacing the former highway safety infrastructure program. As the financial allocation changes, the US DOT requires the state and MPO/RPC to reassess each project's likelihood to be funded by the US DOT.

Another method for each state to secure federal funding for a "high priority" project is through specific allocations. An allocation is a distribution of funds with no formula described in the laws. The US DOT can grant an allocation to the specific state at any time during the fiscal year. Congress allocates specified funds to particular projects through legislative earmarks, or can include these in committee report's statements of congressional intent. The funding for these projects is large but limited; therefore not every state receives allocations. If the state fails to use an allocation within a specified time, the funding can be redirected to other states or other federal priorities.

Since the US Congress enacted ISTEA in 1991, transportation legislation reduced the focus on building new highways, and instead emphasizes the preservation of existing infrastructure. This new framework establishes a more complex and intertwined federal, state, and local planning hierarchy by strengthening the planning roles of the MPO and encouraging coordination between states and local municipalities. Currently, the MPO must develop a twenty-year LRP, consistent with a short term TIP, in coordination with the municipalities, state transportation agency, FHWA representatives, and interested public citizens. The state transportation agency incorporates the approved LRP and TIP, in collaboration with the MPO and local government officials, into the SLRP and STIP. The SLRP and STIP are presented to the state legislator and Governor, who finalize the STIP's prioritization list through a political voting process. Finally, the US DOT apportions funds, in accordance with the STIP, to the state's transportation agency. Then, depending on state regulations, the state completes federally funded projects by either distributing money to a municipality or completing the project themselves.

RESEARCH METHODS

Interviews and Focus Groups

A series of in-depth interviews and two focus groups with multiple stakeholders, including local government officials, CCMPO staff and board members, VTRANS officials, FHWA and USDOT officials, federal and state senator office representatives, and local NGOs were conducted in the fall of 2010 to enhance our understanding about the intergovernmental project prioritization process. Figure 1 shows a schematic representation of this process for the state of Vermont. Decisions concerning which transportation projects to scope, plan and build are made by a complex latticework of intergovernmental networks of organizations ranging from the USDOT, State DOTs, MPOs and Rural Planning Organizations (RPOs), and municipal governments, as shown and explained in Figure 1.

From the focus groups and interviews with MPO and RPC officials, we elicited the current institutional rule structure, i.e., criteria for project prioritization that are followed by the state and MPOs/RPCs in prioritizing transportation projects. Table 1 shows these rules for six project classes, including roadways, paving, bridges, bike/pedestrian, traffic operations and park & ride. Currently, the agent based model is limited to simulating the project class of roadways.

Analysis of Project Prioritization Data

We have analyzed CCMPO and VTRANS project scoring data from 2006 to 2010 to elicit their numerical decision heuristics. The remaining 9 RPCs in the state

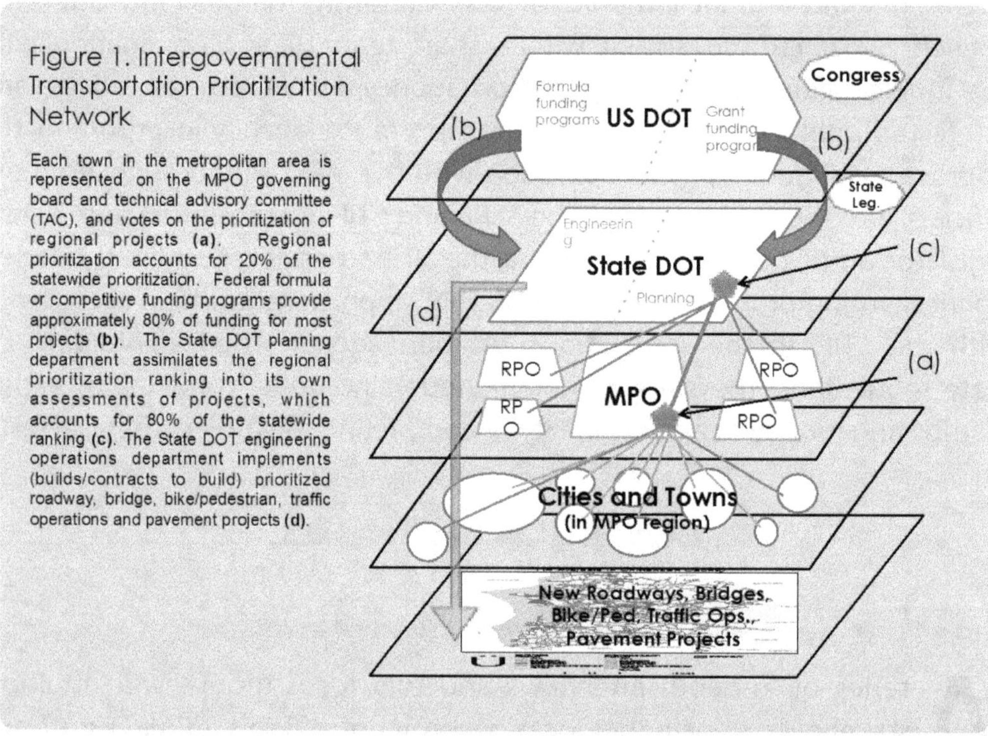

Figure 1 *Intergovernmental Project Prioritization Network*

of Vermont merely rank the projects without formal Multi Criteria Analysis and provide their rankings to the state agent. For each of the six project types, as shown in Table 1 and equation 1, CCMPO calculates an expected value function (CCMPO_EV) for each project included on TIP, as shown in Equation 1:

$$\text{CCMPO\_EV}(\text{Project}_i) = \Sigma_{j=1}\, w_j . X_j \qquad \text{[Equation 1]}$$

Where, there are i=1,2...N projects and j=1,2....M criteria. Each project is evaluated qualitatively on seven criteria, denoted by *Xj*. Further, all seven criteria are weighted equally, i.e., $\Sigma_{j=1} w_j = 1$ and $w_1=w_2=w_3=w_4=w_5=w_6=w_7$. Table 2 below shows descriptive statistics for CCMPO_EV, its natural log transformation (Ln_CCMPO), and seven criteria scores (Xj) from 2007 to 2010 TIPs for Roadway projects. Standard errors are shown in brackets.

	State Level		Regional Level			
	SDOT Criteria		Chittenden County		Other Counties	
Project Class		Wt.	MPO Criteria (applied across all classes)	Wt.	RPC Criteria since 2006	Wt.
Roadway	Highway system*	.40	Economic vitality Safety and security Accessibility, Mobility and connectivity Environment, Energy and Quality of life Preservation of Existing System Efficient System Management	.166 @	The impact of the project on congestion and mobility conditions in the region The availability, accessibility and usability of alternative routes The functional importance of the highway or bridge as a link in the local, regional or state economy The functional importance of the facility in the social and cultural life of the surrounding communities. Conformance to the local and regional plans Local support for the project.	Ranked by priority from 1 (being the highest) to 5 (being the lowest)
	Cost per vehicle mile*	.20				
	Regional priority	**.20**				
	Project momentum	.20				
Paving	Pavement condition index*	.20				
	Benefit/cost*	.60				
	Regional priority	**.20**				
Bridges	Bridge condition*	.30				
	Remaining life*	.10				
	Functionality*	.05				
	Load capacity and use	.15				
	Waterway adequacy & scour suscept.	.10				
	Project momentum	.05				
	Regional input and priority	**.15**				
	Asset-benefit cost factor	.10				
Bike/ Pedestrian	Land use density	.20				
	Connectivity to larger bike/ped network	.10				
	Multi-modal access	.05				
	Designated downtown/village center	.05				
	Project cost	.20				
	Regional priority	**.20**				
	Project momentum	.20				
Traffic operations	Intersection capacity*	.40				
	Accident rate	.20				
	Cost per intersection volume*	.20				
	Regional input and priority	**.20**				
	Project momentum	.10				
Park and ride	Total highway and location*	.40				
	Cost/parking space	.20				
	Regional Input priority	**.20**				
	Project momentum	.20				

***denotes Asset Management System**

Table 1 *Decision Making Criteria For Transport Project Prioritization At State And Regional Levels*

	2007 N= 24	2008 N=15	2009 N=15	2010 N=18
Expected Value	46.08 (5.41)	49.06 (6.01)	50.53 (6.20)	47.22 (9.21)
Economic Vitality	7.41 (2.48)	8.53 (1.95)	9.06 (1.66)	8.38 (2.17)
Safety and Security	6.37 (2.22)	8.00 (1.46)	8.00 (1.46)	7.72 (1.52)
Accessibility, Mobility and Connectivity	6.70 (2.40)	7.60 (2.41)	7.86 (2.19)	7.27 (2.53)
Environment, Energy and Quality of Life	6.25 (2.57)	5.80 (2.75)	5.80 (2.75)	5.72 (2.86)
Preservation of Existing System	4.62 (2.42)	5.06 (2.81)	5.20 (2.75)	5.05 (2.55)
Efficient System Management	6.29 (1.80)	7.40 (1.80)	7.40 (1.80)	7.00 (1.87)
TIP Status	8.41 (2.56)	6.66 (3.90)	7.20 (4.05)	6.55 (4.43)

Table 2 *CCMPO Prioritization Scores: Descriptive Statistics For Roadway Projects*

The observed patterns from the project prioritization data shared by CCMPO within its jurisdictions present deterministically chaotic trends in terms of allocation of funding within the 21 jurisdictions of CCMPO that include 18 local governments and interstate highway, transit and regional projects. Figure 2 shows the observed chaotic trends of allocated funds for these 21 jurisdictions. Similar chaotic trends are observed for other local towns in the state of Vermont. When we scale up to the regional commission level, V-Trans funding allocation data (2008 to 2010) presents similar patterns of deterministic chaos in terms of annualized funding allocations.

Table 3 below shows descriptive statistics for roadway project expected values from 2007 to 2010 that were derived from V-Trans project prioritization data.

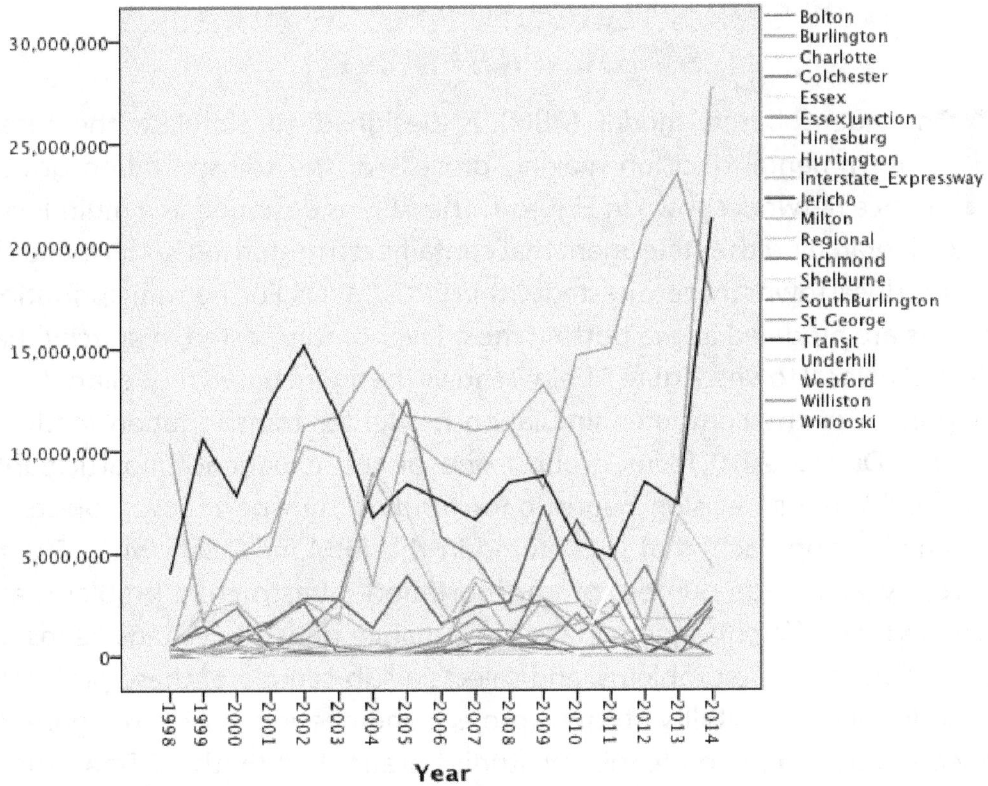

Figure 2 *Deterministically Chaotic Patterns Of Observed Funding Allocations Among Local Towns In CCMPO, 1998-2014.*

	2007 N=73	2008 N=61	2009 N=60	2010 N=61	2011 N=57
Expected Value	48.15 (15.864)	50.07 (13.103)	49.82 (13.662)	50.36 (14.709)	49.33 (15.399)
Highways System	20.808 (9.984)	20.721 (9.510)	21.050 (9.826)	20.820 (10.044)	21.474 (9.838)
Cost/vehicle mile	11.041 (5.397)	11.508 (5.793)	10.900 (5.780)	11.016 (5.895)	10.316 (5.938)
Regional priority	9.643 (6.947)	10.721 (6.802)	9.633 (7.192)	10.557 (6.714)	9.263 (7.889)
Momentum	6.658 (4.808)	7.115 (5.410)	7.066 (5.590)	6.820 (6.412)	7.228 (6.598)
Designated Downtown			1.166 (3.237)	1.148 (3.213)	1.053 (3.096)

Table 3 *V-Trans Scores for Roadway Projects from 2007-2011: Descriptive Statistics*

THE STRUCTURE OF THE AGENT BASED SIMULATION MODEL

The agent based model (ABM) is designed to simulate the inter-governmental decision making process of the transportation governance network shown in Figure 1. The ABM is designed as a multi-level, nested hierarchy with a state agent that contains ten nested RPCs/MPOs and six hundred local towns that are nested within RPCs/MPOs. Further, transportation projects are modeled at the bottom most layer of this nested hierarchy that arise out of local towns. Figure 3 below shows the agent based modeling (ABM) component of the computer simulation model for transportation roadway projects. During 2010 focus groups, one of the experienced participants described V-Trans decision heuristic for funding transportation projects as a "funneling approach" that is captured in the ABM model shown in Figure 3. Every year, a large number of transportation infrastructure problems are identified by different agents in the governance network. V-Trans keeps an updated list of these problems, and selects a sub-sample of these problems for undertaking feasibility studies (project engineering phase in Figure 3). Projects with completed feasibility studies are ranked by the V-Trans agent according to multiple criteria shown in Table 1 above. Figure 3 shows this only for roadway projects.

If a roadway project is prioritized at this stage, it goes through right of way (ROW) approval and then, if funding is available from relevant federal programs (STP and IM primarily fund roadway projects) and if state and local government matching funds are available, the project is approved for construction phase. Overall, the ABM module of the computer simulation model for roadway (shown in Figure 3) and other project classes captures this "funneling" effect of a sub-sample of initially identified transportation projects that go through three project phases (feasibility, ROW and construction). Further, this ABM loop also captures the delays and the formation of old project queues that are empirically observed in the V-Trans STIP databases.

The six hundred local town agents in the ABM contain the environment of the projects and keep track of the transportation projects that arise over time in a random basis (modeled as a uniform probabilistic function). The ten RPC/MPO agents keep track of their local town agents, as well as rank the transportation projects that are submitted to them every year based upon their multiple criteria expected value functions (shown in Table 1). The parameters

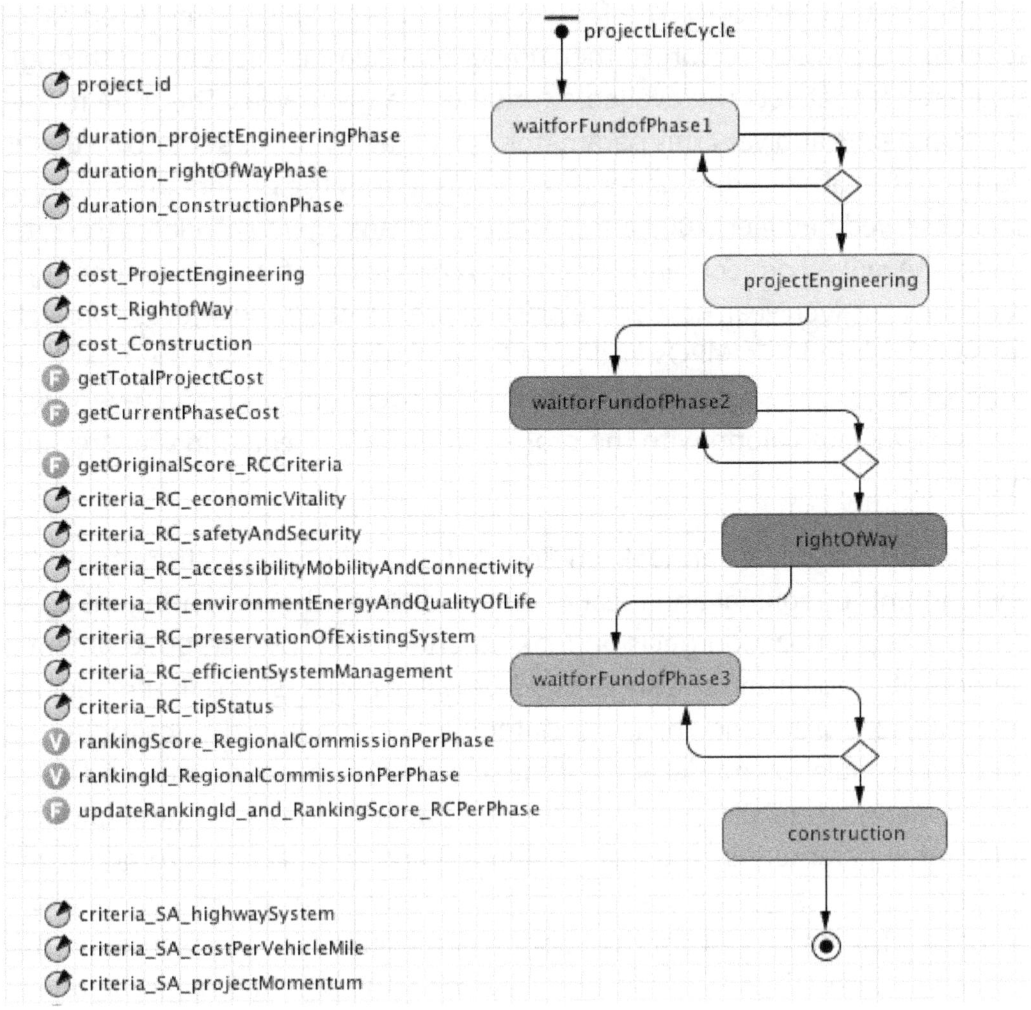

Figure 3 *Agent Based Module Of Current Computer Simulation Model For Simulating Transportation Roadway Project Prioritization Implemented By Governance Network Of State, Regional And Local Level Agents*

for the seven RPC/MPO criteria are derived from the observed probability distributions shown in Table 2. The RPC/MPO agents rank the roadway projects and send them to the state agent. Under the default Business as Usual scenario (Scenario #1), the state agent assigns 20% weight to the regional priority. So a project that is ranked number 1 by RPC/MPO is assigned 20 points by the state agent, project ranked number 2 is assigned 18 points and so on until a project that is assigned a rank of 10 or lower is given zero point by the state agent. Further, the state agent calculates its expected value for each submitted project according to its default criteria (40% highway system, 20% cost per vehicle mile and 20% to the project momentum) and ranks the projects by assigning rank no. 1 to the project with the highest expected value, 2 to the next highest and so forth. The federal agent is not explicitly modeled as an agent in the current version. Rather, the federal agent is represented through a parameter that sets the available funding in a given year. The ABM model user can define and vary this parameter. The default value is set to 10% that is closest to the observed value for V-Trans agent from the project prioritization data observed between 2008 and 2010. The V-trans agent calculates the total cost of the ranked projects and selects top ranked projects for which funding is available in a given year. Rest of the projects are rejected. The rejected projects are sent back to RPCs/MPOs and local towns for evaluation in the next time period. New projects are added (though a user defined parameter) every year that are evaluated along with the projects that were rejected in t he previous cycle.

The ABM calculates annualized flow of financial resources from the state government to regional and local town jurisdictions that is contingent upon the project prioritization decision making undertaken by intergovernmental network of regional and state agents. Approval of projects for different regions on annualized basis and their associated costs can be used to calibrate the computer simulation model against the observed data for Vermont and other similar states. The ABM presented in this paper is thus very generalizable and model parameters could be re-defined and re-calculated to adapt the model design to any intergovernmental transportation governance network in the US. The findings from the experimental simulations presented in this paper, however, are generated from the model that is calibrated to reflect the institutional structure of intergovernmental decision making in the state of Vermont and its ten RPCs/MPO. Figure 4 shows the user interface for the ABM. The users (e.g., policy makers, managers, other stakeholders) could run business as usual and alternate institutional design scenarios by defining the parameters for different scenario runs.

FINDINGS FROM EXPERIMENTAL SIMULATIONS OF ALTERNATE INSTITUTIONAL DESIGNS

First, we present experimental simulations for four scenario runs for the parametric values as shown in Table 4 below. Scenario 1 shows a baseline scenario, whereby regional agents' rankings on transport projects are only given 20% points while the weight on highway system, cost per vehicle mile and project momentum is defined at 40%, 20% and 20% respectively. This scenario assumes that 30 new roadway projects are added statewide

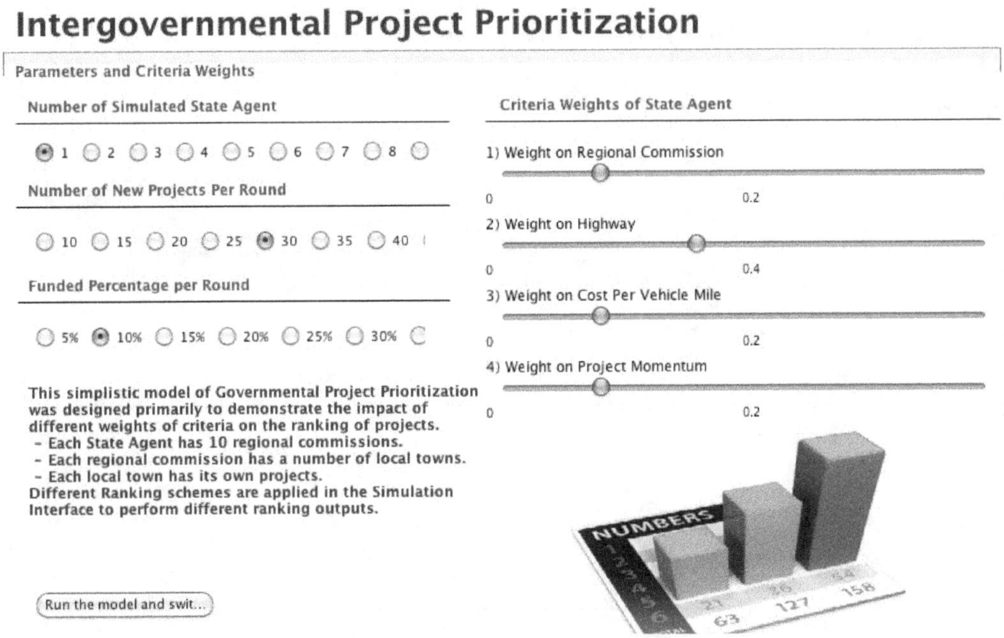

Figure 4 *User Interface of the Computer Simulation Model*

	Scenario 1 (Baseline)	Scenario 2	Scenario 3	Scenario 4
Weight On Regional Commission Priority	0.2	0.4	0.0	0.0
Weight on Highway	0.4	0.2	0.4	0.4
Weight on Cost Per Vehicle Mile	0.2	0.2	0.4	0.4
Weight on Project Momentum	0.2	0.2	0.2	0.2
Number of New Projects Per Round	30	30	30	30
Percentage of Funded Projects Per Round	10%	10%	10%	30%

Table 4 *User-Defined Parametric Values For Four Scenario Runs*

every year and that federal government only makes funding available for the top 10% ranked projects.

Figure 5 shows the emergent patters of funding allocation over a period of 25 years as well as the % of total funded projects across ten RPCs/MPOs under the baseline scenario 1. Broadly speaking, we are able to reproduce the observed deterministic chaos in terms of annualized funding allocations

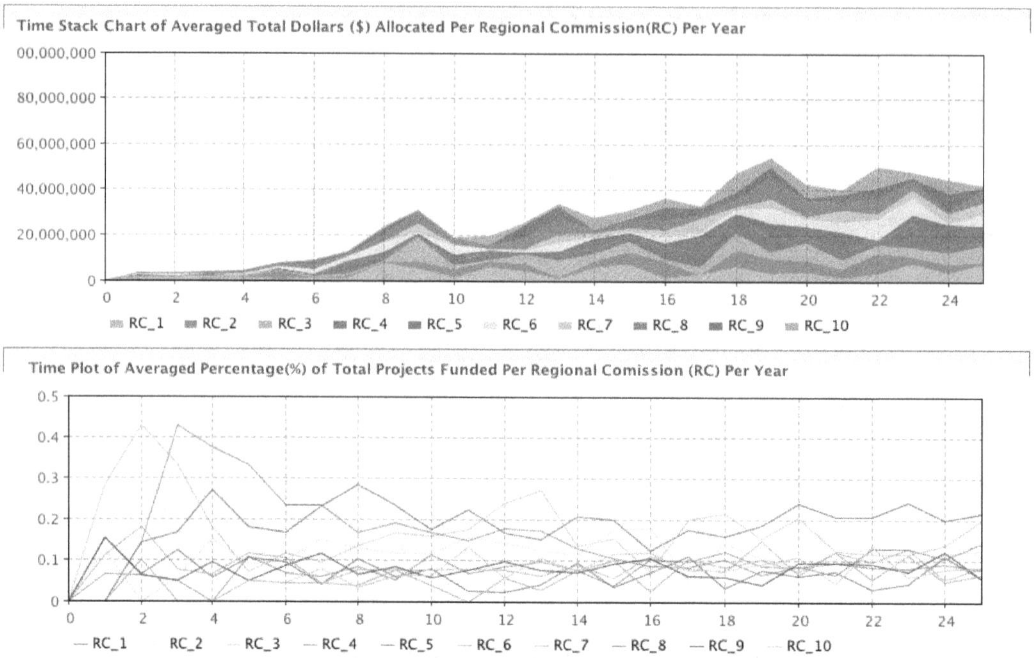

Figure 5 *The Emergent Patterns Of Funding Allocation And % Of Total Projects Funded Under The Baseline Institutional Design Scenario 1*

by modeling the existing institutional design structures for this inter-governmental transportation governance network as defined in scenario 1.

Figure 6 below shows the emergent patterns from running scenario 2. One of the focus group participants (who currently happens to be a high ranking official in the state agency) proposed changes in the institutional design structure that is modeled in scenario 2. Under this scenario, the rankings of regional governments are given 40% weight (20% more weight than the baseline scenario), while the asset management generated values for the criterion of highway system are weighed only 20% (reduced by half as compared with the baseline scenario). Noticeably, the emergent patterns in scenario 2 (Figure 6) show a reduced variability across the % of funded projects for RPCs/MPOs over time.

Figures 7 and 8 below, respectively show the emergent patterns by running the institutional design scenarios 3 and 4 that are defined to reflect complete ascendancy of state agent and its asset management system. Under both scenarios 3 and 4, the regional level agents' ranking are assigned 0% weight (which is equivalent to the assumption that the federal/state governments eliminate funding for RPCs/MPOs), while the asset management derived

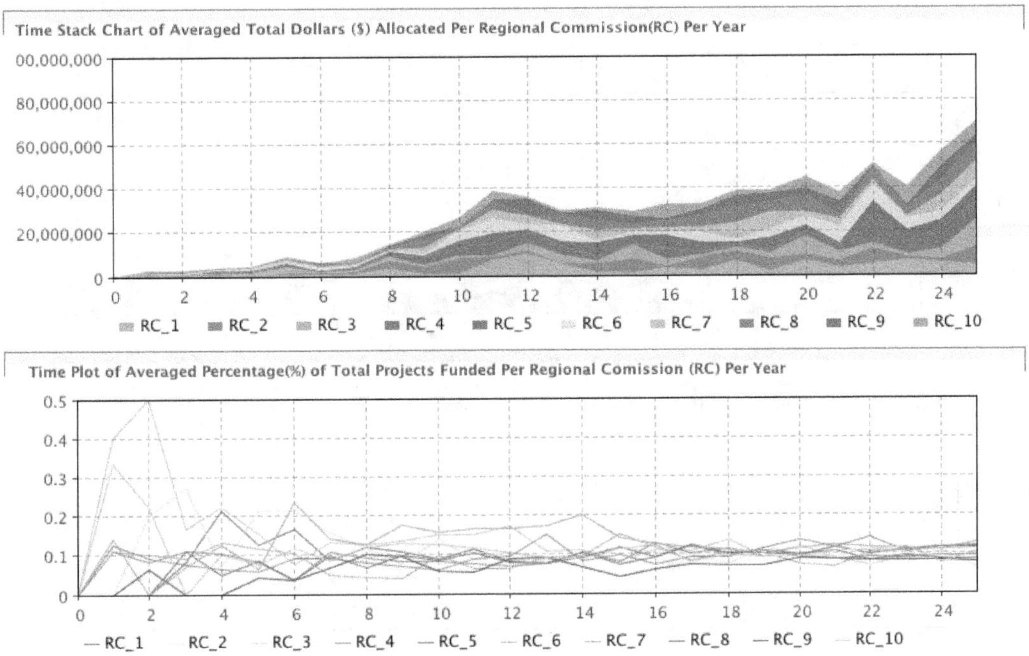

Figure 6 *The Emergent Patterns Of Funding Allocation And % Of Total Projects Funded Under Alternate Institutional Design Scenario 2*

criteria are assigned relatively higher weights (shown in Table 4). The only difference between scenario 3 and 4 is that scenario 3 funds top 10% ranked projects while scenario 4 funds top 30% of the ranked projects (i.e., more money is made available by the federal government in scenario 4 under the assumption that is made available by the elimination of RPCs/MPOs from the intergovernmental system). The deterministic chaos of funding allocations across RPCs/MPOs increases drastically under both scenarios 3 and 4, while the parametric difference in funding 10% versus 30% top ranked projects does not appear to cause much differences except the expected increase in the allocation of funds over time for scenario 4.

LIMITATIONS OF THE CURRENT ABM AND NEXT STEPS

All models are abstractions of reality. It is the function and purpose of designing computer simulation models that broadly defines the boundaries of the dynamic policy and governance system that is abstracted out of the complex reality. The main purpose of designing the ABM presented in this paper was to simulate the operational decision making

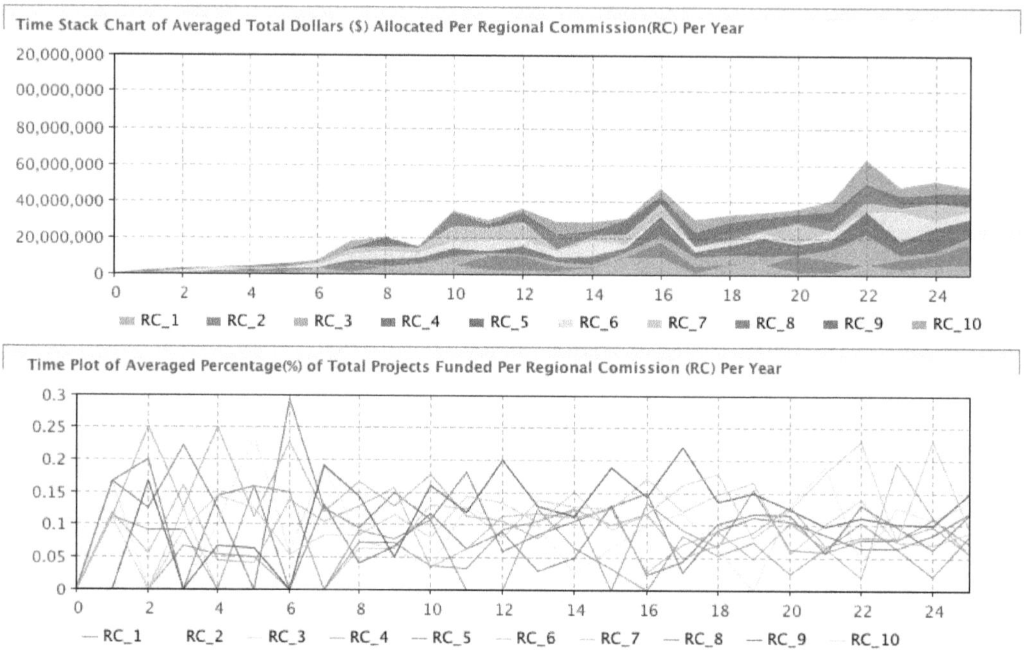

Figure 7 *The Emergent Patterns Of Funding Allocation And % Of Total Projects Funded Under Alternate Institutional Design Scenario 3*

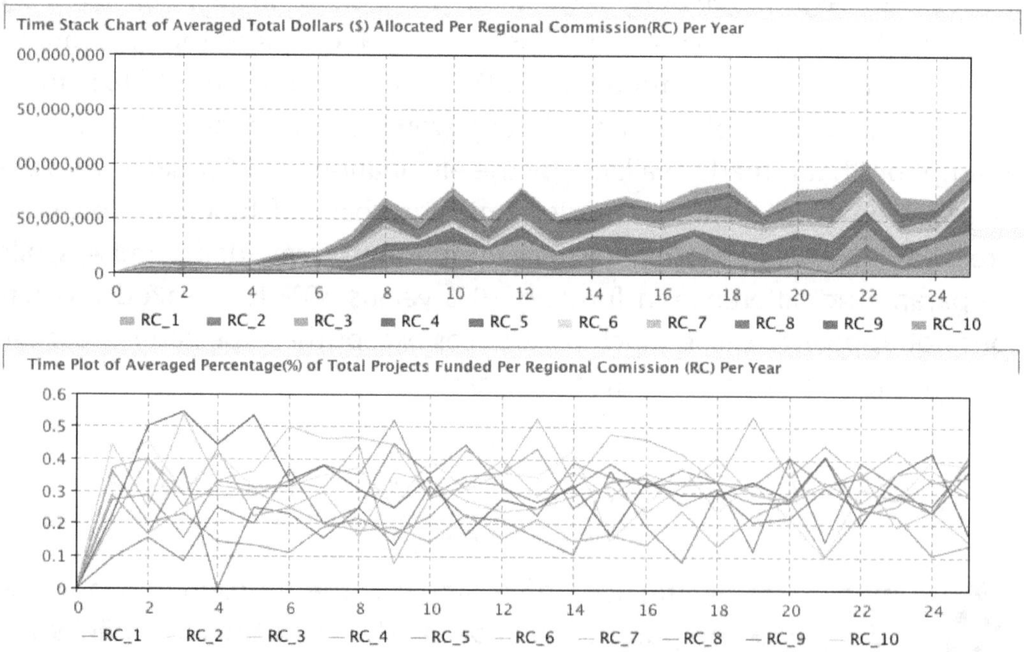

Figure 8 *The Emergent Patterns Of Funding Allocation And % Of Total Projects Funded Under Alternate Institutional Design Scenario 4*

dynamics of project prioritization processes among multi-level governments. The baseline institutional designs and rules that govern the interactions of intergovernmental governance dynamics are the focus of this study, yet the system boundaries of the ABM presented in this study led us to simplify and in some cases even ignore the institutional rules that might have direct bearing on the model outcomes presented in this study. The geophysical boundaries of ABM are limited to Vermont. We intentionally excluded five other project classes and just focused on roadways. While this simplification allowed us to focus on modeling the institutional rules across different levels of government, the competing dynamics that occur in terms of allocation of funding across different project classes (i.e., between bridges and roadway and/or bike/ped and park & rides) have been ignored in setting up the system boundaries. Similarly social boundaries have been scaled up to represent governmental organizations as agents, while in practice the networks of individual and group social actors comprise the governance networks.

There are many possible ways, both vertically and horizontally, to extend and generalize the ABM presented in this study. Within the current transport policy domain, the ABM could be made spatially explicit by adding GIS layers. Further, the ABM could be extended to all six transport project classes. Explicit rules of federal programs, such as STP, IM and others, could be incorporated in an upward expansion of the model. Similarly, in a downward expansion of the model, the complex dynamics that occur in local towns and their planning commissions/boards could be explicitly captured. Further, even at a finer grain, the outcomes of the funded projects could be ascertained by coupling integrated land-use transportation models such as UrbanSim with the governance network simulation model presented in this paper. Finally, as well known in the policy implementation literature since Pressman and Wildavsky (1973), financial capital flows that ensue from federal legislative actions to states and local governments typically entail development and execution of discrete projects. Since projects are the fundamental unit of analysis in the ABM presented in this study, similar context specific ABMs could also be developed in other policy domains, e.g., energy, health, education, environment and international development.

IMPLICATIONS OF THE FINDINGS

The computer simulation model presented in this paper could enable policy actors and relevant decision makers to visualize the effects of alternate institutional designs that govern the inter-organizational and intergovernmental network of actors implicated in implementing federal and state transportation and land-use policies through the prioritization of roadway projects. With the addition of complex policy problems such as mitigation and adaptation to climate change, prioritization processes of transport projects would need to be fundamentally reconsidered and governing policies re-designed for explicitly steering federal and state tax dollars towards the development of sustainable transportation systems and sustainable communities. The computer simulation model presented in this paper could aid decision makers in visualizing alternate institutional designs that require addition or re-weighting of project prioritization criteria to accommodate GHG emission accounting and other adaptation considerations. Further, the governance of the transportation governance network could be improved through the governance network analysis that is proposed in this study. The computational modeling of governance networks can improve our understanding about the apparently chaotic emergent patterns that arise due to overlapping and sometimes competing interests among different levels of the government, vertically, and among governments, markets and societies, horizontally.

CONCLUSIONS

In this paper, we introduced "governance network analysis" as a policy analytical tool to compare alternate institutional designs of intergovernmental decision-making processes that are aimed at addressing complex public policy problems. A prototypical Agent Based Model (ABM) of transportation governance network in the state of Vermont has been presented to demonstrate an application of governance network analysis in real world public policy settings. This ABM enables simulation of the dynamics of transportation project prioritization processes under alternate inter-governmental institutional rule structures. Further, this ABM enables decision makers to visualize the impacts of alternate inter-governmental institutional rules on the emergent patterns of financial investment flows from federal to state, regional and local scale governments. The results from experimental simulations suggest that the current institutional configuration between state and regional level governments might not be the only way to fund and prioritize roadway projects. The ABM presented in this study could be extended to test system-wide effects of alternate institutional designs on the differential emergence of project prioritization patterns by adding and/or re-weighting transport project prioritization criteria. Such computational models could also be developed to simulate public policy implementation processes in other policy domains for informing the governance of governance networks.

ACKNOWLEDGMENTS

We acknowledge USDOT funding through UVM-TRC as well as a seed grant provided by UVM's James Jeffords Center for Policy Research. Findings reported in this paper are solely the responsibility of the authors and not the funding agencies.

REFERENCES

Betsill, M.M. and Bulkeley, H. (2004). "Transnational networks and global environmental governance: the cities for climate protection program." international studies quarterly, 48:471-493.

Bogason, P. and Musso, J.A. (2006). "The democratic prospects of network governance." the american review of public administration, 36(1):3-18.

Coen, D. and Thatcher, M. (2008). "Network governance and mlti-level delegation: european networks of regulatory agencies." Journal of Public Policy, 28(1): 49-71.

Haas, P.M. (2004). "Addressing the global governance deficit." Global Environmental Politics, 4:1-15.

Ingram, P., Robinson, J. and Busch, M.L.. (2005). "The intergovernmental network of world trade: IGO connectedness, governance and embeddedness." American Journal of Sociology, 111(3): 824-858.

Johnston, E. and Kim, Y. (2011). Introduction to the special issue on policy informatics. The Innovation Journal. 16(1), http://www.innovation.cc/volumes-issues/vol16-no1.htm.

Jones, C., Hesterly, W.S. and Borgatti, S.P. (1997). "A general theory of network governance: exchange conditions and social mechanisms." Academy of management review, 22(4): 911-945.

Kickert, W.J.M., Klijn, E.-H. and Koppenjan, J.F.M. (1997). "Introduction: a management perspective on policy networks." In Managing complex networks, eds. Walter J. M. Kickert, Erik-Hans Klijn and Joop F. M. Koppenjan. London: Sage.

Klijn, E.-H. (1996). "Analyzing and managing policy processes in complex networks." administration and society, 28: 90-119.

Klijn, E.-H. and Skelcher, C. (2007). "Democracy and governance networks: compatible or not?" Public administration, 85(3): 587-608.

Koliba, C., Meek, J. and Zia, A. (2010). Governance Networks in Public Administration and Public Policy. Boca Raton, FL: CRC Press/Taylor & Francis.

Koliba, C. and Zia, A. (2009). "Dispelling the Myth of the Invisible Hand: An Argument for Democratically Legitimate Inter-Organizational Governance Networks." Administrative Theory & Praxis, 31(2): 417-423.

Koliba, C., A. Zia and Lee, B. H.Y. (2001). "Governance Informatics: Managing the Performance of Inter-Organizational Governance Networks." The Innovation Journal, 16(1):Article 3.

Koliba, C., Zia, A. (2011) Theory Testing Using Complex Systems Modeling in Public Administration and Policy Studies: Challenges and Opportunities for a Meta-Theoretical Research Program. *The American Society for Public Administration, Section on Complexity and Network Studies (SCNS), and Erasmus University Rotterdam, research group Governance of Complex Systems (GOCS) Joint Conference on "Challenges of Making Public Administration and Complexity Theory Work" (COMPACT).* Rotterdam, The Netherlands, June 2011

Lowndes, V. and Skelcher, C. (1998). "The dynamics of multi-organizational parternships: an analysis of changing modes of governance." Public administration, 76: 313-333.

Meyer, W. and Baltes, K. (2004). "Network Failures -- How realistic is durable cooperation in global governance?" In Governance for Industrial Transformation. Proceedings of the 2003 Berlin conference on the human dimensions of global environmental change, eds. Klaus Jacob, Manfred Binder and Anna Wieczorek. Berlin: Environmental policy research center.

O'Toole, L.J. (1997). "Treating Networks seriously: practical and research-based agendas in public administration." Public Administration Review, 57: 45-52.

Park, S.H. (1996). "Managing an interorganizational network: a framework of the institutional mechanism for network control." organization studies, 17(5): 795-824.

Pressman, J. and Wildavsky, A. (1973). Implementation: How great expectations in Washington are dashed in Oakland. Los Angeles, University of California Press.

Provan, K.G. and Kenis, P. (2007). "Modes of Network Governance: Structure, Management, and Effectiveness." Journal of Public Administration Research and Theory.

Skelcher, C. (2005). "Jurisdictiional integrity, polycentrism, and the design of democratic governance." Governance: an international journal of policy, administration, and institutions, 18(1): 89-110.

Sorensen, Eva. (2002). "Democratic theory and network governance." Administrative theory and praxis, 24(4): 693-720.

Sorensen, Eva and Jacob Torfing. (2005). "Network governance and post-liberal democracy." Administrative theory and praxis, 27(2): 197-237.

Torfing, Jacob. (2005). "governance network theory: towards a second generation." european political science, 4: 305-315.

White, Leroy. (2001). "'Effective Governance' through complexity thinking and management science." systems research and behavioral science, 18: 241-257.

Zia, A., and Koliba, C. (2011) Accountable Climate Governance: Dilemmas of Performance Management Across Complex Governance Networks. *Journal of Comparative Policy Analysis,* 13(5): 479-497.

Asim Zia is currently serving as Assistant Professor in the Department of Community Development and Applied Economics at the University of Vermont. He has a Ph.D. in Public Policy from the Georgia Institute of Technology. His research is focused on the development of computational and complex-systems based approaches for Policy Analysis, Governance Informatics and Adaptive Management. His recent peer-reviewed publications have appeared in journals such as Ecology and Society, Sustainability, Public Administration Review, Conservation Biology, Journal of Comparative Policy Analysis, Public Understanding of Science, The Innovation Journal, Transportation Research Part D, and Public Performance and Management Review. Most recently, he has co-authored with Chris Koliba and Jack Meek a book entitled "Governance Networks in Public Administration and Public Policy", published by CRC Press in 2010.

Christopher Koliba is an Associate Professor in the Community Development and Applied Economics Department at the University of Vermont (UVM) and the Director of the Master of Public Administration (MPA) Program. He possesses a Ph.D. and an MPA from Syracuse University's Maxwell School of Citizenship and Public Affairs. His research interests include governance networks and complex adaptive systems, organizational learning and development, action research methods, civic education, and educational policy. His current research program focuses on the development of complex adaptive systems models of regional planning, watershed governance, food systems, and transportation planning, and smart grid energy networks. Chris is the lead author of Governance Network in Public Administration and Public Policy published in the fall of 2010 by Taylor & Francis and has published over 30 articles in such prominent peer reviewed journals as Public Administration Review, the Journal of Comparative Policy Analysis, International Journal of Public Administration, Administration & Society, American Journal of Evaluation, Ecological Economics, Educational Policy, and the American Behavioral Scientist. He, along co-authors Russell Mills and Asim Zia, is the 2011 Marshall E. Dimock Award winner for best lead article in the Public Administration Review. He has served as principal investigator or co-principal investigator in research and program development grants from the National Science Foundation, the United States Department of Transportation, the United States Department of Agriculture, the United States Department of Education, the Corporation for National and Community Service, and the Spencer Foundation. He is chair of the Complexity and Network Studies section of the American Society of Public Administration.

Yuan Tian is currently a Research Officer at the University of Saskatchewan, conducting modeling and analytical work on stroke prevention and control, working closely with the director in University of North Carolina. Her research work is primarily focused on building both population-level and individual-level computational models of stroke and exploring the cost-effectiveness of interventions via sensitivity and uncertainty analysis. Yuan holds a Master degree in Computer Science from University of Saskatchewan, where she focused on individual-based and System Dynamics modeling of infectious diseases, and a Bachelor degree in Information Management and Information System.

SECTION II

COMPLEXITY IN THE REAL WORLD

Chapter 8

THE APPLICATION OF A COMPLEXITY ANALYTICAL LENS TO EXPLAIN AND UNDERSTAND PUBLIC POLICY PROCESSES

Elizabeth Eppel (Victoria University of Wellington, NZL)

An analytical lens developed from complexity concepts is used to create a useful explanation of policy processes. This lens is demonstrated through an empirical example based on a qualitative, interpretative study of tertiary education policy processes in New Zealand (Eppel, 2010). The lens, incorporating the concepts of interdependent actors, interacting complex systems, feedback loops, adaptation, self-organization and emergence is used to create a holistic understanding of the socially interactive, reflexive dynamism of policy processes. Implications for the development and implementation of public policy and the practice of policy analysis are identified. Rather than the conventional notion of thinking of policy processes as achieving optimal or best solutions to problems, complexity theory indicates that we need to think of them as facilitating the achievement of self-consistent, stable patterns which allow for adaptation, coevolution and emergence to occur in local contexts.

INTRODUCTION

While many scholars refer to the 'policy process' as an entity, assumptions about these processes differ. For example, scholars who adopt a stepped approach to analyzing policy processes (such as Bardach, 2005; Dunn, 2008) assume (however tacitly) that the sum of the components reveals the whole. Moreover, conceiving public policy processes as 'wholes' implies there is a boundary between them and an external environment. Many existing theories and approaches towards public policy analysis, such as Kingdon's multiple streams (1995), Sabatier's advocacy coalition framework (Sabatier & Jenkins-Smith, 1993), or Hill and Hupe's multiple governance framework (2006), refer to external environments as fixtures to be accommodated by the policy process. Moreover, causal factors are perceived to arise from an external environment which is generally perceived as a given, that is, unaffected by the policy processes (Howlett & Ramesh, 2003; Kickert, Klijn, & Koppenjan, 1997; Sabatier, 2007). This chapter proposes an alternative, complexity theory-informed, framework for understanding and explaining policy processes. It begins by first reviewing a number of strands scholars have followed in recent decades in the pursuit of theories which takes more account of the socially interactive and dynamic nature of policy processes. These developments are outlined in the next section. The review leads to the conclusion that there is a gap in existing theories of policy processes created by the challenges of dynamism in these processes, especially in their attempts to deal with complex societal problems, and the need to explain them in a holistic way (Budd, Charlesworth, & Paton, 2006; Goodin, Rein, & Moran, 2006; Morcol, 2002; Sabatier, 2007).

Thus the review of existing theories in the next section is used to set the scene for consideration of the contribution complexity theory might have to make to the understanding of policy processes. Complexity theory has been proposed recently as a potential source of insights in the quest for theories to understand public policy processes that target complex societal problems (Dennard, Richardson, & Morçöl, 2008; Klijn, 2008; Koppenjan & Klijn, 2004; Morçöl, 2010; Richardson, 2007). Some public management scholars have explored the usefulness of complexity concepts for understanding and explaining the management of complex public policy processes (Klijn, 2008; Meek, De Ladurantey, & Newell, 2007; Morçöl, 2010; van Buuren & Gerrits, 2008; Wagenaar, 2007; Teisman, van Buuren, & Gerrits, 2009). So far, complexity concepts have been applied mainly selectively, and for analyzing case studies in infrastructure policy and physical systems (e.g., Butler & Allen, 2008; Gerrits, 2010; Klijn & Teisman, 2006; Meek *et al.*, 2007; van Buuren & Gerrits, 2008). While it has been acknowledged the complexity theory has enormous potential for improving understanding of public policy processes and their management, a well developed conceptual model of a complexity-informed approach to public management is absent, and much work still remains to be done in the operationalization of these concepts to the social sciences, and their application to empirical phenomena in public policy and management (e.g., Klijn, 2008; Meek, 2010; Morçöl, 2010). The task of achieving a comprehensive complexity-informed theoretical lens is complicated by the fact that there is no single integrated complexity theory but rather a set of ideas and concepts which have their origins in the natural sciences. Research using concepts from complexity theory in 'softer', social policy domains is scarcer. Additionally, an open question concerns whether complexity theory concepts are useful only at the level of metaphor (e.g., Morçöl, 2002) or whether these concepts may contribute to better theory and a better understanding of dynamic public policy processes and their management. In the third section of the paper "Holistic Complexity Lens", complexity theory is revisited to propose a framework for applying complexity theory to public policy.

The final section illustrates the usefulness of a complexity-informed conceptual model, or 'lens', to analyze public policy processes and their management through an empirical case in a social policy domain—industry skills training policy processes in New Zealand. The exploration draws on part of a larger interpretative study of tertiary education policy processes in New Zealand (Eppel, 2010). The paper concludes in section five with implications for theory and practice of public policy analysis, built on the illustrative case.

POLICY PROCESSES

Public policy scholars' search for better explanations of policy processes and more appropriate concepts to apply to interventions targeted at wicked problems have followed a number of strands over the years. These are summarized as an introduction to the current state of thinking.

One development is the acknowledgement that there are multiple actors beyond the formal power holders and decision makers in policy process (e.g., Kingdon, 1995; Lipsky, 1980; Sabatier & Jenkins-Smith, 1993; Wildavsky, 1987).

Moreover, these actors are constrained in the rationality of their decision making (e.g., Lindblom, 1979), are bounded (March & Olsen, 1984), and become subject to cognitive overload (Jones & Baumgartner, 2005) which can bring about punctuations in policy processes. Acting as 'policy entrepreneurs', individuals can have a significant influence on policy processes (Kingdon, 1995).

A second development acknowledges the importance of institutional actors and processes and the importance of the interplay between the constitutional, governance and operational levels of these institutions in shaping policy processes (e.g., Hill & Hupe, 2006; Ostrom, 2007). Policy processes consist of sets of activity clusters, actors and action situations which take place at the level of the individual, the organization and the system (Hill & Hupe, 2006) which provides explanations for stability and resistance to change.

A third development focuses on the horizontal, nonlinear social interactions between actors, which are theorized as policy communities (Benson, 1975) and networks (Kickert, Klijn, & Koppenjan, 1997). Policy networks are a changing pattern of social relationships between interdependent actors which take shape around policy problems or policy programmes which are formed, reproduced and changed by an ecology games between these actors (Klijn, 1996). The policy processes 'game' is an ongoing sequence of strategic actions between policy process participants, governed by formal and informal rules that develop around issues or decisions in which the actors are interested (Klijn & Teisman, 1997). Much attention has been focused on understanding the internal dynamics of networks and how they might best be governed (for example, Agranoff & McGuire, 2001; Klijn, 2005). Policy processes are perceived to be influenced by 'horizontal' interactions between policy participants as well as participants' interactions with 'vertical' institutional structures (Kickert *et al.*, 1997; Rhodes, 1984). Furthermore, policy making is considered more or less unpredictable because many and varying interdependent actors are trying to influence public policy processes according to their own goals (Klijn, 1996). The advocacy coalition framework was developed to take account of multiple, interdependent, goal-conflicted policy process actors involved in the solution of wicked problems and takes account of the social interactions involved (Sabatier & Jenkins-Smith, 1993).

A fourth development acknowledges temporal dynamics in policy processes. The more or less linear and predictable notion of policy processes: problem identification leading to analysis of options, followed by decision making, implementation and review, is replaced by arguments that the actions of individual policy participants are contingent upon the actions of others; hence outcomes can be unpredictable (e.g., Bardach, 2006; Kickert *et al.*, 1997; Sabatier & Weible, 2007). Policy processes are shaped by more or less accidental couplings of problems and solutions when a window of opportunity provides conducive circumstances (Kingdon, 1995). Policy processes are treated as ongoing with no definite beginning and end. Information attention/ inattention and cognitive processing have a powerful role in shaping which issues are attended to and can create punctuations in the policy process, shifting the focus of the policy process (Jones & Baumgartner, 2005). Policy design and policy implementation are viewed as more contiguous than sequential (Hill, 1997). Also noted for shaping future policy processes are the roles of history and path dependency (e.g., Bennett & Elman, 2006; Cashore & Howlett, 2007; Kay, 2005), feedback (Easton, 1965; Pierson, 1993) and time (Pollitt, 2008).

In a fifth development, concepts of systems and complexity ideas are enlisted in various ways to focus on the interaction between the endogenous machinery of public policy processes and the policy problems involved. Scholars have rejected a closed systems' view of policy processes (for example, Easton, 1965, 1971), in which a distinction is made between the external environment, including policy problems, and the internal machinery of the public sector. Recently, some scholars have turned towards complexity theory to explore the possibility of advanced understanding of dynamics and causality in public policy. For instance, policy processes are considered as social systems, with participants, both as individuals and as members of institutions, interacting in nonlinear and interdependent ways (Chapman, 2004; John, 2003; Klijn, 2008). Interactions entail the exchange of resources, ideas and information both within policy processes, and between them and the environment.

Scholars have yet to advance a complexity-informed lens for public policy analysis. Some scholars talk about nonlinear dynamism in policy processes, applying a variety of metaphors without considering the concepts of complexity theory (e.g., Jones & Baumgartner, 2005; Bardach, 2006; Klijn & Teisman, 1997, Mayer *et al.* 2004; Wildavsky 1987). Jones and Baumgartner (2005), for example, examine 'information attention' and 'disproportionate information processing' in policy processes, resulting in a 'punctuated equilibrium' in policy events. In their view, policy processes are periodically interrupted by changes brought about by some disturbance originating externally to the policy-in-action system, however they do not consider the endogenous dynamics in policy processes, such as the role of interactions between participants. Moreover, scholars who have adopted complexity concepts, such as 'feedback loops', 'adaptation', 'coevolution', and 'emergence' with explicit reference to their origins in complexity theory, have done so selectively and independently, applied to specific parts of public policy and management. Examples include the use of 'adaptive systems' and 'self-organization' to analyze policy implementation (Butler & Allen 2008); the application of 'self-organization' and 'emergence' to the management of administrative networks (Meek *et al.*, 2007; Gerrits, 2009); and the use of 'coevolution' to explain the mutual influencing of different subsystems (Kemp *et al.*, 2007). Other examples are the application of complexity concepts to change management in the public sector (Kiel, 1994; Wallace *et al.*, 2007), and the application of complex adaptive system terminology to policy analysis (Bankes, 2002; Dennard, *et al.*, 2008; Medd, 2002; Midgley & Richardson, 2007), stability and instability in policy systems (Haynes 2008), public decision making (Gerrits, 2010), and the governance of complex systems (Teisman *et al.*, 2009).

In sum, the shifts towards the acceptance of complexity ideas in public policy and management identified above suggest an opportunity to propose a complexity-informed lens for public policy analysis. However, because of the lack of a complexity-informed conceptual model in current public policy theory (Klijn, 2008; Mörçol, 2010), it is appropriate to call on the broader spectrum of complexity literature related to biological and human social systems and organizations (e.g., social sciences, economics and human behavior) because of the commonalities they share with policy processes. The next section provides a brief introduction to the origins of complexity theory as a prelude to identification of a complexity informed lens for public policy analysis.

HOLISTIC COMPLEXITY LENS

Concepts and ideas, often labelled 'complexity theory', draw from varying disciplines. Mounting evidence since the 1950s about the relative and sometimes paradoxical nature of the physical universe implied by the quantum and relativity theories (Bohm & Hiley, 1993) has led to new insights about the physical and biological world. These new perspectives transcend the simple linear causality proposed in Newtonian theories: matter can behave unpredictably, and reorganize itself anew, without external cause. Gradually, these ideas from the natural sciences have been further developed through their use in the social sciences, economics, human behavior and organizational studies (Anderson, 1999; Waldrop, 1992). For instance, attempts to understand changes in biological communities and biological evolution subsequently broadened understanding of the concepts of adaptation, feedback loops, coevolution and emergence (Kauffman, 1995; Waldrop, 1992), and led to wider acceptance of the validity of these concepts in understanding the behavior of *all* complex adaptive systems (Kauffman, 1995; Prigogine, 1987).

Based on an extensive academic literature review of complexity theory-informed insights from different disciplines (Eppel, 2010), the following eight interconnected complexity themes can be identified that could be expected to be found in complex social processes such as public policy processes:

1. **Many interacting actors making a system 'whole'**—Many interacting actors, with interdependent dynamism between them, make up a complex system 'whole' that cannot be understood as the sum of its parts, or reduced to its parts to assist understanding (e.g., Bohm & Hiley, 1993; Kauffman, 1993; Waldrop, 1992). Social systems are made up of living, conscious organisms interacting with each other and comprise wholes that change and are changed by related systems they interact with (Allen *et al.*, 2006; Byrne, 1998; Cilliers, 1998; Midgley, 2000). Organizations are particular examples of human complex systems (e.g., Mitleton-Kelly, 2003; Richardson, 2008), and can be seen as processes for joint action (Stacey, 2003). Organizations are open to a flow of energy, actors, information, and ideas (e.g., Boisot & Child, 1999; Morgan, 1997; Stacey, 2003). While individuals can plan their own actions, they cannot plan the actions of others or the interplay of plans and actions, and therefore cannot know the whole.

2. **Nested, interacting and interdependent systems**—Complex systems can be nested within larger complex systems (Kauffman, 1995). Such nested systems show 'self-similarity' because the characteristics identified at one level of the system are also present in the 'whole'. In order to understand the interdependence of these nested, interacting systems, a holistic view is needed (Byrne, 1998, Chapman, 2004).
3. **Multiple interactive systems, creating feedback mechanisms within and between systems**—Feedback loops are reflexive influence patterns, which arise from interaction between system parts (actors). Negative feedback undoes or compensates for changes elsewhere, resulting in macro-stability. Internal features of organizations—such as structures, hierarchies, rules, controls, cultures, defensive routines, and power relations—are held in place by feedback loops locking-in particular stable patterns (Morgan, 1997). Positive feedback loops amplify changes by reinforcing the direction of change. They can also cause sudden, unpredictable and destabilizing effects.
4. **Adaptation and coevolution within and between systems**—Over time, reflexive changes in one system lead to adaptations in interacting systems, coevolution between biological, social (Kauffman, 1995), and organizational systems (Boisot & Child, 1999; Stacey, 2003). From a complexity perspective, the 'external' environment is an interacting system (Midgley, 2000). Therefore, changes in the environment may not only stimulate system change, but also external change in response. As a result, there is coevolution and adaptation between a system and its environment.
5. **Change through self-organization and emergence**—Self-organization and emergence occur through ongoing changes in feedback loops, and are integral features of biological and social systems (Byrne, 1998; Kauffman, 1993, 1995; Capra, 1996, 2002). Every change produces the stimulus for further change by self-reference to the internal state and adjustment to compensate internally and externally. Change in socio-economic systems occurs as a result of forces which act on the micro-diversity that emerges within the system as a result of feedback loops, adaptation, and emergence (Allen *et al.*, 2006; MacIntosh & MacLean, 1999). Emergence of new levels of order occurs through self-organization of systems around 'attractors' (Richardson, 2008).

Consciousness, learning, and language arise from social interactions (Blackmore, 2005; Watzlawick, 1984). Human beings are well-adapted to reflect internally on their experiences and recognise patterns that help them make sense of their environment (Weick, 1995, 2001). People respond to the complexity of their environment by either: side-stepping complexity through the codification of responses; or, by holding multiple and sometimes conflicting representations of environmental variety, and therefore retaining a repertoire of responses. Pattern recognition can also lead to unpredictable effects when a novel stimulus is set against previous learning (Weick, 1995).

6. **Open systems and socially constructed boundaries**—Social systems are open to their environments (Kauffman, 1993, 1995; Cilliers, 1998), with neither constant nor fixed boundaries. Boundaries are social constructs—artificial or socially imposed reference points—which define the limits of knowledge considered pertinent to the system and to the human agent who generates that knowledge (Midgley, 2000). References to fuzzy boundaries (Darwin, 2001), actors' bounded rationality (Simon, 1991), and the need for frame reflection (Schön & Rein, 1994), acknowledge the difficulties inherent in fluid, socially constructed boundaries.
7. **Stability is not equilibrium**—Despite a sometimes stable macro-appearance, complex systems are in fact 'far-from-equilibrium' (Byrne, 1998; Cilliers, 1998; Eve *et al.*, 1997; Kauffman, 1993, 1995; Richardson, 2008). Such systems can suddenly and unpredictably undergo changes at some critical 'bifurcation point' (Prigogine, 1987). Far-from-equilibrium is often marked by tensions or paradoxes, as changes in the feedback loops destabilise the system (Stacey, 2003). Change or instability can be disproportionate to the stimulus (Kauffman, 1995; Prigogine & Stengers, 1984). New patterns self-organise and emerge from the seemingly chaotic without external intervention. Social systems are both chaotic and stable depending on when one observes them (Eve *et al.*, 1997).
8. **The history of the system influences its starting point for change**—System histories, starting points and feedback loops can create 'path dependencies'(Prigogine & Stengers, 1984). Stabilizing path dependencies arise when negative feedback loops undo externally imposed change. Stable systems are more likely when there is a single, strong attractor influencing feedback loops. Less stable systems are characterized by multiple, weak attractors (Kauffman, 1995). Furthermore, the size, precise

timing and nature of change in a complex system cannot be predicted in advance because of the sensitivity of the system to its initial starting position and contingency of the interactions between the system parts.

USING THE COMPLEXITY LENS: AN ILLUSTRATIVE CASE

In this section, an illustrative empirical case is used to demonstrate the usefulness of all eight aspects of the complexity lens to gain deep and holistic insights about public policy processes related to industry training in the context of tertiary education policy. The case concerning the New Zealand polytechnics and industry training was part of a larger investigation of the dynamics and interdependencies of public policy processes (Eppel, 2010). The study used conversational interviews or 'narratives' (Allison, 1971; Lincoln & Guba, 1985; Patton, 2002; Riessman, 1993), based on the experiences of 65 policy process participants with different roles and involvement in New Zealand tertiary education policy. Research participants were asked about their experiences of tertiary education policy processes during the time-period under study, beginning their 'narrative' from any point they considered relevant to their understanding of what occurred. A meta-narrative of the dynamics of New Zealand tertiary education policy processes was constructed on the basis of these individual narratives and industry skills training is part of this meta-narrative. Background information, including an introduction of the main policy actors, institutions and policy processes in this section introduces the complexity analysis which then follows.

In 2003, the New Zealand government established a new permanent statutory Tertiary Education Commission (TEC), to fund and oversee all tertiary education including foundation-level adult education, industry training, polytechnics, universities and private training establishments. The tertiary sector included eight public universities, 20 public polytechnics, three wānanga[1], 40 industry-training organizations (ITOs), over 400 publicly funded private tertiary education providers (PTEs), and about 400,000 students. Creating TEC was seen as a step towards a new government's vision of linking knowledge creation, economic development and tertiary education (New Zealand Labour Party, 1999). Among the impediments to this vision, the government noted little cooperation across the sector; competition for students; increasing costs to taxpayers; homogeneity in offerings tied to

1. Indigenous Māori tertiary institutions.

popular courses; funding based on student enrolments encouraging growth in numbers taught not necessarily related to industry and society needs; and uneconomic trade and technical education, contributing to skill shortages and viability concerns in regional polytechnics.

Prior to TEC's creation, government agencies involved in the tertiary sector included the Ministry of Education (MoE), Skill New Zealand (SNZ), and the New Zealand Qualifications Authority (NZQA). MoE and SNZ funded different parts of the tertiary education sector. NZQA oversaw quality assurance of New Zealand qualifications. The new TEC was a Crown entity, controlled by a governing board appointed by the Minister, and required by statute to 'give effect to' government policy (New Zealand Government, 2004). It had a full time chair and deputy chair. On establishment, thirty staff positions, about NZ$2 million in operational funding, and grant funding of NZ$1.5 billion for research and teaching were transferred from MoE to TEC. Also, all SNZ offices, 250 staff, an operational budget of about NZ$14 million, and administered grant funding of NZ$300,000, became part of TEC. From 2003, the TEC board was responsible for funding and giving effect to the government's 2002–2007 Tertiary Education Strategy and priorities and ensuring that each tertiary education entity (polytechnic, PTE, ITO, for example) had a 'charter' approved by the Minister, and a 'profile', which set out the entity's tertiary education offerings for public funding. TEC was to decide which courses to fund, guided by the government's statement of priorities. The government made no other changes to existing funding policies. The MoE remained responsible for all education policy and advice on the financial viability of public providers such as polytechnics and universities.

The establishment of TEC coincided with a period of strong economic growth and the New Zealand economy was experiencing skill shortages. The government looked to its industry training policy and TEC to increase skills training in the areas of scarcity. Prior to TEC's establishment, industry training policy was implemented by the MoE, and SNZ. MoE funded the polytechnics to offer trades-related courses, and SNZ contracted the 40 ITOs to develop qualifications where they did not exist and arrange on and off-job training for people employed in the industries they represented. ITOs could arrange training through any suitable registered provider, but mostly favoured private providers over public polytechnics. NZQA worked with ITOs on the development and quality assurance of qualifications and recorded the credits earned by learners towards qualifications on the National Qualifications Framework.

Understanding the policy process interactions which occurred between the government agencies, and polytechnics, private providers and ITOs is assisted by taking a complex adaptive systems view using the lens outlined in section 3. There are at least groups of actors interacting at the macro level: the policy system made up of the people in government agencies involved in industry training policy —Ministers, TEC, MoE, NZQA; the industry training system —employers, ITOs, polytechnics and students/trainees; and the policy processes system through which the two former two groups of actors and organizations interacted.

Significantly, the policy processes system had an existing dynamic shaped by the policy history which had preceded the establishment of TEC. Feedback loops were seen at work in the interaction between the policy system and the industry training system, and these feedback loops were influenced by the history of prior policy processes. The ITOs had been formed in 1992, to bring some unity and industry ownership of skills development to meet industry needs. Prior to this, industry believed the polytechnics controlled the industry skills curriculum and qualifications. Industry found the former system costly and not sufficiently responsive to changing industry needs. From 1992-2002, there had been development of a systematic approach to qualifications for all industries facilitated by NZQA, SNZ and the ITOs concerned. New industry-relevant qualifications, developed through interactions between the policy system and the industry training system, encouraged firms to train more staff to increase productivity. When training was purchased by ITOs using funding from SNZ and industry, the ITOs favoured private providers because they were generally more flexible and lower cost. The ITOs were tightly contract managed by SNZ using simple metrics such as number of credits on the Qualifications Framework earned by trainees through training offered by the ITOs. The initial effect was loss of industry trainees from the polytechnic sector, loss of income by the polytechnic sector and in some cases, threat to the viability of the polytechnic, especially in the case of smaller, regional polytechnics.

Polytechnics responded to the ITO decisions and loss of income in a range of ways intended to offset the changes caused by ITOs and restore lost income —demonstrating another set of feedback loops leading to adaptation in the polytechnic system.

Over the ten years between the creation of ITOs and the creation of TEC, there was coevolution between the polytechnic system and the industry training policy processes system, and emergence of some new phenomena. The polytechnics were funded through MoE for all students enrolled in courses leading to registered qualifications. Polytechnics mounted new courses in competition for learners, forming alliances with parts of industries poorly served by ITOs, and purchasing PTEs which they operated as wholly-owned subsidiaries. The polytechnics also shifted the overall balance of their business away from the trade training areas into business courses and degree-level programmes. Polytechnics in some cases severely curtailed or ceased their industry training offerings because they were costly to run with low student numbers. The pre-employment courses in polytechnics counteracted two aspects of the intended industry training policy: first, it avoided the intended industry control; second, the industry co-payment was also sidestepped, bringing about an increased cost to government and the trainees involved because the students were charged fees and could draw on student allowances and loans.

The ITO performance measures used by SNZ also led to emergence in the industry training system. The ITOs clubbed together and created a new body, the Industry Training Federation (ITF), to represent them collectively. Individually and collectively, the ITOs were developing strategies, working closely with their industries and employers. The ITOs considered they belonged to industry and were becoming increasingly resistant to micro-management by SNZ. The strategic thinking of individuals in ITOs flowed into the ITF and helped to build a stronger ITO-owned sense of the industry training strategy and future strategy. The contracting relationship between the ITOs and the funding agency did not allow for interaction around strategic thinking and forward looking matters because of the way the boundaries had been defined by their co-funder, SNZ. So the ITF began to look for other ways to influence the policy processes system, initiating engagements with other parts of that system such as the MoE and ministers over policy, and even other policy systems such as the labour market and employment policy systems. Industry wanted more influence over policy and the decisions about the relevance to industry skills and knowledge needs of what was funded. The ITF wanted industry training to become a core part of tertiary education, thus they lobbied for its inclusion in the ambit of TEC, and the non continuation of SNZ. They were only partially successful because SNZ was moved into TEC unchanged and because of its size and the existing operational feedback loops which shaped organizational culture and behavior inside TEC, continued to influence TEC strongly in its early operation which tended to be very bureaucratic and SNZ-like, instead of the 'light-handed' steering body the government had intended.

In 1999–2000 a change in polytechnic funding policy made under a previous government came into effect. The cap which had existed on the number of funded student enrolments was lifted and thereafter all enrolments in courses leading to qualifications above level 4 on the Qualifications Framework were eligible for funding. The effect of this change in the funding policy was that polytechnics accelerated the moves that they had begun making into alternative areas (a positive feedback loop begun by the creation of ITOs further reinforced by positive feedback through funding). The funding also stimulated other responses within the industry training system which amounted to new phenomena which had not been anticipated. Emergence occurred in the form of new PTEs entering the field, some created directly by industry or ITOs to claim funding for training previously done without government funding or through the industry training fund contracted through SNZ and existing PTEs increased their activity to take effect of the policy change. The policy system had never been intended that ITOs would become direct providers of training, in competition with the polytechnics, however many ITOs became involved in direct provision to maximise the share of funding to their industry and avoid the performance management arrangements developed by SNZ.

In response to industry and ITF lobbying, the policy system initiated an Industry Training Review in 2002, later agreeing to a stronger strategic role for ITOs. At the time of writing there was still little movement on how the strategic functions would be operationalized within the performance management approach adopted by TEC. Thus by the time TEC was trying to address skill shortages, polytechnics were no longer the predominant provider of skills training in their regions, instead moving into short courses and degree programmes, which was not the outcome the policy system had wanted. There were in effect two industry training systems. One operated through the ITOs and co-funded by employers and TEC, and the other funded directly through providers of industry training and funded directly by TEC and through loans and allowances to trainees/students.

The policy system (minister) also wanted to reduce funding of PTEs to only strategically important areas which could not be met by the public sector and placed a cap on funding to that sector. The feedback loops between the PTEs and the polytechnics were further affected by these changes in the policy system affecting PTE funding. Instead of PTEs simply folding up shop as the policy system intended, there was adaptation, coevolution, self-organization, and emergence in the industry skills training system. Enrolments in polytechnics had previously been negatively affected by competition from PTEs which generally had smaller overheads. In the face of funding losses in both the PTE and polytechnic sectors, PTEs and polytechnics formed new alliances. Some PTEs were bought by polytechnics and operated as subsidiaries, keeping the students and courses already running and claiming funding from government through the polytechnic which was unaffected by the policy change. In other cases a joint venture was formed, where the polytechnic claimed funding and subcontracted the PTE to do the recruitment and training.

Space available here limits how fully the dynamics of the policy processes which occurred through the interactions between the policy and industry training systems can be described. However, with the assistance of the complexity lens in this section, it is possible to explain and understand the interdependent and ongoing iterative response and counter response between the policy system, the industry training system and the policy processes system a way that demonstrates the endogenous origin and ongoing nature of the dynamics. In the next section, the wider implications of the policy processes picture seen with the help of the complexity-informed lens are discussed.

TOWARDS A HOLISTIC VIEW OF POLICY PROCESSES

The eight complexity themes described above, which make up the complexity lens used in the analysis of the industry training policy processes are well-developed in areas of the social sciences that overlap with current thinking about public policy and management. Indeed, organizational and human interactions viewed as complex social systems have many commonalities with institutional and participants' interactions in public policy processes. Moreover, recent developments in public policy and management literature suggest that the eight complexity themes and related concepts are ripe for adoption in the study of public policy processes. Consequently, this integrated set of complexity themes and concepts is put forward as a complexity-informed lens for public policy analysis.

Messy, unordered and chaotic systems of public policy processes, such as the case described in this paper, are only partially understood through the application of traditional lenses outlined above. Adopting the complexity-informed lens creates new perspectives which have the potential to bring together existing perspectives. Understanding and explaining public policy processes requires a focus on at least three interacting complex systems: the policy-in-action system, such as industry skills training; the policy system (which is the domain of government and its agencies; and the policy processes system, where the policy-in-action system and the policy system come together. As examples of nested systems, institutional theory points towards individuals, organizations and constitutional/legislative systems having distinctive, interacting roles in policy processes (e.g., Ostrom, 2007; Hill and Hupe, 2006). Interdependent interactions occur as one institution responds to changes made in another system, or self-organizes to form a new level of organization. Activities across one level of organization, such as changes in rule-setting, spontaneously generate behaviors at another level of organization, such as decision making by public officials.

Acknowledging multiple policy actors, institutions and processes as parts of a complex social system, and interdependent dynamism between these many interactive parts, the system 'whole' cannot be reduced to its parts to assist understanding. A complexity-informed lens not only focuses on the dynamics of policy processes, but also recognizes the direction, contents, speed, and intensity of changes in a system. For instance, interactions between actors, institutions and processes, result in reflexive actions which undo change (negative feedback loops) and positive feedback loops, which amplify changes by reinforcing the direction of change in a system 'whole'. Feedback loops can result in stability around an attractor (e.g., a new policy document, institution, funding instrument), or in destabilizing and chaotic disturbance leading a system to suddenly and unpredictably re-organise itself around a new attractor (e.g., an unintended effect of policy change). The size of a response (e.g., budget cuts) is not necessarily proportional to the initial stimulus. The self-organization which occurs during adaptation and coevolution of far-from-equilibrium policy processes systems can result in emergence of new phenomena. Policy processes systems are also sensitive to their history, which affects the starting point for any change.

Acceptance of the complexity-informed view of policy processes means recognizing that it is impossible to control the evolution of policy processes in conventional ways. Moreover, the social construction of system boundaries in policy processes is important for understanding the goal conflicts that impede more effective exploration of policy problems and the identification of workable solutions. It follows that understanding policy problems and solutions will lack coherence if boundaries as identified and maintained by different policy participants are ignored, with consequent difficulties for managing policy processes. Taking an open systems' perspective, a complexity-informed lens sees complexity operating within nested system levels that comprise a whole. In the empirical case, PTEs, polytechnics and ITOs are all separate subsystems operating as part of a complex whole industry skills system and are affected by changes in the policy system in ways that were not predictable. Adaptation occurred in response to changes in the interacting systems leading to coevolution between the systems over time. These systems are open and therefore subject to ongoing changes in composition, resources and ideas. Their boundaries are social constructions and not necessarily fixed: what policy processes actors regard as 'environment' and 'internal' are outcomes of socially negotiated reference points.

Further empirical use in real-time policy processes will allow further refinement of this complexity informed lens, along with articulation of its implications for the design and management of policy processes.

REFERENCES

Agranoff, R. and McGuire, M. (2001). Big questions in public network management research. *Journal of Public Administration Research and Theory*, 11(3): 295-326.

Bardach, E. (2005). *A Practical Guide for Policy Analysis: the eightfold path to more effective problem solving* (2nd ed.). Washington DC: CQ Press.

Benson, J.K. (1975). The interorganizational network as a political economy. *Administrative Science Quarterly*, 20, 229-249.

Budd, L., Charlesworth, J., and Paton, R. (eds.). (2006). *Making policy happen.* Abingdon UK: Routledge.

Butler, M.J.R., and Allen, P.M. (2008). Understanding policy implementation processes as self-organizing systems. *Public Management Review,* 10(3): 421-440.

Dennard, L., Richardson, K.A., and Morcol, G. (eds.) (2008). *Complexity and policy analysis: Tools and concepts for designing robust policies in a complex world.* Goodyear, AZ: ICSE Publishing.

Dunn, W.N. (2008). *Public policy analysis: An introduction* (4th ed.). Upper Saddle River, NJ: Pearson Education.

Eppel, E. (2010). *The contribution of complexity theory to understanding and explaining policy processes: A study of tertiary education policy processes in New Zealand.* Unpublished PhD Thesis, Victoria University of Wellington, Wellington, NZ. http://researcharchive.vuw.ac.nz/handle/10063/1202

Gerrits, L. (2010). Public decision-making as coevolution. *Emergence: Complexity and Organization,* 12(1): 19-28.

Goodin, R. E., Rein, M., & Moran, M. (2006). The public and its policies. In R. E. Goodin, M. Rein & M. Moran (Eds.), *Oxford handbook of public policy* (pp. 3-38). Oxford UK: Oxford University Press.

Hill, M.J. (1997). *The public policy process* (4th ed.). Essex, UK: Pearson Education.

Hill, M.J., & Hupe, P. L. (2006). Analyzing policy processes as multiple governance: accountability in social policy. *Policy and Politics,* 34(3): 557–573.

Howlett, M., and Ramesh, M. (2003). *Studying public policy* (2nd ed.). Ontario: Oxford University Press.

Jones, B.D., & Baumgartner, F.R. (2005). *The politics of attention. How government prioritises problems.* Chicago IL, London: University of Chicago Press.

Kickert, W.J.M., Klijn, E.-H., & Koppenjan, J.F.M. (eds.). (1997). *Managing complex networks: Strategies for the public sector.* London, Thousand Oaks, New Delhi: Sage Publications.

Kingdon, J.W. (1995). *Agendas, alternatives and public policies* (2nd ed.). New York: Addison-Wesley Educational Publishers.

Klijn, E.-H. (1996). Analyzing and managing policy processes in complex networks: A theoretical examination of the concept of policy network and its problems. *Administration and Society, 28*(1): 90-199.

Klijn, E.-H. (2005). Networks and inter-organisational management: Challenging, steering, evaluation and the role of public actors in public management. In E. Ferlie, L. E. J. Lynn & C. Pollitt (Eds.), *Oxford handbook of public management* (pp. 257-281). Oxford UK: Oxford University Press.

Klijn, E.-H. (2008). Complexity theory and public administration: What's new? *Public Management Review,* 10(3): 299-317.

Klijn, E.-H., and Teisman, G. (1997). Strategies and games in networks. In W. J. M. Kickert, E.-H. Klijn & J. F. M. Koppenjan (Eds.), *Managing complex networks: Strategies for the public sector* (pp. 98-118). London, Thousand Oaks, New Delhi: Sage Publications.

Klijn, E.-H., and Teisman, G. (2006, 2-4 April). *Complexity theory and public administration: A critical appraisal.* Paper presented at the XI International Research Symposium of Public Management, University of Potsdam, 2–4 April 2006.

Koppenjan, J.F.M., and Klijn, E.-H. (2004). *Managing uncertainities in networks: A network approach to problems and decision making.* London: Routledge.

Meek, J.W. (2010). Complexity theory for public administration and theory. *Emergence: Complexity and Organization,* 12(1): 1-4.

Meek, J.W., De Ladurantey, J., and Newell, W.H. (2007). Complex systems, governance and policy administration consequences. *Emergence: Complexity and Organization,* 9(1-2): 24-36.

Mitleton-Kelly, E. (Ed.). (2003). *Complex systems and evolutionary perspectives on organizations: The application of complexity theory to organizations.* Oxford UK: Elsevier Science.

Morçöl, G. (2002). *A new mind for policy analysis : Toward a post-Newtonian and postpositivist epistemology and methodology* Westport CN: Praeger.

Morçöl, G. (2010). Issues in reconceptualising public policy from the perspective of complexity theory. *Emergence: Complexity and Organization,* 12(1): 52-60.

Ostrom, E. (2007). Institutional rational choice: An assessment of the institutional analysis and development framework. In P. A. Sabatier (Ed.), *Theories of the policy process* (pp. 21–64). Boulder CO: Westview Press.

Richardson, K.A. (2005). Complexity thinking and policy analysis. Retrieved from http://kurtrichardson.com

Richardson, K.A. (2007). Complex systems thinking and its implications for policy analysis. In G. Morcol (Ed.), *Handbook of decision making* (pp. 189-222). Boca Raton FL: CRC Press.

Sabatier, P.A. (ed.). (2007). *Theories of the policy process.* Cambridge MA: Westview Press.

Sabatier, P.A., and Jenkins-Smith, H. (1993). *Policy change and learning: An advocacy coalition approach.* Boulder, Co: Westview Press.

Sabatier, P.A., and Weible, C.M. (2007). The advocacy coalition framework. In P. A. Sabatier (ed.), *Theories of the policy process.* Cambridge MA: Westview Press.

Teisman, G., van Buuren, A., and Gerrits, L. (eds.). (2009). *Managing complex governance systems: Dynamics, self-organization, and coevolution in public investments.* New York: Routledge.

Teisman, G., Westerveld, E., and Hertogh, M. (2009). Appearances and sources of process dynamics: The case of infrastructure development in the UK and the Netherlands. In G. Teisman, A. van Buuren & L. Gerrits (Eds.), *Managing complex governance systems: Dynamics, self-organization, and coevolution in public investments* (pp. 56-75). New York: Routledge.

Teisman, G., and Klijn, E.-H. (2008). Complexity theory and public management: An introduction. *Public Management Review,* 10(3): 287-297.

van Buuren, A., and Gerrits, L. (2008). Decisions as dynamic equilibriums in erratic policy processes. *Public Management Review,* 10(3): 381-399.

Dr. **Elizabeth Eppel** is a Research Fellow in the School of Government, Victoria University of Wellington, New Zealand. Her research interests include the application of complexity theory to understanding policy processes, collaborative network and governance, and new media in governance.

Chapter 9

PUBLIC SERVICES BETWEEN LAISSEZ FAIRE ECONOMIC AND PREDATORY FINANCIAL CAPITALISM

Alexander Dawoody (Marywood University, USA)
Peter Marks (Erasmus University Rotterdam, NLD)

With globalization and the morphing of multi-national corporations, Laissez-Faire economic is abandoning its impartial spectator and gradually disappearing for the sake of the newly-emerging predatory financial capitalism. The result is the shrinking of the middle class, the undermining of governmental authorities and the rule of public administrators, and the increase in threats to economic security. We address a complexity response to the ill-logic of artificial engineering in bailing out predatory financial capitalism by tax-dollars and emphasizes on complexity solutions in allowing for Pareto-optimal move through self-organization and emergence of new dynamic system that is better capable of dealing with changes in interconnected global economy without artificial engineering. We illustrate the importance of coordination, autonomy, networking and decentralization in the materialization of expertise and shared-resources. Understanding public policy's role in the economy through complexity lens enables us to capsulate the bigger picture in change dynamics and have better appreciation of the multilayered dynamics that interplay during their display.

INTRODUCTION

Capitalism continually renews and reinvents itself

Karl Marx

After the financial crisis of 2007-8, many people though that predicting the subprime meltdown would have helped. It would not have, since it was a symptom of the crisis, not its underlying cause (Taleb & Blyth, 2011). Life is not predictable (Barabasi, 2003). A key distinction between most natural sciences and social sciences, like politics, public administration, and economics, is that natural sciences are linear and social sciences are complex (Miller & Page, 2007). Yet, the complex nature of these social sciences, even to most of their observers, is misunderstood. This is true because we, as human beings, inherit our knowledge linearly, especially in understanding causal relations, and it is difficult transferring it to complex domain (Taleb & Blyth, 2011). Nevertheless, we live in both worlds simultaneously. Our linear domain is characterized by predictability and the low degree of interaction among its components. This allows us to use mathematical methods to make forecasts (Guastello, 2002). In the complex domain, we are devoid of visible causal links between elements and rely, instead, on interdependence and extremely low predictability (Kauffman, 1993).

One of the errors we make when we are in the linear domain is that we have an urge to control (Capra, Juarrero, & Uden, 2007). We do this in our daily routine interactions, or in public and economic policies (Harrison, 2006). Although all indicators point to the contrary and results demonstrate the fatality of such behavior, we, nevertheless, persist on maintaining this trait (Buchanan, 2003). In addition to control, we also exhibit another fatal tendency that we inherit from the linear domain, and that is the propensity to predict (Brown, 1995). Yet, the most fatal and dangerous element we inherit from the linear domain is our tendency to prevent systemic volatility and persisting on the illusion of maintaining "stability" through artificial engineering (Goldstein, 2007). This type of error, often adapted by policymakers, is the recipe for disaster and often results in catastrophe (Brown, 1995). To help explain this dangerous, yet often practiced method, we will use the observation of capitalism as a system that had transformed through artificial engineering from a volatile, yet dynamic system as the one explained by Adam Smith, into a predatory system of amalgamation of capital and industrial global octopus as the one we are witnessing today. Although such morphology was an innovative process to escape the rise in sophistication of organized labor and a phase shift from national bourgeoisie-oriented market into a global capital, its consequence is severe and impacts the entire human race.

ADAM SMITH AND ARGUMENT AGAINST STABILITY

A market economy is sometimes said to be self-organizing (Krugman, 1996; Redmond, 2010). Self-organization is related to the question how (labor) markets come to being by consumptive and productive activities by actors. The notion of the invisible hand (Smith, 2010) is a predecessor of the self-organization notion and holds that interactions lead to coordination as an unintended consequence of the self-interested behavior. Or as Hayek (1967) coins his catallaxy that states the spontaneous order of the free market as a "self-organizing system of voluntary cooperation". Traditionally, the self-regulating properties of the price mechanism are emphasized (negative feedback) which stabilizes the equilibrium. However, stable equilibrium cannot explain the driving forces of economic development, and in a changing environment markets hardly ever reach a state of perfect coordination due to problems in information processing (Witt, 1997). New profit opportunities are opened when individuals successfully innovate, opening up new knowledge that diffuses through the economy, creating new products, technologies and / or markets. These self-augmenting features of the market process create positive feedback away from the stable equilibrium and have a de-coordinating effect. Due to the self-regulating and self-augmenting features production, trading and division of labor take place, but there are surprises, misallocations, backlashes and losses (Witt, 1997).

According to Marx, capitalism is the combination of capital and industrial (later the service sector was added) operations that monopolizes labor in turning raw materials into products and generate surplus (profit) in the process (Marx, 2007). The market used for the introduction (sale) of goods is national, hence creating a national bourgeoisie, antagonistic among itself because of competition for market niche but united in safeguarding its interests against labor (Marx & Engels, 2011). Democracies emerged to protect these means of production through enacted laws and public institutions (Przeworski, 1986).

With the rise in labor demand and unity, the national boundaries of capitalism market gradually started to blur and fuse with each other (Ricardo,

2010). The national bourgeois and the socio-political and economic system that it had represented at one point morphed into global capitalism (Frieden, 2007). Public entities also morphed to either restrict or support this growth under the guise of providing stability (Gilpin, 2001). Global organizations, such as the World Trade Organization, the World Monetary funds, and the World Bank emerged in order to support these new developments (Eichengreen, 2008).

While many economists hold that planning disturbs the self-organizing system, from a socialist economic interpretation the state should direct production and pricing due to considerable market failures, for instance to protect against severe deprivation. However, this leads to distortions of the pricing mechanism. The public restraint on competition has a lasting hindering effect on the process of self-control of the catallaxy (Streit, 1993). The problems created by the interference are not recognized as related, which triggers even more interventions to neutralize the ill side-effects. Hence, the intervening institutions become locked-in (North, 1990; Pierson, 2000), which is even more affected by the lobbying of interest groups (Olson, 1982).

It is here that classic capitalism as professed and its dynamics of booms and busts explained by Adam Smith (2010) evolved to predatory capitalism. Public sector's constant interference in the affairs of predatory capitalism, either supporting it through subsidies, tax cut, tariff regulations, and anti-labor measures, or preventing its collapse through bailouts, are exercises of suppressing volatility in the name of stability (Acharya & Richardson, 2009). In doing so, the public sector had tied its fate with the fate of predatory capitalism, choking in the process windows for natural evolution and for the complex system to self-organize and adapt to changes in the environment (Ott, 2003). These forms of artificial engineering are dangerous elements in the public sector meddling through, and a departure from Adam Smith' premise to allow the invisible hand in the market to work its magic.

Companies that survive in the competitive market do so by luck or design, coming close to maximizing profits (Alchian, 1950). The magic of the market is that through individual consumptive and productive self-interest the actors enforce self-regulating features (Witt, 1997). Ecologist Holling (1973) introduced the term resilience to understand the characteristics of change and the interactions between human and natural systems. Resilience is a measure of the persistence of systems and of their ability to absorb change and disturbance and still maintain the same relationships between populations or state variables (*ibid.*). Economic resilience is the ability of an entity or system to maintain function (e.g., continue producing) when shocked (Rose, 2007). There are two alternative interpretations (Bergh, 2007). From a more linear interpretation economic resilience is how far the system has moved and how much time was necessary for the system to return to its single equilibrium steady-state after a perturbation, i.e., this type is associated with 'shock counteraction' (Briguglio 2009). From a complexity interpretation economic resilience is the amount of disturbances or shocks that can be neutralized or rendered negligible before flipping over into another regime of behavior, also referred to as 'shock absorption' (*ibid*). In the latter multiple stables states exist. Adam Smith called for the market to heal itself without government interference and artificial engineering. Adam Smith, for example, argued against price control and called instead for the market's invisible hand to work the fluctuations (Smith, 2010).

Today, however, and because of the morphing of classic nation-state capitalism (bourgeoisie oriented market) to global predatory capitalism, price controls by government are designed to limit volatility on the basis that stable prices are better for the consumer/citizen (Stiglitz, 2010). The behavior of the economy is thus heavily influenced by events of the past and the inherited institutional structure. These institutional structures determine overall economic performance, and the pattern of the evolution of institutions may cause that efficient institutional forms do not survive and perpetuate suboptimal institutions (Poirot, 2001). These suboptimal institutions reinforce the pattern of institutional evolution undermining the efforts to transform it. Changes will then likely be discontinuous, because they need to cross a certain threshold. The long-term consequence of price controls by governments is costly explosion that exceed the benefits gained (Taleb & Blyth, 2011).

DEPARTING FROM ADAM SMITH'S BOOMS AND BUSTS

In the development of Western market economies institutions were designed to regulate the relationship between capital and labor, or to protect vital interests of economic agents. These institutions tune economic fluctuations to a certain socially acceptable range (Matutinovic, 2006). Opposite of these self-regulating processes (negative feedback) there are self-augmenting processes (positive feedback). The self-augmenting processes (Witt, 1997) exist because of the continuous drive of profit making, growth preference and cost cutting. Successful innovations and profit opportunities and the diffusion and imitation of this knowledge create new products, technologies and/or markets, i.e., it represents an endogenously caused de-coordinating effect (Witt, 1977). Hence we have both goal-oriented tuning and dynamic drivers balancing in the capitalist market (Matutinovic, 2006). This balancing between coordinating and de-coordinating tendencies in the markets by and large do not allow for a viable coordination to be reached. This makes the evolutionary path of the economic system unpredictable and nonlinear. Nonlinearity gives rise to uncertainty about the future behavior of economic variables and the rules that govern behavior (Poirot, 2001).

The morphing of national capitalism to predatory capitalism is grounded in the rise of geopolitical and global economic interdependence. There are three general responses to shifting stability and crises within such interdependence. A first is to test the resilience of a system by doing nothing, but wait to see if the system will return to an acceptable state. The second is actively trying to get the system back to the desired stability, and third is to accept that the system will irreversibly change and adapt to this new altered state. These complex systems, according to Taleb and Blyth (2011), if experiencing artificially suppressed volatility by the public sector, become extremely fragile (even if they do not exhibit visible risks at the beginning or appear to be calm while silent risks are accumulating under that calm). It is, therefore, dangerous not to examine low probability risks and remove it from observation (*ibid.*).

Yet, the public sector seems bent on adapting policies to eliminate fluctuation and prevent booms and busts in the market in order to sustain the illusion of stability (Stiglitz, 2010). During most of the 20th century this command and control approach focuses on controlling a certain target variable of interest (replacing uncertainty of nature with certainty of control), but then slowly changes other parts of the system. These policies can weaken both the market and the public sector. Examples of these policies can be illustrated in measures such as over regulations, reducing velocity through lowering interest rates, providing government loans, enforcing price control, and sustaining stability.

These measures often come in two types: conforming to the market mechanism or regulating it to ease-up its side-effects (Krugman, Obstfeld, & Melitz, 2011). Both types, however, are suppressive of volatility and have unintended and unforeseen consequences (Nowak, 2008). It is better to tolerate these systems' fluctuations and allow them to absorb their bumps in order for true organic change to take place from within these systems, including their collapse (Guastello, 2002). Suppressing volatility not only fails to avoid these bumps but it will also increase the probability of leading these systems toward major explosions (Juarrero & Rubino, 2008). Nurturing and preserving the elements that enable the system to renew and reorganize itself ought to be encouraged (Allison & Hobbs, 2004). The public sector, as such, must allow market failures in order to allow for new and more salient orders to emerge (Miller & Page, 2007).

These systems, regardless of public sector interventions (Gruber, 2009), will eventually experience explosions, undoing years of illusionary government-desired stability (Elliot & Kiel, 1996). Often, they will end up far worse than they were in their initial volatile state. The longer it takes for these systems to explode and collapse the worse the outcome will be for both the public and private sectors (Sandri, 2008). These crises typically end in default and restructuring with high levels of debt that take years to work off, and have a lasting effect on housing, output and employment as businesses and consumers rebuild their balance sheets (Reinhart & Rogoff, 2011). Though sector reform will not by itself overcome hysteresis, doing so will not be possible without effective regulatory design in the sector (Poirot, 2001).

LEARNING FROM COMPLEXITY

Complexity sciences teach us many things that we can apply to public policy and economic development. An important lens that we can learn and adapt from complexity sciences is the notion of self-organization. Self organization is the idea that living systems are capable of self-organize themselves in ways that all their components and processes can jointly produce the same components and processes as autonomous agents (Vesterby, 2008). This concept is also known as autopoiesis (Maturana & Varela, 1991). A key notion of this concept is self-referentiality (Sandri, 2008). The idea of self-reference designates the unity that a dynamic system is for itself, and that unity can be produced through relational operations (Little, 1999).

The interplay between a dynamic system and its environment is an on-going enterprise of continuous relationships of interplaying and interconnected networks (Johnson, 2002). Changes that take place in the system's environment act as "kicks" to generate changes within the system's internal structure (Dawoody, 2011). Communications between the system and its environment is based on positive and negative feedback (Morgan, 2006). Environmental kicks are received by the system through its sensory receptors which act as strange attractors in order to prepare the system internally to reshuffle its internal structure and change its older order to correspond to changes in the environment. If the internal order is incapable of change, then the system's entire structural order must collapse in order to allow for new structural order emerge and deal with the new environmental changes (Prigogine and Stengers, 1984). Sustaining the older structures through artificial engineering may buy the system some time, but it will not prevent its ultimate collapse, at which point it may be accompanied with catastrophe (Brown, 1995; Dawoody, 2011).

Without the collapse of older structure there will be no birth of a new order. This concept is also referred to as bifurcation (Kuznetsov, 2010), and translated in phase shifts in the order of the system's dynamics (Wheatley, 2006). As the self-organizing order emerges out of the interaction of elements within the system, the system's own parameters become unstable and the older order starts to collapse (Brem, 1999). Public policy must be understood according to this interpretation in order to safeguard it from costly errors of resisting change or attempting artificial engineering (Dawoody, 2011; Richardson & Goldstein, 2007).

The relationship between the system and its environment operates on feedback that is either positive or negative (Morgan, 2006). Feedback as stimuli is retransmitted by the environment and cause random changes in the agent's internal processes (Wheatley, 2006). This behavior contains the agent's morphology from static equilibrium to a state of chaos and disorder. Disorder then leads to new structures and practices (Prigogine, 1996). The phase-shifts from equilibrium to disequilibrium to equilibrium are self-organizing and irreducible, and unpredictable (Nicolis & Prigogine, 1989).

Understanding public policy's role in the economy through complexity lens enable us capsulate the bigger picture in change dynamics and have better appreciation of the multilayered dynamics that interplay during their display (Richardson & Goldstein, 2007).

IMPLICATIONS OF A COMPLEX OBSERVATION OF PUBLIC SERVICE AND THE ECONOMY

Should the economy not be in the realm of linear disciplines? Does the public sector's role is to observe how the private sector is operating and understand when not to intervene? Complexity sciences do not specify where a specific discipline ought to operate. The specifics of each discipline are the subject and responsibility of its agents. Complexity, however, give us good recommendations on what to avoid in order to better understand our world and the complex, interconnected and interweaving dynamics of our function regardless of the discipline that we are involved with. Complexity, for example, advises us to avoid control, prediction, long-term planning, and artificial engineering. It encourages minimizing uncertainties without necessarily presenting it as a formula for staring at a crisis. Limiting control over policies does not produce 'wise' and self-regulating systems. However, allowing the system to collapse when reaches atrophy and dissipation in its natural cycle, and avoiding the prolonging of outdated order through interjected public policies can be beneficial in sustaining a healthy interaction between any system and its environment, bypass catastrophes, and allow for the nature of things to exist, mature, go through changes, collapse, and emerge in newer structures naturally and without artificial engineering. If government did not interfere in bailing out Wall street, for example, the system would had exhibited its natural cycle of decay, collapse and the rebirth of a new structure that better fits in responding to changes in the environment. However, by interjecting itself in the economy to prevent the natural cycle of progression in the financial industry, government had only delayed the inevitable through the artificial engineering of the bailout. In doing so, it had invited catastrophe that is much more harmful when takes place at the eventual collapse of this decaying system that if it allowed to collapse and then self-organize naturally. The case of the European Union's interjection in Greece is another example.

The world is a world of participatory collusion among particles in which entities separated by space and possess no mechanism for communicating with one another can exhibit correlations in their behavior (Overman & Loraine, 1996). Structures collapse and evolve because of consistently small reasons that grow larger and become more nonlinear (Brem, 1999; Dawoody, 2011).

We are no longer constrained by a single ontological model. Truth can now be seen not as an attribute inherent in a system or event but as the meaning we attribute to that system (Buchanan, 2003). This kind of ontological liberation is evident in the paradigm shift to the world of nonlinear sciences (Wheatley, 2006). These types of shifts include moving from structure to process, from objective to epistemic science, and from building block to network (Evans, 1999). Nonlinearity and its model free us from the burden that comes from needing to control rather than to evoke process and relationship (Overman and Loraine, 1996). This understanding forces us to examine public policy and its role in the economy not through the isolated observation of its building-blocks, but in relationship of these particles with themselves and the environment of the system as a whole (Dawoody, 2011; Johnson, 2002).

CONCLUSIONS AND LESSONS

We, as public administrators, need to learn from a complexity observation of the transformation of the market to predatory capitalism and from observing governmental intervention in the market through various forms of artificial engineering in the name of maintaining stability and preventing collapse. This will also help us to apply complexity sciences into public policy areas at various capacities despite the current (and persistent) linear applications in public policy.

It is uncomforting to let go of long application of linear causality and conceptual models. We need to get comfortable in simultaneously inhabiting both the linear and complex domains and offer complexity analysis and solutions prior to crisis. Off course, this mean stepping out of a comfort zone, so policymakers, researchers, analysts, public administrators and academics should be assisted in transferring their inherent sophisticated linear knowledge to the complex domain, since this is problematic. First step is to accept and get comfortable with the absence of visible causal links between elements masking a high degree of interdependence and extremely low predictability. This is despite human nature in begging for clear causal relation and continuing with patterns that seek visible causality. No matter how much time we spend on devising models and instruments for predictability, we will never be able to trace chance (Capra, 2004). Policymakers and public administrators ought to be trained avoiding control, predictability, the use of catalyst as cause, or explaining systems through events (especially last events).

Once we accept that predictability is fraught with difficulties, we need to overcome the fear of chance and randomness (Juarrero & Rubino, 2008). When accepting that stability is an illusion that we should not constantly try to maintain, randomness and variation are not feared, but rather welcomed since they are the source of variation and innovation. That is, when we allow for complexity to govern our analyses and observations we can rescue ourselves from control and prediction. This of course means that low probability risks are no longer neglected of pushed away into end tails statistical probabilities. In fact, they need to be visible to allow for volatility. This volatility is necessary in order for the complex system to self-organize.

This understanding will shift the emphasis from control to enhanced flexibility, in which systems are allowed their natural booms and busts, where collapse is a natural consequence in system morphology, instead of massive blowups. Systems would be able to absorb our imperfection rather than that we seek to change them.

As such, a typical dynamic system can exhibit a variety of temporal behavior. When the behavioral history of a system is examined, the nature of change becomes the core of its inquiry (Brown, 1996). If a system becomes unstable, it will move first into a period of oscillation, swinging back and forth between two different states. After this oscillation stage the next state is chaos, and it is then the wild gyrations begin (Dawoody, 2011; Wheatley, 2006).

If we look at public policy as a dynamic system and examine the nature of changes within it, especially in relation to the economy, we can see these changes requiring oscillation, chaos and the birth of new order. However, often these changes are artificially engineered in form of reforms in order to stop the systemic collapse and prolong its decaying structure beyond its natural time.

We need to realize that fluctuations can take place (Kendall, Schaffer, Tidd & Olsen, 1997). Fluctuations are initiated by changes in the environment and lead to corresponding changes within the system through positive and negative feedback. Positive feedback translates changes in the environment to more changes in the system's internal dynamics, and fewer changes in the environment will lead to fewer changes within the system. Negative feedback, on the other hand, is when more changes in the environment lead to fewer changes within the system while fewer changes in the environment lead to more changes within the system (Morgan, 2006). This environmental stochasticity increases the probability of some policies of program extinction. Policies and programs that evolve are those who are selected against (Dawoody, 2011; Kendall, Schaffer, Tidd & Olsen, 1997).

The resulting configuration within the system's internal order is emergent, allowing for new structures, patterns and processes to emerge through self-organization in order to fit best with the changing dynamics in the environment (Vesterby, 2008). The relationship between the system and its environment is an active relationship that benefits from feedback and translates into systemic morphology (Ruelle, 1993). Stimuli from the environment and the system's response are based on short or long-term transitions and corresponding changes in the system's internal dynamics are irreducible, unpredictable, and nonlinear (Dawoody, 2011).

Newton's laws in physics and the subsequent linear interpretation that came out of it were tremendous step in human progress that unveiled new frontiers, including in public policy. These interpretations helped shaping modern decision-making processes, analytical interpretations of the world (Kauffman, 1993). However, nonlinearity has uncovered an entire new frontier (Kiel, 1999) that transcends singular causality, certainty, predictability, hierarchy, centralization, and control (Wheatley, 2006). It presents another way into looking to the world (Dawoody, 2011; Kaplan & Glass, 1997).

The linear approach to public policy reduces process to the simplistic notions of accountability versus non-accountability. Citizenry interests are simplified into public opinion, which triggers the political system to irritate the administrative system, reducing responsiveness and causing the alienation of citizenry itself (Little, 1999). It is this drive toward nonlinearity that places demands on public policy to develop new interpretations for dealing with challenges (Kiel, 1999).

In order for public policy live within a continually changing environment of the economy, it must act as a dynamically adapting system. This nonlinear system must be composed of many interacting and autonomous agents on global, national, state, and local levels. Each agent has to be associating with the other as partners. In such a scheme, no single agent can be poised in controlling position. Instead, the autonomy and network nature of the associations among the agents must enable continuous transformation of the economy by meshing more suitably with changes in the external structures of the economic system than to trap it within an outdated mode or order (Dawoody, 2011).

We need to shift our focus from control to influence and from prediction to anticipation. We need to deemphasize prediction, and long-term planning. Instead, we need to acknowledge the uncertain outcome of deterministic systems, accept the unexpected and random consequences, and engage in processes and pattern observation. This orbital decomposition will enable us be part of self-transcendency constructs, appreciate patterns of similarities and differences, welcome change, and better prepare for emergence of the new (*ibid.*).

REFERENCES

Acharya, V. and M. Richardson. (2009). *Restoring financial stability: how to repair a failed system*. Indianapolis, IN: Wiley.

Alchian, A. (1950). "Uncertainty, evolution and economic theory." *Journal of Political Economy* 58 (June): 211-221.

Allison, H. and R. Hobbs (2004). "Resilience, adaptive capacity, and the 'Lock-in Trap' of the western Australian agricultural region." *Ecology and Society*, 9(1).

Barabasi, A. (2003). *Linked*. NY: Plume.

Bergh, C.J.M. van den (2007) "Evolutionary thinking in environmental economics." *Journal of evolutionary economics,* 17: 521-549.

Brem, R.J. (1999). The Cassandra complex. In G. Morcol & L. Dennard (ed), *New Sciences for Public policy and Policy, Connections and Reflections,* 125-150. Burke, VA: Chatelaine Press.

Briguglio, L, Cordina, G. Farrugia, N. and Vella, S (2009) "Economic Vulnerability and Resilience: concepts and measurements." *Oxford development studies,* 37(3): 229-247.

Brown, C. (1995). *Chaos and catastrophe theories.* Thousands Oaks, CA: Sage.

Buchanan, M. (2003). *Nexus*. NY: W.W. Norton and Company.

Capra, F. (2004). *The hidden connections*. NY: Anchor Books.

Capra, F., A. Juarrero, and J. Van Uden. (2007). *Reframing complexity*. Mansfield, MA: ISCE Publishing.

Dawoody, A. (2011). "Teaching public policy as a nonlinear system." *Journal of US-China Public Administration,* 8(4): 372-386.

Guastello, S. (2002). *Managing emergent phenomena*. Mahwah, NJ: Lawrence Erlbaum Associates, Publishers.

Elliot, E. & D. Kiel. (1996). *Chaos theory in the social sciences*. Ann Arbor: The University of Michigan Press.

Eichengreen, B. (2008). *Globalizing capital: a history of the international monetary system*. Princeton, NJ: Princeton University Press.

Evans, K. (1999). "Imagining anticipatory government." In G. Morcol & L. Dennard (eds.), *New Sciences for Public policy and Policy: connections and reflections*, 195-220. Burke, VA: Chatelaine Press.

Frieden, J. (2007). *Global capitalism: its fall and rise in the twentieth century*. NY: W. W. Norton & Company.

Gilpin, R. (2010). *Global political economy: understanding the international economic order*. Princeton, NJ: Princeton University Press.

Goldstein, J. (2007). "A new model for emergence and its leadership implications." In J. K. Hazy, J. Goldstein and B. B. Lichtenstein, (Ed.), *Complex systems leadership theory*, 61-92. Mansfield, MA: ISCE Publishing Company.

Gruber, J. (2009). *Public finance and public policy*. NY: Worth Publishers.

Harrison, N. (2006). *Complexity in world politics: concepts and methods of a new paradigm*. Albany, NY: State University of New York Press.

Holling, C.S. (1973) "Resilience and stability of ecological systems." *Annual Review of Ecology and Systematics*, 4: 1-23.

Kaplan, D. and L. Glass. (1997). *Understanding nonlinear dynamics*. NY: Springer.

Kauffman, S. (1993). *The origin of order*. NY: Oxford University Press.

Kendall, B., W. Schaffer, C. Tidd, and L. Olsen. (1997). The impact of chaos on biology: promising directions for research. C. Grebogi and J. Yorke (ed), *The impact of chaos on science and society*, 190-218. Tokyo: United Nations University Press.

Kiel, D. (1999). "The science of complexity and public policy." In G. Morcol and L. Dennard. (ed), *New Sciences for Public policy and Policy*, 63-80. Burke, VA: Chatelaine Press.

Krugman, P. (1996). *The self-organizing economy*. Oxford: Blackwell Publishers.

Krugman, P., M. Obstfeld, and M. Melitz. (2011). *International economics*, 9th edition. Upper Saddle River, NJ: Prentice Hall.

Kuznetsov, Y. (2010). *Elements of Applied Bifurcation Theory*. NY: Springer.

Juarrero, A. and C. Rubino. (2008). *Emergence, complexity, and self-organization*. Goodyear, AZ: ISCE Publishing.

Johnson, S. (2002). *Emergence*. NY: Touchstone.

Little, J. (1999). "Governing the government." In G. Morcol and L. Dennard (eds.), *New sciences for public policy and policy: connections and reflections*. Burke, VA: Chatelaine Press.

Marx, K. (2007). *Das capital*. Fairlawn, Ohio: Synergy International of the Americas, Ltd.

Marx, K. and F. Engels. (2011). *The communist manifesto*. Watertown, WI: Tribeca Books.

Maturana, H.R. and F.J. Varela. (1993). *Autopoiesis and cognition: the realization of the living*. NY: Springer.

Matutinovic, I. (2006). "Self-organization and design in capitalist economies." *Journal of Economic Issues*, XL(3): 575-601

Miller, J. and S. Page. (2007). *Complex adaptive systems*. Princeton: Princeton University Press.

Morgan, G. (2006). *Images of organization*. Beverly Hills, CA: Sage Publications.

Nicolis, G. and I. Prigogine. (1989). *Exploring complexity*. NY: W.H. Freeman and Company.

North, D. (1990). *Institutions, institutional change and economic performance*. Cambridge, UK: Cambridge University Press.

Nowak, M. (2006). *Evolutionary dynamics*. Cambridge, MA: Harvard University Press.

Ott, A. (2003). *The public sector in the global economy: from the driver's seat to the back seat*. Northampton, MA: Edward Elgar Publishing.

Olson, M. (1982). *The rise and decline of nations: economic growth, stagflation, and social rigidities*. New Haven: Yale University Press.

Overman, S. and T. Loraine. (1996). "The new sciences of administration: chaos and quantum theory." *Public Policy Review*, 56(5): 487-491.

Pierson, P. (2000). "Increasing returns, path dependency, and the study of politics." *American Political Science Review*, 94: 251-267.

Poirot, C. (2001). "Financial integration under conditions of chaotic hysteresis: the Russian financial crisis of 1998." *Journal of Post Keynesian Economics*, 23(3): 485-507.

Prigogine, I. (1996). *The end of certainty*. NY: The Free Press.

Prigogine, I. and I. Stengers. (1984). *Order out of chaos*. NY: Bantam Books

Przeworski, A. (1986). *Capitalism and social democracy*. Cambridge, UK: Cambridge University Press.

Redmond, W. (2010). "Rules and roles in the marketplace: self-organization of the market." *Journal of Economic Issues*, XLIV(2): 337-344.

Reinhart, C. and K. Rogoff. (2011). *This time is different*. Princeton, NJ: Princeton University Press.

Ricardo, D. (2010). *On the principles of political economy, and taxation*. Highlands Ranch, CO: General Books LLC.

Richardson, K. and J. Goldstein. (2007). *Classic complexity: from the abstract to the concrete*. Mansfield, MA: ISCE Publishing.

Ruelle, D. (1993). *Chance and chaos*. Princeton: Princeton University Press.

Rose, A. (2007) "Economic resilience to natural and man-made disasters: Multidisciplinary origins and contextual dimensions." *Environmental Hazards*, 7: 383-398.

Sandri, S. (2008). *Reflectivity in economics: an experimental examination on the self-referentiality of economic theories*. Physica-Verlag HD.

Smith, A. (2010). *The wealth of nations*. Hollywood, FL: Simon & Brown.

Stiglitz, J. (2010). *Freefall: America, free markets, and the sinking of the world economy*. NY: W. W. Norton & Company.

Streit, M. (1993). "Cognition, competition, and catallaxy: in memory of Friedrich August van Hayek." *Constitutional Political Economy*, 4(2): 223-262.

Taleb, N. and M. Blyth. (2011). "The black swan of Cairo." *Foreign Affairs*, V(90)3: 33-39.

Vesterby, V. (2008). *Origins of self-organization, emergence, and cause*. Goodyear, AZ: ISCE Publishing.

Wheatley, M. (2006). *Leadership and the new science*. San Francisco: Berrett-Koehler Publishers.

Witt, U. (1997). "Self-organization and economics: What is new?" *Structural Change and Economic Dynamics*, 8: 489-507.

Alexander Dawoody is an Assistant Professor of Public Policy and Administration and the Adviser of the Master of Public Administration at Marywood University in Pennsylvania, USA and a Visiting Professor at the University of Electronic Science and Technology in Chengdu, China. His researches are in the areas of complexity and systems and Middle Eastern studies. He is the founder of the Section on Complexity and System Studies Network (CSN), as well the Section on Effective and Sound Administration in the Middle East (SESAME) at the American Society for Public Administration (ASPA). He is also the founder and president of the Association for Middle Eastern Public Policy and Administration (AMEPPA).

Dr. **Peter Marks** works at the department of Public Administration at the Erasmus University Rotterdam. Having studied economics and philosophy his research mainly focuses on complex decision making processes from an evolutionary (economics) perspective. The target domain in his research is public safety policy issues. He also teaches courses in systems and evolutionary economics.

Chapter 10

DESIGN, DEVELOPMENT, AND SUSTAINABILITY IN MULTILEVEL GOVERNANCE IN MANAGING EMERGENCIES AND CRISES

Naim Kapucu (University of Central Florida, USA)
Vener Garayev (Gediz University, TKY)

Organizations are restructured into collaborative systems in order to address complex problems by combining expertise distributed across functions, knowledge specialties, and geographic locations. Often times these systems face complex and multifaceted goals requiring distinct teams of public managers to coordinate their efforts and compile crucial information distributed across a network. This article renders an understanding of the requisite structural and relational dynamics that enable network organizations to function effectively in preparedness, mitigation, response to and recovery from disasters. A network of public, private, and nonprofit organizations with stronger and denser ties is critical for the success of disaster preparedness, response, and, recovery networks. Networks not only need to be designed, but also developed and sustained effectively. Information communication technologies are a cornerstone to achieve these goals. Findings from this study contribute insights to the literature on networks and multi-team systems (e.g., intra-organizational and inter-organizational level). In particular, this knowledge informs scholors about needed alterations to the design, leadership, training, policy, and feedback systems in such organizations.

INTRODUCTION

Organizations are restructuring into collaborative governance systems in order to address complex problems by combining expertise distributed across functions, knowledge specialties, and geographic locations. It has become normal for organizations to deal with multiple goals and requirements to achieve end results (Koliba, Meek, & Zia, 2010). This prompts for organizational and managerial restructuring that would address and encompass multifaceted functions, knowledge, skills and expertise. This type of design has come to require participation and involvement of several actors, from different entities, striving to achieve a common goal by combining efforts, resources, and information sharing.

Virtual networks are non-spatial structures composed of organizations collaboratively working toward a common goal and dependent on information technologies for communication purposes (Ahuja & Carley, 1999). The underlining characteristic of such systems is their relatively ad-hoc structure that is viable only through continuous efforts to sustain the existing relationships and communication among the actors involved. Network sustainability, in this regard, becomes crucial especially for such short-run structures as disaster response and recovery networks. While disasters are relatively non-frequent (low probability high consequence events), continuation of inherently ad-hoc network systems is essential for their future successes. This is especially important because disaster networks are very much dependent on communication and coordination of structurally, culturally, and cognitively separate entities (Drabek *et al.*, 1981; Kapucu & Ozerdem, 2011; Kapucu & Van Wart, 2006; Kapucu, 2009). In this article, specific emphasis is placed on how these entities with distinct functions and tasks cooperate with each other to produce a unified and network structure.

This article analyzes how functionally collaborative networks (FCN)—a special case of virtual organizations—in an emergency management context are designed, developed, and sustained. Functionally collaborative networks are joint virtual structures composed of organizations with shared goals and efforts working to produce a collective benefit. The following research questions are examined in the context of emergency management networks: What are the key characteristics of functionally collaborative networks? What are the challenges to sustain effective collaboratively functional networks? Are the network relationships solely informal or formalized so that they are sustainable over time? How can the functionally collaborative networks function in response to disasters in metropolitan regions? These research questions are examined using content analysis of the archival data in response to the four hurricanes of the 2004 hurricane season, and an open-ended survey administered to the officials of four counties in the State of Florida. The findings from this research should benefit network organizations operating in the complex environments of disasters, as well as contribute to the literature on network organizations. In particular, this study is expected to improve the design, development, and implementation aspects of inter-organizational emergency management networks in metropolitan regions.

LITERATURE REVIEW AND CONCEPTUAL FRAMEWORK

This section discusses functionally collaborative networks in light of the literature on networks. It also specifies the main aspects of FCNs, setting a stage for the analysis of networks of local agencies chosen for this study.

The ever-increasing severity and scope of disasters over the past decades, complexity, and the dynamic nature of disasters make it impossible for public agencies to cope with the consequences on their own. Specific emphasis and importance is placed on the need to collaborate across different sectors and with other levels of government for more effective and successful results (Callahan & Holzer, 1994; Vangen & Huxham, 2003; Bryson, Crosby, & Stone, 2006; Kapucu & Van Wart, 2006). Such an approach requires that agencies with disaster management responsibilities collaborate through networks (Weber, 2003). According to Snow, Lipnack, and Stamps (1999), network organizations comprise three factors: they are multisite, multi-organizational, and dynamic. They are entities comprising several geographically distributed actors that are temporarily tied to each other for competitive advantage (common interests) and share common values and operations supported and maintained through information technology (Grabowski & Roberts, 1999; Paula *et al.*, 2004). Such organizations allow for dispersed information sharing, decision-making and resource allocation without being physically present in a certain location (Davis, 2002). Therefore, the main goal of a network is to establish a platform for the agencies representing different values and norms with the purpose to accomplish a common goal using information management and communication technologies.

This article uses virtual networks interchangeably with networks. A virtual organization is an entity with or arrangement of non-spatial inter-organizational relationships that do not comprise a formal organization on its own. As with any approach, networks have their pros and cons (Fedorowicz, Gogan, & Williams, 2007). The advantageous part of networks is that they provide an opportunity for flexibility in terms of the structure, the decision-making and speedy processes. Since networks/virtual organizations are non-spatial in physical terms, they allow for organizations to be part of the system independently and regardless of the physical boundaries that traditional organizations might be imposing otherwise. In addition, the network actors enjoy dispersed decision-making allowing for responding to different conditions, situations and requirements despite frequent changes and volatility in terms of the goals and operations of the actors in the system (Grabowski & Roberts, 1999). Lastly, networks are tools for constant communication with the purpose of synchronization, which is generally a cumbersome process in traditional organizations due to the usual trend to discuss such issues in frequent long face-to-face meeting environments. Such flexibility is an opportunity for more frequent and more streamlined communication and information sharing among actors, minimizing possible delays and maximizing process efficiency (Snow, Lipnack, & Stamps, 1999).

The first disadvantage of the networks when compared to traditional organizations, on the other hand, is the heterogeneity and difference of values, ideas and norms of network actors. Such a gap may result in a delayed decision-making process, thus in slower operations overall. The second disadvantage is the variety of technological tools and applications used by different actors. Since every single organization's technical background is comparatively different from others' due to disparities in organizational goals, and preferences and capacity, it is difficult, if not impossible, to have a standard set of technologies across organizations involved in networks. The third disadvantage of the networks is the relatively lower level of trust between the actors and entities involved in the system. While trust is a cornerstone of network-based relationships, it is a challenge for networks to achieve an ideal level of trust among the parties especially due to lack of or less face-to-face interaction, different set of ideas and norms, and/or an inability to clearly convey messages or meanings establishing trust through virtual environments (Paula *et al.*, 2004; Kapucu, 2006; Kapucu & Ozerdem, 2011). The final disadvantage of networks is the essential question of how they promote participation, not only in terms of membership but also in terms of decision-making. In other words, who should be included in or excluded from the network is not always clear. What is more, decision-making mechanism in networks is problematic in terms of its inherent capacity to end up with conflicts slowing the overall process and network effectiveness.

According to Fedorowicz, Gogan, and Williams (2007), a functionally collaborative network is a "joint organizational entity, infrastructure, business processes, resources, and relationships which support a shared effort to provide some collective benefit, whether it is a program, service, or a product" (786). Such networks may be governed formally (Milward & Provan, 2006) or informally (Bardach, 2001) and are created if they agree upon information sharing and communication on an ongoing basis, and have their own distinct strategy, governance structure, goals, processes, inter-organizational structure, and technology base. These networks are also "networks of actions" that take place between organizations based on their common-goal interactions (Pentland, 1999). The organizational actions described by Pentland (1999) are oriented towards network goals that intend to produce a unified result. Functionally collaborative networks are characterized by intense communication and information exchange, shared goals, understanding and responsibility, and joint decision-making which accelerate and improve the process overall (Jassawalla & Sashittal, 1999). A functionally collaborative network, thus, is a set of two or more coordinated and streamlined cliques of activities/actions (with actors centralized around those activities) serving a common and more superior goal of the overall network, whether formally or informally, by focusing on the delivery of services in the most efficient way. Such network governance is unique in terms of its emphasis on the efficient collaborative accomplishment of specific functions/activities by related (both goal-wise and objective-wise) expertise which would be impossible otherwise. Perhaps the most important distinction between a functionally collaborative network and other types of networks is that this network comprises several sub-networks of organizations around specific functions, in which organizations of a specific sub-network can be also a part of other sub-networks. In a sense, this is a network placing importance on functions, not actors, when establishing an agenda.

Jones, Hesterly, and Borgatti (1997) establish a framework to analyze the governance network, which is composed of four main elements: (1) demand uncertainty; (2) human asset specificity; (3) task complexity; and (4) exchange frequency. When this framework is applied to a disaster setting, it presents a good fit for functionally collaborative networks with each element observed at the extreme end of the spectrum. That is to say, in terms of demand uncertainty, emergency events are characterized by extreme levels of uncertainty and unpredictability in terms of (a) timing of the events, (b) scale of risk involved, (3) level of need for responses, and (4) possible consequences. In terms of human asset specificity, emergency events require a very specific skill set of knowledge, expertise, and information and processes, which is only possible through a coordinated network governance effort. In terms of task complexity, emergency events are perhaps the most complex tasks to be accomplished under time pressure and with a multiplicity of actors working toward one common goal. Lastly, in terms of exchange frequency, emergency situations create an environment of increased, enhanced and more frequent interaction among functionally related actors. A functionally collaborative network in emergencies, therefore, is a unique case of network governance with increased emphasis on both structural and relationship factors. The main implication then is that investment in, as well as development and sustainability of such networks is critical. This paper identifies the following set of features of functionally collaborative networks that make them unique in conceptual terms.

Interdependence of Networks

Networks are created in the form of inter-organizational and intra-organizational relationships and interactions (Snow, Lipnack, & Stamps, 1999; Vangen & Huxham, 2003; Provan *et al.*, 2005; Bryson, Crosby, & Stone, 2006). Networks are human and non-human entities operating across various sectors (Kapucu, 2009), and such entities would combine into a socio-technical network system (Kelly & Stark, 2002). Broadly defined, network is "a group of three or more organizations connected in ways that facilitate achievement of a common goal" (Provan, Fish, & Sydow, 2007: 4). Several factors are involved within this definition. First, it means that organizations are managed horizontally rather than vertically; meaning that complete alignment of goals, values, norms and preferences is difficult, if not impossible. In other words, the actors/organizations of the networks are tied enough to accomplish a task at a micro level, their network-related duty, but are separate enough to follow their own agenda at a macro level, pertaining to their organizational mission and goals. In addition, being formally separate organizations, actors in networks establish and maintain their own organizational goals, values and norms, as well as enjoy distinct preferences. In a sense, networks are characterized by interdependencies with different goals, values, norms, and preferences (DeSanctis, Staudenmayer, & Wong, 1999).

Organizations mainly form networks due to structural, resource, and cost-minimization interdependencies as well as network embededness (the level and quality of informal relationships) (Gulati & Gargiulo, 1999), which would result in organizational success (Galaskiewicz, 1989; Provan *et al.*, 2005; Ansell & Gash, 2007) and overall increase the productivity of assets and resources (Fiszben & Lowden, 1999). The idea of shared goals is central to the functionally collaborative network perspective. It is for this reason that networks incorporate multi-stakeholder perspectives (Davies, 2003) and trans-jurisdictional approaches to complex public policies in general, and to complex emergency management issues in particular (Thurmaier & Wood, 2002). Disaster response is multi-organizational and emergent (Drabek *et al.*, 1981; Jordan, 2010). It is important to note, however, that networks are dynamic systems based on the evolution of organizational structures and actions (Gulati & Gargiulo, 1999). The following proposition was developed based on the literature:

Proposition 1: *Networks comprise interdependent organizations distinctive in their structural aspects but with a common and shared goal.*

Complexity of Identity and Interactions

The complexity of network organizations is characterized by two factors. The first factor is that it entails a variety of ideas, goals, values, and preferences with each organization tending to preserve the most of their identity. Such a combination of different identities adds to the overall complexity of operations as well as their results. Above all, it adds to the second-nature complexity ensued by a variety of network interactions in general, and interactions in response to disasters in particular. The increased formal and informal relationships are assumed as a basis for complexity of networks (Kapucu, 2009). Networks, in this regard, present an example of complex adaptive systems (ever-changing, inter-dependent, inter-organizational relationships) comprising several actors and extensive interactions among those actors. Collaborative behaviors are complex by nature (Jassawalla & Sashittal, 1999), and they are impractical in traditional hierarchies making them dysfunctional (Kapucu, 2008). DeSanctis,

Staudenmayer, and Wong (1999) argue that complexity is the result of a multiplicity of actors and identities that prevent standardization. Emergency management is a field that extensively relies on the involvement of inter-governmental and cross-sector collaborations during multiple phases of emergency management (Donahue & Joyce, 2001; Jordan, 2010), which makes it difficult to produce a clearly unified action. Scholars have called for the need to analyze emergency management from the complexity perspective (McEntire, 2004). The second proposition developed based on the literature is as follows:

Proposition 2: *Networks comprise organizations with various identities and a certain level of interactions resulting in complexity.*

Utilization of Information Communication Technology (ICT)

The socio-technical changes and innovations (globalization, social media, technological innovations, etc.) are critical to the realization and maintenance of the networks (Mowshowitz, 1997a; Snow, Lipnack, & Stamps, 1999). They are very much dependent on the use of technology that would facilitate communication and information exchange (Dawes & Eglene, 2004). Today networks are essentially practicable because of this very reliance on technology, which minimizes costs, increases efficiency, and increases the speed of operations as well as adjustments and changes occurring during the process (Mowshowitz, 1997b; den Hengst & Sol, 2001). Remote communication is generally considered a barrier to trust, security and culture (Walsh & Maloney, 2003). Nevertheless, technology-supported networks enjoy more effective collaborative practices because technology is the facilitating and enhancing, though not sufficient, factor when collaboration is considered (Cohen & Mankin, 1999).

Technology has become a tool undermining traditional hierarchies, short-circuiting the red-tape and vertical communication, and promoting horizontal relationships; thus, it is an unprecedented opportunity for geographically dispersed and highly collaborative organizations to enhance their operations and actions (Kelly & Stark, 2002). Knowledge management is just one, and perhaps the most important example of how ICT utilization in networks may be beneficial (Katz & Lazer, 2002). A widely used ICT mechanism, WebEOC, is just one example of effective utilization of ICT in emergency management. The system is a web-based crisis information and communication management tool enabling real-time information sharing and collaborative decision-making. The following proposition flows out of the literature above:

Proposition 3: *Information technologies are imperative for networks for continuity and sustainability of operations*

Sustainability

Sustainability of network relationships is a critical factor affecting their performance and the success (Weber, 2003). This is especially true for networks emerging during emergency situations or disasters; the networks developed during these stages are rarely sustained because of the role an emergency situation plays as a bridging function. Since, networks comprise entities that are geographically dispersed, structurally different and operate in different time periods, it is important to focus on the design, development, and sustainability of these different and multi-faceted networks (Trotter *et al.*, 2008). Stability of solid network relationships over time, especially between disasters in emergency management networks is crucial; this is especially crucial to reduce resistance to change and to institutionalize collaborative practices projected over time (Milward & Provan, 2000; Ansell & Gash, 2007).

According to Katz & Lazer (2002), sustainability is partly dependent on the success of, as well as the level of trust developed during previous collaborative practices. The establishment of trust and shared visions is also a factor influencing the level of interaction and involvement across time (Tsai & Ghoshal, 1998). It is also a function of node connectivity and ICT preferences (Balakrishnan, Magnanti, & Prakash, 1998). Milward & Provan (2001) claim that multiplexity, which is defined as the strength of ties between agencies based on the number of connections, is also a strong determinant of network sustainability. According to Gillespie *et al.* (1993), network relationships are sustained if (1) there is an active problem to be addressed; (2) the interactions are sustained; (3) interactions are conducted in a professional cooperative manner; and (4) organizational representatives have good personal relationships. Barriers to sustainability, on the other hand, are due to goal differences and fear of liability. It is also a problem of motivating factors as well as a lack of standardized technology that would prevent network sustainability. The following proposition was developed based on the related literature:

Proposition 4: *For effective collaborative response during emergencies, network relationships need to be sustained before and after disasters.*

In the previous section of the article, the characteristics of networks are defined broadly, and each network should be analyzed separately to find out case-specific differences. Networks are established with various purposes and functioning across several phases of emergency management (Hansen, Mors, & Lovas, 2005). The functionally collaborative networks in emergency situations are also established for specific purposes and missions. The table below specifies possible combinations of networks based on expertise, resources, tasks, and functions.

Pentland (1999) describes networks as patterns of actions that are to be discovered through network analysis. He characterizes networks by temporally and spatially distributed sequences of interactions towards a specific goal. The matrix establishes a framework for network-based relationships that are exercised by various entities on daily basis. Functionally collaborative networks are represented by the intersection of agents/actors and functions, which identifies which organization is responsible for which function. The function-based categorization is one of the most important aspects of the emergency management in the U.S. Before 1978, all emergency management functions were handled by seven inter-related agencies. The functions were then a classification of activities and operations in line with their commonalities. This approach was systematized with the creation of the Federal Emergency Management Agency (FEMA) in 1978 (formally authorized in 1979), which envisioned the coordination of dispersed functions of federal agencies. The Federal Response Plan (FRP) created in 1992 was a further attempt to standardize and categorize functions, tasks, duties and roles for better coordination of federal activities with specific focus on their mission, policies and responsibilities. The six sections of the FRP outlined respective primary and support federal agencies and the Emergency Support Functions (ESFs) they would execute.

While the FRP had 12 ESFs, the National Response Plan (NRP) of 2003 had 15 ESFs. The National Response Framework (NRF) that was created in 2008 also has 15 ESFs. The main theme of the latter is inter-sector and inter-government coordination and collaboration that would increase emergency response effectiveness (Kapucu, 2009). Today the NRP/NRF, which provides structure and mechanisms for nation-wide policies concerning incident management,

	Agents	Knowledge	Resources	Tasks	Functions
Agents Relation	**Communication Network** *Who knows who*	**Knowledge Network** *Who knows what*	**Capabilities Network** *Who has what resource*	**Assignment Network** *Who does what*	**Role Network** *Who is responsible for what function*
Knowledge Relation		**Information Network** *What informs what*	**Skills Network** *What knowledge is needed to use what resource*	**Needs Network** *What is needed to do that task*	**Competency Network** *What knowledge is where*
Resources Relation			**Substitution Network** *What resources can be substituted for which*	**Requirements Network** *What resources are needed to do that task*	**Capital Network** *What resources are where*
Tasks Relation				**Precedence Network** *Which task must be done before which*	**Sectoral (Responsibility) Network** *What tasks are done where*
Functions Relation					**Affiliation / dependence Network** *Which organization is affiliated with which function*

Table 1 *Meta Matrix for Functionally Collaborative Networks* **(***Adapted from Kapucu 2005***)**

and the National Incident Management System (NIMS), which provides a common template for jurisdictions concerning incident management, serves as a philosophical as well as operational guide to cope with disasters from all-hazards perspective.

Functionally collaborative networks serve as an example of affiliation networks, which has widely been studied in the organizational context. The application and analysis of this phenomenon in the context of emergency management, which has different structural and relational requirements, is relatively a novelty. Therefore, this paper aims to link network attributes to organizational policies and plans. The following sections further explain how this system fits into the framework of functionally collaborative networks.

METHODOLOGY

The data for this research was derived from content analysis of publicly available documents, namely State Emergency Response Team (SERT) situation reports, county situation reports provided by emergency managers of the four counties studied, and newspaper news reports available through the academic Lexus-Nexus database that were related to the six-week period, with four consecutive hurricanes (Charley, Frances, Ivan, and Jeanne), of 2004 in Florida. The sources were analyzed to identify which organizations specified in County Comprehensive Emergency Management Plans (CEMP) performed what function during the hurricane season. Emergency response and recovery agencies are responsible for certain operations that fall within a category of certain ESFs—a mechanism of classification and organization of the agencies in terms of their functions. Each organization in the CEMP has its own responsibilities summoned under the ESFs. The operations and activities identified in the sources were coded in the agency/ESF matrix. There was assumed to be an affiliation if an agency was specified under a specific ESF and was involved in related operations. The functional network of primary and support agencies within the CEMPs (design) at the local level (Orange, Miami-Dade, Hillsborough, and Duval counties) were compared, first, with other counties' plans on a theoretical base, and, second, against the actual network (implementation) in light of responses to the 2004 hurricanes. The latter comparison was conducted only for two counties, namely Orange County and Hillsborough County, due to the fact that not enough information was collected through the content analysis. The network perspective was utilized in this research to analyze the structure of relationships among organizations (Milward & Provan, 1998). The social network analysis software UCINET (Borgatti, Everett, and Freeman 2002) was utilized to analyze and depict inter-organizational relationships in the form of diagrams. Network analysis allows making "invisible" informal relationships across boundaries and functions "visible" (Cross, Borgatti, & Parker, 2003).

In addition to the archival data, a survey was designed and e-mailed to 67 county emergency managers in the State of Florida in 2008 with 36 (53.7%) returned responses. The survey was designed in order to get additional feedback from real-world experiences of practitioners to support the results of the content analysis. It included open-ended questions with the purpose to derive qualitative information and insight about issues pertaining to their jurisdictions. The survey provided a qualitative observation of the emergency managers on the four key characteristics of the functionally collaborative networks that was used to support the discussion around the concepts. The following four questions were asked: 1) What are the benefits of having ESF based partnerships/networks? How do you develop these partnerships? 2) How do ESF-based networks/partnerships increase the complexity of the working relationships? 3) How do you utilize information technology in your communication with ESF agencies besides your regular face to face meetings? What are the benefits of using information technologies? 4) How do you maintain the relationships with ESF agencies in the absence of disasters/ emergencies? What are the most critical factors that help you maintain the relationships?

The survey was distributed separately through e-mail addressing each emergency manager. The questions targeted to identify respective emergency management operation centers' structure and nature of operations in line with the research goals. The questions were related to how emergency responders consider ESF-based partnerships/networks, the level of complexity coming with those networks, the use of information technology for network purposes, and the issue of network sustainability in the absence of disasters/ emergencies?

FUNCTIONARY COLLABORATIVE NETWORKS IN EMERGENCY MANAGEMENT

While networks are practiced across a range of fields and disciplines, they are quite popular in and eccentric to the emergency management field. Emergency management networks characterized by flexibility primarily focus on the common issue that brings those actors together. In a sense, such networks are project or actor/task teams that are tied around a certain issue often on an ad-hoc basis. Davies (2003) calls these networks "common interest networks," through which actors are connected by ideas, concerns, events, and organizations, which is similar to the meta matrix developed by Kapucu (2005) mentioned earlier in this article. Functionally collaborative networks are established around certain functions that need to be executed in complex environments. In other words, they comprise entities that have proximities based on their interest profiles (Laumann & Knoke,

1986). The capacity of a network to adapt is as important as the initial design for sustainability and survival (Aldrich, 1999; Boin *et al.*, 2010)

While the National Response Framework in the US provides a framework for the federal government to arrange functional goals and responsibilities among related entities, not all governments adopt the mechanism in the same form. In analyzing the case of the metropolitan area emergency management systems in the State of Florida, this article focused on four metropolitan areas (Orange, Miami-Dade, Hillsborough and Duval counties). The rationale behind specifically choosing these counties is that this study targeted only four metropolitan areas specified in the grant received for this study[1]. Each of these counties has a CEMP entailing and outlining the entire emergency management system of the counties and how it would work in case of emergency situations or disasters. The plans also provide a list of ESFs and the primary as well as support agencies responsible for those ESFs during emergency and response operations. The primary agencies are those responsible for coordination of related field agencies, while support agencies are secondary agencies coordinated by the primary agencies if/when needed. For example, American Red Cross is the primary agency responsible for coordinating mass care organizations in most counties. According to the plans, Orange County has 20, Miami-Dade County has 18, Hillsborough County has 17, and Duval County has 18 ESFs with minimal differences in the titles of those ESFs (see Appendix A for Hillsborough County as an example).

Several inferences can be made in terms of ESF-based collaborative networks at the local level. Figure 1 below depicts the anticipated function-based affiliation networks of the four counties in Florida. Accordingly, it is ESFs, not organizations, that are the focus of the networks, and specific organizations are theoretically expected to be tied to specific ESFs. For example, all mass care organizations are tied around ESF #7. Inter-organizational ties are not depicted in the diagram; the diagram only shows how organizations should be structured in terms of ESFs.

The picture shows little multiplexity in the CEMPs established by the counties. The level of collaboration, thus, is expected at a relatively moderate level. This expectation is seen when one analyzes the cliques of organizations around specific ESFs. In other words, county emergency management plans are

1. This research is funded by National Science Foundation (Award #: 0943208). Title: VOSS: Creating Functionally Collaborative Infrastructure in Virtual Organizations.

cautious about assigning ESFs to multiple organizations. Another explanatory factor is the number of organizations expected to be involved in a possible emergency situation. The complexity and difficulty inherent in coordination and collaboration between large numbers of organizations leads county emergency management practitioners to use a more balanced approach. While there is no specific standard for this, the counties rely on a collaborative approach involving at least one or two external agencies. However, in case of catastrophic disasters, involvement of state and federal level entities makes the situation more complicated.

When a visual comparison is made between counties, the more centralized ESF-based affiliation network is Orange County's network. While the other three counties are relatively decentralized, Miami-Dade County's affiliation network is characterized by more rigid functional boundaries, which is followed by Hillsborough and Duval Counties. Such rigidity may be an advantage in that it would increase specialization and focus on the functions the agencies are assigned to, thus, leading to the possibility of sustained relationships. Distraction and loss of focus is inevitable when an organization is assigned to multiple functions and tasks. It is, therefore, for this reason that Orange County's plan presents a more complex, and thus more vulnerable and less

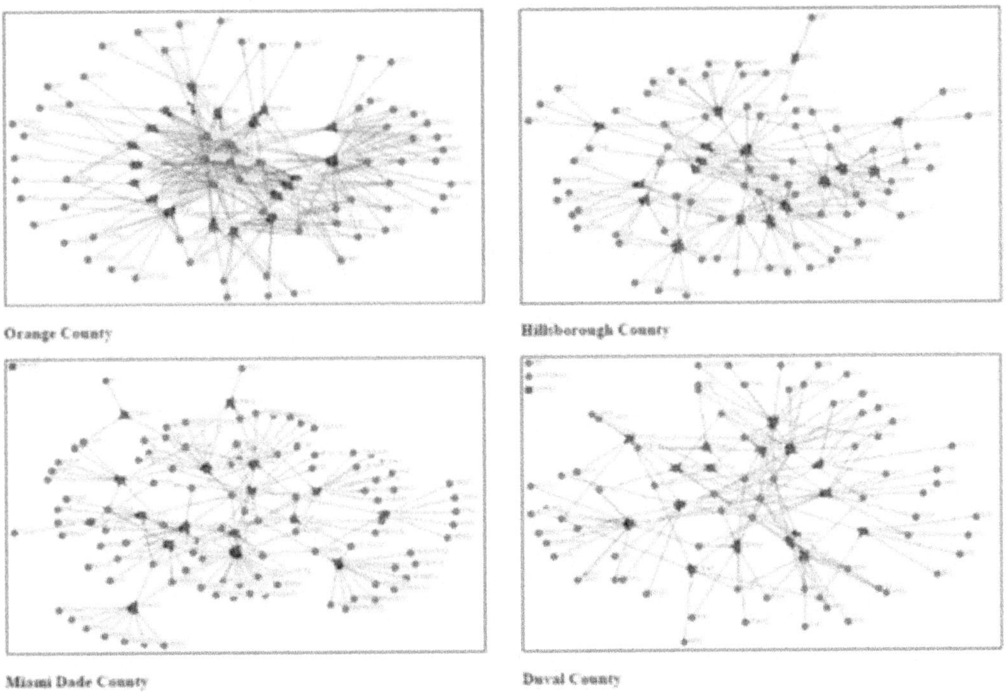

Figure 1 *Function-Based Affiliation Networks of Four Counties in the State of Florida*

fragile system. Having a system resembling Miami-Dade County's plan with cliques formed around functions would guarantee a focused approach, hence a more specialized and flexible group of entities.

The above-mentioned comparison so far has been a comparison of the CEMPs of the four counties. The content analysis of the actual affiliation network relationships allowed identification of the actual networks of Orange County and Hillsborough County, which are compared in the Figures 2 and 3 below:

Orange County's affiliation network comparison presents a relatively big gap between the design in the CEMP and actual networks. While no ESF was left outside of the network, many organizations seem to be uninvolved during the hurricane season. First of all, the content analysis might have limited data that would reflect un-captured relationships. Second, the County's emergency response agencies' involvement might be limited in accordance with the

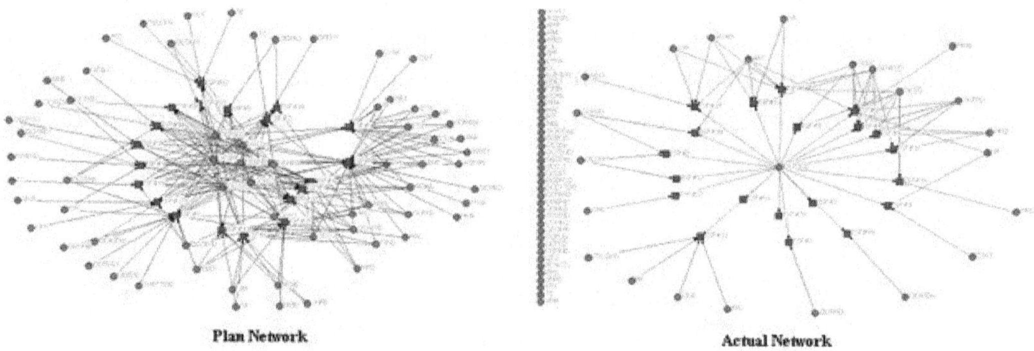

Figure 2 *Orange County Function-Based Plan vs. Actual Affiliation Networks*

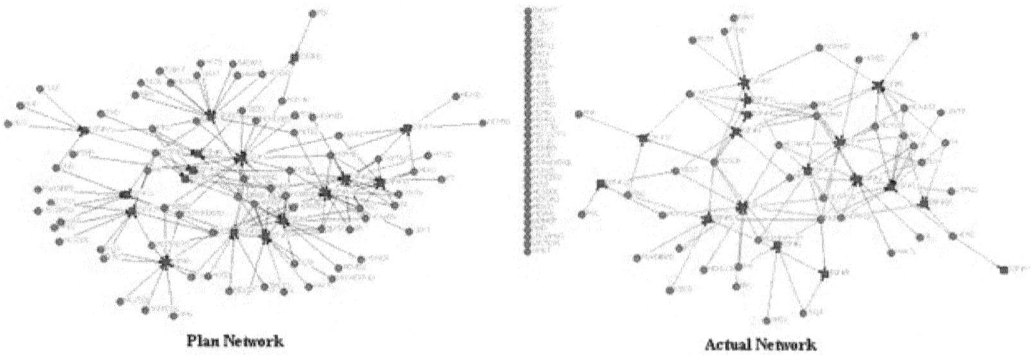

Figure 3 *Hillsborough County Function-Based Plan vs. Actual Affiliation Networks*

need for other agencies during the hurricane season. It is essential, however, to identify the isolate organizations and determine the reason behind their being passive in the network. The isolates are still organizations that possess the necessary resources, knowledge and human capital that is crucial during any emergency and disaster situation. Therefore, their minimal involvement is imperative regardless of the level of disaster or capacity of other already involved organizations.

The Hillsborough County's actual response network, though, presents a relatively smaller gap when compared to the plan. In addition, as in Orange County's case, all of the ESFs were activated during the hurricane season, which also shows the severity of the disaster for both cases. Many organizations seem to be called upon during the response, but the number of inactive organizations or agencies is smaller when compared to Orange County's case. Again, one should note the limitations of content analysis as well as the severity of emergency situations, both of which might be determining factors in how many organizations are involved in response operations.

DISCUSSION

CEMPs are documents pre-determining the structure, size, scope and content of functionally collaborative networks. County plans can be analyzed on intra-functional and inter-functional basis. The former means how individual organizations structured around a function interact with and relate to each other. The later, on the other hand, looks at how organizations tied to one function interact with and relate to organizations tied to another function. The four assumptions identified based on the literature review are analyzed below in terms of these two dimensions. Analysis was supported by the results of the survey, with open-ended questions on the four main assumptions, that was sent to emergency managers in the State of Florida.

When the above-mentioned design networks of the respective counties are analyzed in terms of their *network* dimension, it would ideally be more appropriate to have networks of organizations around specific ESFs. In other words, intra-functional networks are preferred to inter-functional networks, at least in the context of emergency management. Such an approach would and should secure concentration on the task as well as a more streamlined and effective collaborative response. Miami-Dade County's plan is an example of such a plan. The County's approach seems to be aligned with the idea that multiplexity, in the sense that individual organizations have multiple connections across functions, is not favorable. On the other hand, Orange County has a plan with an inter-functional focus, which assigns multiple tasks to single organizations.

The ideal network structure for an emergency management plan, however, is the one in which organizations are involved only in intra-functional networks, extensively interacting with other organizations involved at an intra-functional level. This is required to ensure a focused, coordinated and eased response by organizations affiliated with each other on organizational goal lines. It should also be noted, however, that it is an organizations' tasks to determine, based on their organizational goals and capacity, the verticality (intra-functionality) or the horizontality (inter-functionality) of the network they are to be involved in. This was also the main theme of the responders to the survey: they claimed that ESF-based partnerships or networks are a tool to "create an order out of chaos when disasters or large-scale emergencies threaten" (Sallade, 2010) as well as a mechanism to bring together subject-matter experts (Faulkner, 2010; McNally, 2010; Shepard, 2010). The main idea behind the claim is to bring together as many related expertise from the outside as possible to produce a coordinated, well-informed and professional response to disasters and emergencies.

As far as *complexity* is concerned, it is common sense to have a balanced level of multiplexity that would not be detrimental to organizational capacity and inter-organizational relationships. However, complexity needs to be considered from both, the intra-functional and inter-functional perspectives. While complexity at the intra-functional level is necessary and beneficial, complexity at the inter-functional level is detrimental, and thus to be avoided. The former approach is characterized by multiplexity among organizations related to a function, and is essential even if redundancy is in place. Redundancy in this sense provides for resource availability and increased response capacity. The latter approach, on the other hand, is characterized by inter-organizational interaction across functions, and is detrimental due to dispersed organizational capacity that can be utilized for one function/task otherwise. Such complexity stemming from multiple organizational goals is beneficial only if the organization clearly and professionally coordinates its inter-functional interactions. Again, Miami-Dade County is an example of 'positive' complexity characterized by relative absence of inter-functional complexity (intra-functional complexity is not observable in Figure 1 due to absence of anticipated inter-organizational ties). Orange County, however, stands as an example of inter-functional complexity only to the extent that relatively more organizations are affiliated with multiple functions.

On the other hand, inter-organizational ties at the inter-functional level are only preferred for coordination purposes. In other words, while inter-organizational complexity is beneficial at intra-functional level for comprehensive collaboration, it is preferred at the inter-functional level for coordination purposes only. The survey responses align with the idea that complexity arising from intra-functional networks contributes to a clearer vision, tighter relationships, effectiveness and efficiency (Sallade, 2010; Shepard, 2010). The general idea, thus, is that ESFs do not create complexity; it is rather minimized because the ESF-based structure "has made partnerships/networks work more effectively and efficiently due to the clear scope of responsibility and standard language" (Holman, 2010). Mike Faulkner (2010), the emergency management director of the Okeechobee County, noted that the size of the jurisdiction is also crucial in this regard, which does not create complexity especially at the local levels if jurisdictions are small.

When the plans are analyzed in terms of the *ICT utilization* aspect, one should take into account the extent to which an organization is involved in intra-functional vs. inter-functional network. Generally speaking, ICT dependence is critical for both types of networks because functionally collaborative emergency management networks are effective to the extent they have intra-organizational and inter-organizational interactions. ICT utilization is imperative to coordinate both intra-organizational and inter-organizational networks, while the latter may be problematic due to possible differences in organizational capacity and non-standard technical and technological structures, which might complicate coordination and synchronization processes. There is also a need for related training of staff on the use of those technologies. This situation may be further complicated by the reason specific technological tools are used within an organization. In other words, certain organizations may use technology for knowledge management purposes, while others might be interested in the use of technology for coordination purposes only. The Orange County CEMP, for example, identifies more organizations with multiple goals, which stresses the importance of technology for intra-organizational coordination. Overall, the county emergency managers agreed that ICT utilization is beneficial to save money and time, as well as for coordination purposes (McNally 2010). Charles Craig (2010), the emergency management director of the Volusia County, notes, however, that face-to-face meetings are a best choice when decision-making is concerned.

Finally, in terms of *sustainability*, the networks presented above are sustainable over time as long as they develop task-oriented and trust-based relationships during the incidents they are involved in. Such success is ideally possible through concentrated networks as in the Miami-Dade case especially due to an expectation that extensive intra-functional interactions would lead to long-term relationships and cooperation. In other words, while inter-organizational interactions at inter-functional levels might be short-term, inter-organizational interactions at intra-functional levels would be more trust-based, hence long-term. Continuity of collaboration, cooperation and partnerships before and after a disaster strikes is one of the determining factors of successful network response in emergency management. Nevertheless, there are many organizational as well as contextual factors that need to be taken into account before saying a final word on the issue of network sustainability. It is likewise a matter of organizational goals, intra-organizational and inter-organizational politics, as well as organizational capacity whether intra-functional networks are sustained over time. Periodic communication, face-to-face meetings, collaborative trainings as well as effective leadership are among the factors to foster network sustainability. This issue was strongly supported by county emergency managers as well, which is summoned in Volusia County emergency management director's statement: "You must constantly nurture these relationships especially in fiscally restrained times, thru invitation to train and exercise together and involving each other in activities throughout the year. If you only contact them when you want something, you are doomed. As with any partnership, there must be something in it for everybody, if not they will fold" (Craig, 2010).

The overall analysis of the survey responses provided by emergency managers also presents key points about the four dimensions discussed above. The general theme emerging from the responses to the first question is that ESF-based partnerships bring subject matter experts together, allow for effective and efficient use of resources, provides a better understanding of non-disaster entities, and overall, creates order out of chaos. The relationships based on the ESF structure are formed and developed by periodic organizational/agency meetings, workshops, exercises, and by deliberate and targeted outreach of respective emergency operation centers (EOC). Some of the respondents use the Incident Command System instead of the ESF Structure, arguably due to a more streamlined and easier process as well as better organization and coordination of activities.

The general theme emerging from the responses to the second question is that, while multiple tasks and organizational missions/goals during emergency situations are a reality, the ESF-based networks and partnerships do not increase or result in complexity; they rather facilitate working relationships due to the clearer scope of responsibility and standard language. Complexity seems to not be a problem, especially for the small counties.

In terms of the third question, the emerging theme is that technology has become an indispensible part of emergency management operations. Communication with the purpose of synchronization, information sharing and coordination is a critical component provided by technological tools and facilities. E-mail, web-conferencing, and phones are amongst the most widely used methods of communication. The WebEOC program is also a tool used by some EOCs to effectively and efficiently operate before, during and after emergency situations. The program is a web-enabled crisis information and communication management mechanism enabling emergency management officials and staff to share real-time information and make collaborative decisions. The software was one of the systems developed and enhanced as a reaction to the September 11, 2001 terrorist attacks that signified failure in crisis management. Today, the system is used by not only emergency and crisis management agencies, but also by several government agencies, healthcare associations, corporations and universities. The system is considered user-friendly, easy-to-tailor, easy-to-maintain and complies with the provisions of the Incident Management System (ICS) as well as the ESF (ESI, 2010). All in all, one of the greatest advantages of technology is time-saving and coordination.

In regard to the last question, the main emerging theme is that periodic collaborative activities are the best way to sustain relationships established during emergencies. Such activities, whether drills, trainings, workshops or exercises, develop trust-based relationships that stretch beyond emergency/ disaster periods. Special emphasis is put on the need to develop open dialogue and inclusive participation in order for other agencies to buy in. Face-to-face outreach activities are the best way to establish, develop and sustain relationships, and they should be done between disasters so that a mutual understanding is developed, shared values are established and trust is developed and sustained. Networks are useful not only to the extent they accomplish specific task they are created for, but also to the extent they are sustained to achieve efficiency and effectiveness in the future incidents and cases.

The theoretical contribution of this article is the linkage between structural and relational factors of a network, and specifically an affiliation network. First of all, inter-organizational networks are very much about the structure before any process can take place. In a sense, most inter-organizational networks are functionally collaborative networks in that several organizations come together to collaborate towards a common goal which itself is composed of multiple sub-goals. The actors' organizational affiliation, goals as well as capacity are crucial determinants of their being involved in intra-functional vs. inter-functional relationships, or both. It is for this reason that specifically in the emergency management context it is important to create a balance between the capacity needed to respond to disasters and the level/nature of coordination required for respective response. The ultimate aim should be creating a network of cliques of organizations with maximum capacity concentrated around the main functions, and minimal involvement in other functions. The formula can be summarized as maximum capacity, minimum complexity.

The general implication for practitioners from this article, as supported by the survey responses is that intra-functional relationships, or relationships between organizations tied to one specific function, need to be tight, tense and elaborate. Such multiplexity and complexity would produce expertise-based, focused decisions and well-directed operations, especially due to the aggregated mindset of related practitioners. Inter-functional relationships, on the other hand, should only be preferred to the extent they facilitate inter-functional collaboration, cooperation and coordination. Inter-functional ties bringing no resources or other types of capital should be avoided for the sake of minimal complexity in relationships.

CONCLUSION

Functionally collaborative networks are an essential part of today's emergency management systems. This is especially true for emergency management systems within the US that foster collaboration, coordination and partnerships at all levels of government, and between different sectors. Networks have distinguishing characteristics that have to be considered when designing, developing and sustaining such organizations. The analysis of the CEMPs of four counties as well as the survey responses from 36 county emergency managers in the State of Florida presents different local level approaches in terms of anticipated function-based affiliation network structures and relationships. Generally speaking, county plans envision more fragmented and focused perspectives, especially due to the relatively larger number of organizations expected to be involved in response networks. The network structure and dynamics are the factors that would affect the success as well as the future of the networks. It is with this issue in mind that emergency managers need to pay attention to how emergency management plans are designed, including the issues of network multiplexity, complexity of interactions, correct and sufficient use of technology, and network sustainability.

The practical implication to flow out of this article is that network sustainability is of particular significance to foster sophisticated and well-organized emergency response in the future instances. Future research needs to focus on how network sustainability can be achieved in the absence of disasters leading to those networks. This would be a contribution not only to the field of emergency management, but also to other disciplines affected by network theory and management.

REFERENCES

Ahuja, M.K. and Carley, K.M. (1999), 'Network Structure in Virtual Organizations', *Organization Science*, 10(6): 741-757.

Aldrich, H.E. (1999), *Organizations Evolving,* London, Sage Publications.

Ansell, C. and Gash, A. (2007), 'Collaborative Governance in Theory and Practice', *Journal of Public Administration Research and Theory*, pp. 1-29.

Balakrishnan, A., Magnanti, T.L. and Prakash, M. (1998), 'Designing Hierarchical Survivable Networks', *Operations Research,* 46(1): 116-136.

Bardach, E. (2001), 'Developmental Dynamics: Interagency Collaboration as an Emergent Phenomenon', *Journal of Public Administration Research and Theory,* 11(2): 149-164.

Boin, A., Kuipers, S., and Steenbergen, M. (2010), 'The Life and Death of Public Organizations: a Question of Institutional Design', *Governance: An International Journal of Policy, Administration, and Institutions,* 23(3): 385-410.

Borgatti, S. P., Everett, M.C., and Freeman, L C. (2002). *UCINET for Windows: Software for Social Network Analysis*. Harvard, MA: Analytic Technologies.

Bryson, J.M., Crosby, B.C. and Stone, M.M. (2006), 'The Design and Implementation of Cross-sector Collaborations: Propositions from the Literature', *Public Administration Review*, pp. 44-55.

Callahan, K. and Holzer, M. (1994), 'Rethinking Governmental Change: New Ideas, New Partnerships', *Public Productivity and Management Review,* 17(3): 201-214.

Cohen, S.G. and Mankin, D. (1999), 'Collaboration in the Virtual Organization', In Cooper, C.L. and Rousseau, D.M. (Eds.), *Trend in Organizational Behavior: The Virtual Organization,* Chichester, John Wiley & Sons Ltd, pp. 105-120.

Craig, C. (2010, March 2), *Survey Responses by Volusia County Emergency Management*, Volusia County, Florida.

Cross, R., Borgatti, S.P. and Parker, A. (2003), 'Making Invisible Work Visible: Using Social Network Analysis to Support Strategic Collaboration', In Cross, R.L., Parker A. and Sasson, L. (eds.), *Networks in the Knowledge Economy*, New York, Oxford University Press, pp. 261-282.

Davies, R. (2003, November 24), 'Network Perspectives in the Evaluation of Development Interventions: More Than a Metaphor', www.mande.co.uk/docs/nape.pdf

Davis, S.C. (2002, July), 'Virtual Emergency Operations Centers', *Risk Management*, pp. 46-52.

Dawes, S.S. and Eglene, O. (2004), 'New Models of Collaboration for Delivering Government Services: A Dynamic Model Drawn from Multi-national Research', *2004 Annual National Conference on Digital Government Research,* Seattle, WA, Digital Government Society of North America (pp. 1-11).

den Hengst, M. and Sol, H.G. (2001), 'The Impact of Information and Coordination Technology on Interorganizational Coordination: Guidelines from Theory', *Informing Science,* 4(4): 129-138.

DeSanctis, G., Staudenmayer, N. and Wong, S.S. (1999), 'Interdependence in Virtual Organizations', In Cooper, C.L. and Rousseau, D.M. (Eds.), *Trend in Organizational Behavior: The Virtual Organization*, Chichester, John Wiley & Sons Ltd, pp. 81-104.

Donahue, A.K. and Joyce, P.G. (2001), 'A Framework for Analyzing Emergency Management with an Application to Federal Budgeting', *Public Administration Review,* 61(6): 728-739.

Drabek, T.E., Tamminga, H.L., Kilijanek, T.S. and Adams, C.R. (1981), *Managing Multiorganizational Emergency Responses: Emergent Search and Rescue Networks in Natural Disaster and Remote Area Settings*, Boulder, University of Colorado.

ESI. (2010), *WebEOC*, http://esi911.com/esi/index.php?option=com_content&task=view&id= 14&Itemid=30 (accessed 21 April 2010).

Faulkner, M. (2010, March 2), *Survey Responses by Okeechobee County Emergency Management*, Okeechobee County, Florida.

Fedorowicz, J., Gogan, J.L. and Williams, C.B. (2007), 'A collaborative network for first responders: Lessons from the CapWIN case', *Government Information Quarterly*, 24: 785-807.

Fiszben, A. and Lowden, P. (1999), *Working Together for a Change: Government, Civic and Business Partnerships for Poverty Reduction in Latin America and the Carribean*, Washington, D.C., Economic Development Institute of the World Bank.

Galaskiewicz, J. (1989), 'Interorganizational Networks: Mobilizing Action at the Metropolitan Level', In Perrucci, R. Potter, H.R. (Eds.), *Networks of Power: Organizational Actors at the National, Corporate, and Community Levels*, Hawthorne, Aldine de Gruyter, pp. 81-96.

Gillespie, D.F., Colignon, R.A., Banerjee, M.M., Murty, S.A. and Rogge, M. (1993), *Partnerships for Community Preparedness*, Boudler, University of Colorado.

Grabowski, M. and Roberts, K.H. (1999), 'Risk Mitigation in Virtual Organizations', *Organization Science*', 10(6): 704-721.

Gulati, R. and Gargiulo, M. (1999), 'Where Do Interorganizational Networks Come From?', *American Journal of Sociology*, pp. 177-231.

Hansen, M.T., Mors, M.L. and Lovas, B. (2005), 'Knowledge Sharing in Organizations: Multiple Networks, Multiple Phases', *Academy of Management Journal*, 48(5): 776-793.

Holman, K. (2010, March 2), *Survey Responses by Martin County Emergency Management*, Martin County, Florida.

Jassawalla, A.R. and Sashittal, H.C. (1999), 'Building Collaborative Cross-Functional New Product Teams', *Academy of Management*, 13(3): 50-63.

Jones, C., Hesterly, W.S. and Borgatti, S.P. (1997), 'A General Theory of Network Governance: Exchange Conditions and Social Mechanisms', *Academy of Management Review*, 22(4): 911-945.

Jordan, A.E. (2010), 'Collaborative Relationships Resulting from the Urban Area Security Initiative', *Journal of Homeland Security and Emergency Management*, 7(1): Article 38.

Kapucu, N. (2008), 'Collaborative Emergency Management: Better Community Organising, Better Public Preparedness and Response', *Disasters*, 32(2): 239-262.

Kapucu, N. (2006), 'Interagency Communication Networks During Emergencies: Boundary Spanners in Multiagency Coordination', *The American Review of Public Administration,* 36(2): 207-225.

Kapucu, N. (2009), 'Interorganizational Coordination in Complex Environments of Disasters: The Evolution of Intergovernmental Disaster Response Systems', *Journal of Homeland Security and Emergency Management,* 6(1): 1-26.

Kapucu, N. (2005), 'Interorganizational Coordination in Dynamic Context: Networks in Emergency Response Management', *Connections,* 26(2): 33-48.

Kapucu, N. and Ozerdem, A. (2011), *Managing emergencies and crises.* Boston, MA, Jones & Bartlett Publishers.

Kapucu, N. and Van Wart, M. (2006), 'The Evolving Role of the Public Sector in Managing Catastrophic Disasters: Lessons Learned', *Administration and Society,* 38(3): 279-308.

Katz, N. and Lazer, D. (2002), 'Building Effective Intra-organizational Networks: The Role of Teams', http://www.hks.harvard.edu/leadership/Pdf/KatzLazerWorkingPaper.pdf (accessed 10 February 2010).

Kelly, J. and Stark, D. (2002), *Crisis, recovery, innovation: responsive organization after September 11,* New York, Center on Organizational Innovation.

Koliba, C., Meek, J.W. and Zia, A. (2010), *Governance Networks in Public Administration and Public Policy,* New York, CRC Press.

Laumann, E.O. and Knoke, D. (1986), 'Network Theory: Social Network Theory', In Lindenberg, S., Coleman, J.S. and Nowak, S. (Eds.), *Approaches to Social Theory,* Chicago, Russel Sage Foundation, pp. 83-109.

McEntire, D. A. (2004, June 4), The Status of Emergency Management Theory: Issues, Barriers, and Recommendations for Improved Scholarship, Emmitsburg, MD, Federal Emergency Management Agency.

McNally, P. (2010, March), *Survey Responses by Polk County Emergency Management,* Polk County, Florida.

Milward, H.B. and Provan, K.G. (2006), *A Manager's Guide to Choosing and Using Collaborative Networks,* Washington, DC, IBM Center for the Business of Government.

Milward, H.B. and Provan, K.G. (2000), 'How Networks Are Governed', In Heinrich, C.J. and Lynn L.E. (Eds.), *Governance and Performance: New Perspectives,* Washington, DC, Georgetown University Press, pp. 238-262.

Milward, H.B., & Provan, K.G. (1998), 'Measuring Network Structure', *Public Administration,* 76: 387-407.

Mowshowitz, A. (1997a), 'On the Theory of Virtual Organization', *Systems Research and Behavioral Science,* 14: 373-384.

Mowshowitz, A. (1997b), 'Virtual Organization', *Communications of the ACM,* 40(9): 30-37.

Paula, S., Seetharamanb, P., Samaraha, I. and Mykytyn, P.P. (2004), 'Impact of Heterogeneity and Collaborative Conflict Management Style on the Performance of Synchronous Global Virtual Teams', *Information and Management*, 41: 303-321.

Pentland, B.T. (1999), 'Organizations as Networks of Action', In Baum, J. and McKelvey B. (Eds.), *Variations in Organization Science: In Honor of Donald T. Campbell*, Thousand Oaks, CA: Sage Publications, pp. 237-253.

Provan, K.G. and Milward, H.B. (2001), 'Do Networks Really Work? A Framework for Evaluating Public Sector Organizational Networks', *Public Administration Review*, 61(4): 414-423.

Provan, K.G., Fish, A. and Sydow, J. (2007), 'Interorganizational Networks at the Network Level: A Review of the Empirical Literature on Whole Networks', *Journal of Management*, 33(3): 479-516.

Provan, K.G., Veazie, M.A., Staten, L.K. and Teufel-Shone, N.I. (2005), 'The Use of Network Analysis for Strengthening Community Partnerships in Health and Human Services', *Public Administration Review*, 65: 603-613.

Sallade, W. (2010, March 3), *Survey Responses by Charlotte County Emergency Management*, Charlotte County, Florida.

Shepard, R. (2010, March 2), *Survey Responses by Hardee County Emergency Management*, Hardee County, Florida.

Snow, C.C., Lipnack, J. and Stamps, J. (1999), 'The Virtual Organization: Promises and Payoffs, Large and Small', In Cooper, C.L. Cooper and Rousseau, D.M. (Eds.), *Trend in Organizational Behavior: The Virtual Organization*, Chichester, John Wiley & Sons Ltd, pp. 15-30.

Thurmaier, K. and Wood, C. (2002), 'Interlocal Agreements as Overlapping Social Networks: Picket-Fence Regionalism in Metropolitan Kansas City', *Public Administration Review*, 62(5): 585-698.

Trotter, R.T., Briody, E.K., Sengir, G.H. and Meerwarth, T.L. (2008), 'The Life Cycle of Collaborative Partnerships: Evolution of Structure and Roles in Industry-University Research Networks', *Connections*, 28(1): 40-58.

Tsai, W. and Ghoshal, S. (1998), 'Social Capital and Value Creation: The Role of Intrafirm Networks', *Academy of Management Journal*, 41(4): 464-476.

Vangen, S. and Huxham, C. (2003), 'Nurturing Collaborative Relations', *The Journal of Applied Behavioral Science*, 39(1): 5-31.

Walsh, J.P. and Maloney, N.G. (2003, May 15), *Problems in Scientific Collaboration: Does Email Hinder or Help?*

Weber, E.P. (2003), *Bringing Society Back In: Grassroots Ecosystem Management, Accountability, and Sustainable Communities*, Cambridge, The MIT Press.

APPENDIX A—HILLSBOROUGH COUNTY CEMP MATRIX

	ESF#1 Transportation	ESF#2 Communications	ESF#3 Public Works & Engineering	ESF#4 Firefighting & EMS	ESF#5 Info & Planning	ESF#6 Mass Care	ESF#7 Resource Support	ESF#8 Health & Medical	ESF#9 Search & Rescue	ESF#10 HazMat	ESF#11 Food & Water	ESF#12 Energy	ESF#13 Military Support	ESF#14 Public Information	ESF#15 Volunteers & Donations	ESF#16 Law Enforcement & Security	ESF#17 Veterinary/Animal Protection	ESF#18 Community and Business
ARES		S				S												
ARC						P								S			S	
ACP																		S
ABAC																	S	
ABPD				S					S							S		
BPD																S		
ATT		S	S															
CAP				S					S									
DCECG		S																
DCHD						S		P						S			S	
DCPA					S													
DCSB	S				S	S								S				
EEN																		
ESHC						S												
FCDC						S		S										
FDCF						S												
FDH								S										
FDLE														S		S		
FDOT	S																	
FDEM					S													
FFWCC										S							S	
FHP																S		
BNG							S							S		S		
FPIC								S										
FGBA								S										
JAFD							P											
JAES																	S	
JAA	S																	

JACCD																	P	
JBAC																	S	
JBED	S											S						
JBFD				S					S									
JBPD				S					S							S		
JBJ																		S
JCC																		S
JMA								S										
JEDC																		P
JEPD	P/S	S	P	P	P/S	P		P/S	P	P	P	P		S		P	P/S	P/S
CECD			S					S		S								
JFRD				P	S			S	P	P								
JFM							P											S
JGD					P													
JHND						P/S												
JHS																	S	
JITD		P	P															
JMAPS														P				
JOVS															P		S	
JRCSD	S		S			P/S		S			P				P			
JPDD					S													
JPA	S																	
JPD		P					P				P				P			
JPID														S				
JPWD	S		P	S	P/S				S									
JSO					S	S	S	S		S						P		
JSO911								S										
JTA	P																	
JVMS																	S	
JZG																	S	
JEAU			S		S							P						
JTAC	S					S												
MEO								S								S		
NBAC																	S	
NBPD																S		
NFCRT								S										
NFRC					S													
SHFB						S												
SA						S												
USCGCP	S																	
USCGSM											S							
USNASJCVO																	S	

USNASJROC																		
UWNF										S				S				
HOJ							S							S				
WPSP		S	S															

Naim Kapucu, Ph.D., is professor in the School of Public Administration at the University of Central Florida. His main research interests are emergency and crisis management, decision-making in complex environment, collaborative governance, and organizational learning and design. His work has been published in *Public Administration Review, Journal of Public Administration Research and Theory, Administration & Society, the American Review of Public Administration, Public Administration, and Disasters*. His book *Network governance in response to acts of terrorism: Comparative analyses* is forthcoming in 2012 by Routledge. He teaches public and nonprofit management, emergency and crisis management, and analytic techniques for public administration courses. He can be reached at kapucu@ucf.edu.

Vener Garayev, Ph.D., is an assistant professor in the Department of Management at Gediz University, Izmir, Turkey. His main research interests are emergency and crisis management, decision-making under uncertainty, collaborative governance and networks. His work has been published in *Public Administration Review (PAR), Journal of Emergency Management and Homeland Security (JEMHS), and International Journal of Public Administration (IJPA)*. He can be reached at vener.garayev@gediz.edu.tr.

Chapter 11

URBAN CLIMATE FRAMEWORK: A SYSTEM APPROACH TOWARDS CLIMATE PROOF CITIES

Sonja Döpp (TNO Built Environment and Geosciences, NLD)
Fransje Hooimeijer (TNO Built Environment and Geosciences, and University of Technology Delft, NLD)
Nienke Maas (TNO Built Environment and Geosciences, NLD)

The urgency of climate proofing urban areas is increasingly recognized and various adaptation and mitigation measures are being developed. However, to combine these measures in comprehensive strategies and actually implement them is challenging. The incorporation of climate change issues in day to day urban development requires better understanding of the urban system and the relations between climate and non-climate urban issues. In this paper we describe a theoretical framework and practical method to get a grip on the urban complexity and climate change effects: the Urban Climate Framework (UCF). The UCF approach is based on different system approaches and aims to expose underlying relations in the urban system and takes the complexity perspective. This perspective takes the unpredictable behavior and unforeseen consequences as starting point and enables better understanding of second order effects of climate change impact and the identification of win-win solutions to balance between mitigation and adaptation. The framework is developed in the context of adaptive management and can be used to structure information of different fields of knowledge and stakeholder groups. The UCF serves as a knowledge brokerage method to incorporate climate change in the complex processes of urban development and can be used by researcher, policy makers, urban planners and designers.

INTRODUCTION

The major contribution of cities to the global climate change and the urgency of climate proofing urban areas are increasingly recognized (e.g., VROM 2008). Many cities are struggling with a new task: How to prevent further climate change (by CO_2-reduction) and how to deal with the effects of climate change like heat, water excess and drought and their secondary effects? These effects will influence the physical characteristics and functions of cities (Figure 1) which results in various economic, social and health issues (e.g., Kuypers 2008; Döpp *et al.*, 2009). However, development of comprehensive strategies and the actual implementation of measures contributing to sustainable urban development is still a big challenge, particularly in the dynamic and complex context of cities.

In the traditional practice of urban development and design it is not common to relate interventions that are necessary to tackle the effects of climate change with 'general' urban issues. Successful incorporation of climate proofing elements requires 'mainstreaming' (Klein *et al.*, 2005) of mitigation and adaptation into urban development. In order to find relations between general (non-climate) urban issues and those related to climate change, a good understanding of the urban system and the way the different professionals within this system as a whole is required. Cooperation between urban planners, policy makers, scientists and other actors and combining their knowledge is crucial in finding effective integral solutions. There is a need for new methods to support these processes.

Existing decision support tools and methodologies for developing climate change strategies focus on measuring impacts and the modeling of scenarios (Sheppard & Shaw, 2007). These tools lack the ability to assist actors at different scales in the formulation, implementation and evaluation of climate adaptation and/or mitigation strategies (Laukkonen *et al.*, 2009). Moreover, existing models are not designed to facilitate stakeholder participation. Here the complexity theory can offer additional value because it involves a variety of perceptions and values' regarding what is true, relevant and desirable, and to the dynamics that results from the interactions between al these ingredients (Whitty & Maylor, 2009).

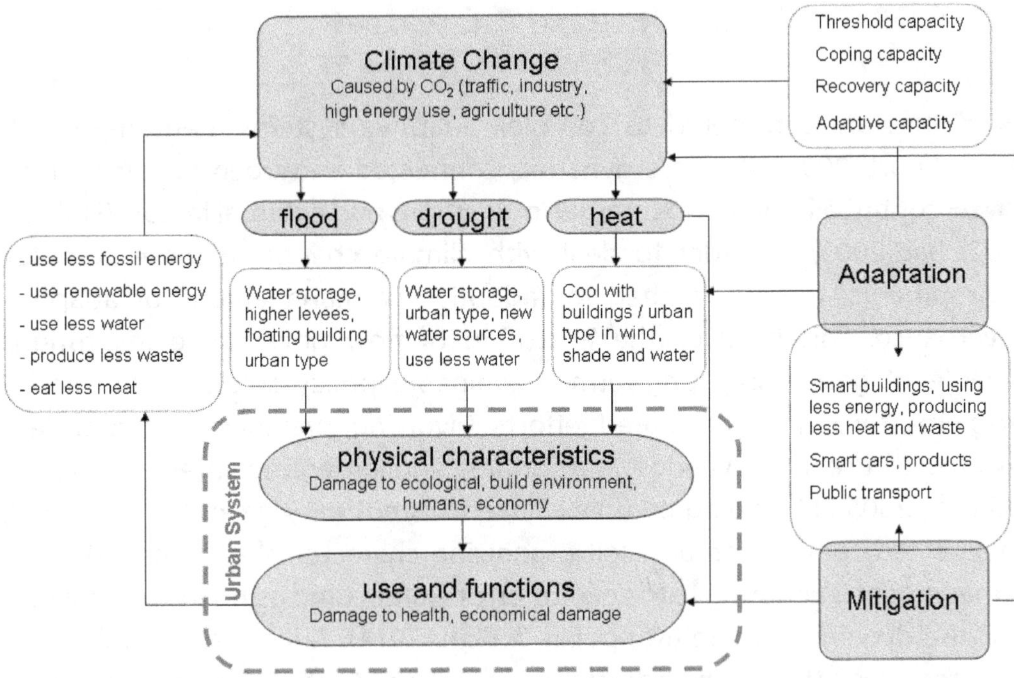

Figure 1 *Schematic representation of climate change impact on the urban system, and (examples of) adaptation and mitigation measures to be taken.*

To support the incorporation of climate proofing in the day to day practice of city building the method Urban Climate Framework (UCF) is developed. The UCF helps understanding the urban and climate system, and to expose underlying relations in the urban system for better understanding of the first and secondary effects of climate change impact. The UCF also acts as a knowledge brokerage instrument to facilitate the dialogue between different stakeholders and find win-win solutions. The UFC is developed on the theoretical background of system approaches taking the perspective of complexity theory by first emphasizes that decision making has a non-linear character. This results from unexpected behaviour and unforeseen consequences of interaction of agents (Koppejan & Klijn, 2011). Second in constructing a model of the landscape the agents are in to grasp onto the continuous change within the system and external influences of changing other systems that are related. Rhodes (2008) calls this the "performance landscape".

This paper explains the construction of the Urban Climate Framework as a theoretical model and practical method, shows the usability and options for improvement for the city of Rotterdam and the working of the method in workshops.

ADAPTIVE GOVERNANCE, COMPLEXITY AND SYSTEM APPROACH

Cities are considered as complex adaptive systems (Benton-Short & Short, 2008); they are dynamic, connected and open systems which continually evolve to internal interactions and external factors (Holling, 2002; Bai, 2003). In order to deal with climate change in these complex environments, there is a need for 'adaptive co-management' or adaptive governance. This involves the 'evolution of new governance institutions capable of generating long-term, sustainable policy solutions to wicked problems through coordinated efforts involving previously independent systems of users, knowledge, authorities, and organized interests' (Scholz & Stiftel, 2005). Managing complex systems is not easy given the dynamics, unexpected coevolution and self organizing character of systems. Adaptive management is adapting to the evolution of systems and using the possibilities that the dynamics create (Koppejan & Klijn, 2011). Adaptive governance in cities requires a thorough understanding of the potential of the "performance landscape" that Rhodes defines: the urban system and input of a multi-disciplinary and multi-perspective group of stakeholders. The complexity perspective focuses on the interactions within the decision making system and with other systems. The UCF is developed to connect problems solutions and choice opportunities in facilitating the process of gathering and combining this information in an effective way to steer decision making.

A system approach makes it possible to understand the city and the issues of climate change in relation to each other. The use of system approaches emerged in the 1950s as a reaction on the reductionist approach where the issues were de-contextualized and reduced to their essence but made complexity too simple (Flood, 2001). As Einstein advised: 'make your object of study simple as possible but no simpler'. The system approach is a method to study phenomena as emergent properties of the interrelated whole with a mutual consistency and in interaction with the surroundings (Heylighen, 2000). This makes the object of study simpler, but still enables to give meaningful understanding of issues that deal with elements of a different nature and coherence. The latter is specifically interesting in the issue of climate and cities, since there is a mix of elements like urban structures and users that are both part of the problem and part of the solution.

The UCF, built upon several system approach methods, aims to investigate the city as a complex adaptive system including the impact of climate change, and to act as a knowledge brokerage method. Complex systems are systems comprised of numerous interacting identities (parts) each of which is behaving in its local context according to some rule(s), law(s) or force(s) (Maguire & McKelvey, 1990: 26). Therefore framework should enable to set together and bring order in information of different nature, from different fields of knowledge and stakeholder groups. It should be simple but comprehensive, easy to understand, offering insights in the relations in the city and possible interventions. Elements were used from four system approach methods:

- **DPSIR framework** (EEA, 1994); causal framework to asses interactions between social and environmental system, based on Driving forces, Pressure, State, Impacts and Response;
- **SCENE-model** (Rotmans, 1998); supporting tool for regional sustainability analyses, based on stocks and flows in the SoCial, Environment and Economic sustainability-capital domains;
- **Layers Approach** (Hoog, Sijmons & Verschuuren, 1998); conceptual framework to guide spatial planning policy;
- **Vulnerability Approach** (e.g., De Graaf 2009); general approach to identify vulnerability of a system and ability to deal with impacts, based on threshold capacity, coping capacity, recovery capacity and adaptive capacity.

URBAN CLIMATE FRAMEWORK

The Urban Climate Framework can be described as a system modulation with a four-step approach that facilitates stakeholder participation. Starting point for the construction of the UCF are the three key urban functions for maintaining healthy and livable cities: living, working and amenities. In order to investigate how and to which extent these functions are subjected to climate change and the influence on the urban system and potential solutions, we identified their main stocks (based on SCENE-model). The functions and stocks for which information can be collected are displayed in Figure 2.

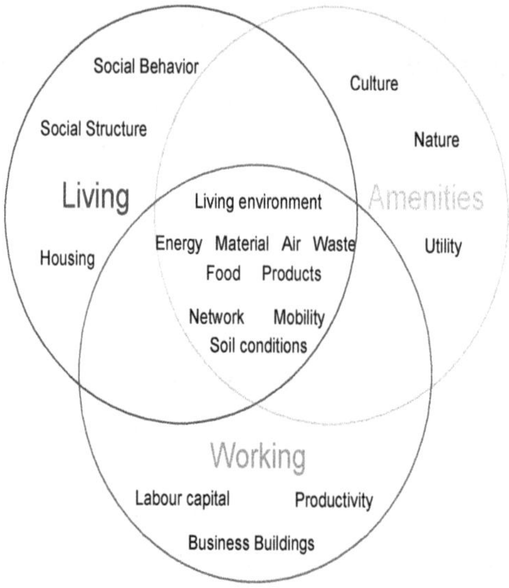

Figure 2 *Essential Urban Functions and Required Stocks.*

The second approach that is used is the Layers Approach (Hoog, Sijmons & Verschuuren, 1998). In the original approach, three types of connected layers can be distinguished that are characterized by different rates and types of spatial development and change: subsoil (soil, water, nature, and landscape), network (infrastructure) and occupation layer (living and working).

In relation to climate change, the original layer approach is a good starting point but not refined enough for the complex urban system. For the purpose of the UCF the original layers approach is extended by the layers public space and users, and the layer metabolism (flows of energy, water, air, goods etc.) based on work of Heeling *et al.* (2002) and Van Schaik & Klaasen (2007) respectively (see Table 1).

Layer/function	Living	Amenities	Working	General conditions (for several functions)
Users High dynamic	Social structure Social behaviour		Labour capital Labour productivity	
Metabolism High dynamic				Facilities (Energy, Water, waste, air, material, food, products)
Occupation Medium dynamic	Housing	Utility Culture	Offices	
Public space Medium/low dynamic		Nature Culture		Living environment
Infra structure Low dynamic				Mobility Network
Underground Very low dynamic				Network (water, energy, waste) Soil conditions

Table 1 *The Ordering of the Stock in the Layers of the City.*

By structuring the stocks (from the SCENE-model) according to the different layers, it is possible to understand the size and impact of climate change for the specific layers. Moreover, it gives an indication on the ease and speed with which changes can be achieved, e.g., by interventions for climate adaptation. For example, adaptation measures like simple changes in behavior (user layer) could be achieved on shorter notice and with less investment than some infrastructural adjustments (infrastructure layer). The stocks ordered by layers make up the y-axis of the UCF supporting table (Table 2).

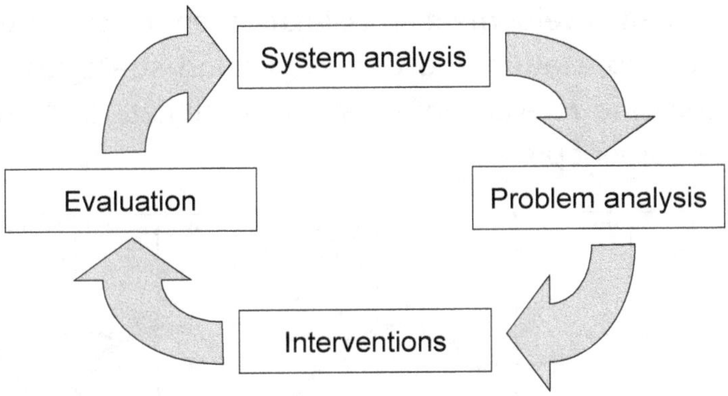

Figure 3 *Scheme of the System Approach of the UCF.*

The elements along the x-axis of the UCF supporting table represents the four-step system approach of the framework (Figure 3) which is derived from DPSIR. Collecting data and information per stock following these four steps supports developing integral climate adaptation strategies.

The **system analysis** aims at a good overview of the local conditions and current state of the urban system, and the (future) pressures that the system is subjected to. In the UCF table, information for the system analyses is placed in the columns *State* and *Pressure*. Relevant questions in the system analysis are: what are the social, physical, chemical and/or biological characteristics and current conditions of the different stocks? The description of the current situation, the state of the different stocks within the layers, can take in existing problems and qualities that characterizes the specific layer.

The second part of the system analyses is a description of the physical and social-economic pressure for each stock. Here the connection between climate impact and contemporary urban development can be made since both pressures are ordered by stocks right next to each other.

Relevant questions in the **Problem Analysis** are: how vulnerable are the stocks for the effects of climate change and/or social-economic pressures, what is the potential impact on the urban system? The impact of climate change in one stock might results in negative results in other stocks as well.

Understanding the vulnerability of the system is useful in the third step of **Interventions**. Options for measures are based on a strategy of lowering vulnerability by increasing the threshold capacity, coping-, recovery- and/or adaptive capacity. More in general: UCF could connect the physical and social component in urban development and climate change issues. Public services, looking for new cooperation and communication strategies, could jointly provide and combine relevant information on climate and environmental aspects in development projects.

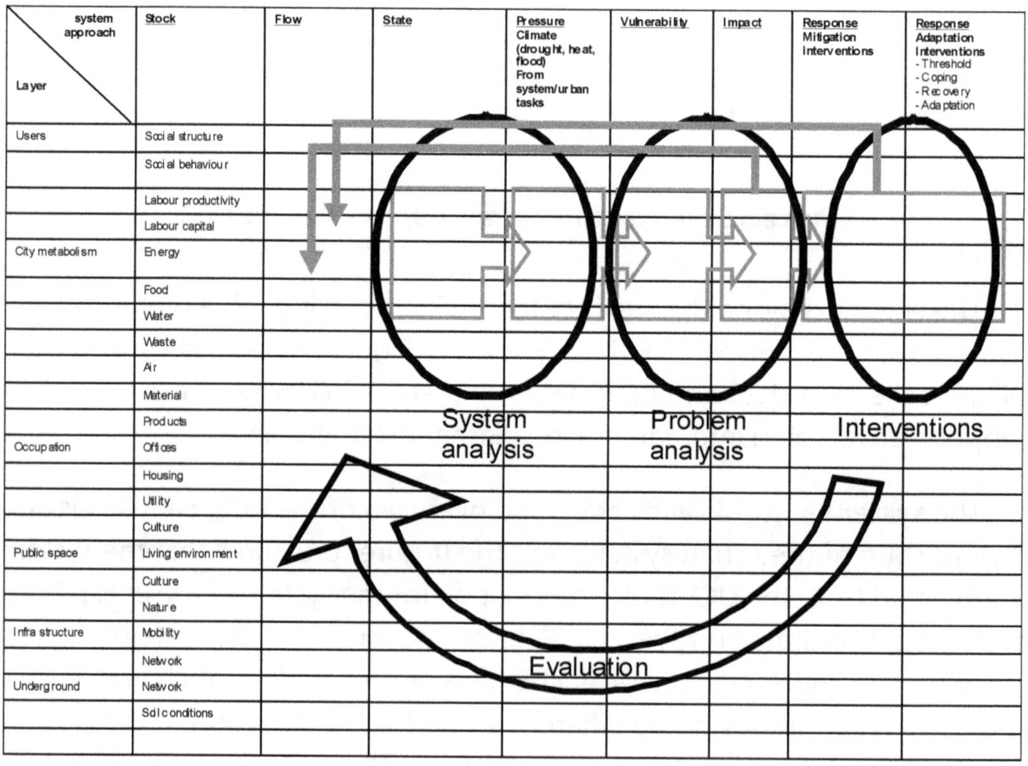

Table 2 *The Urban Climate Framework Knowledge Brokerage Method (source: TNO).*

The fourth step of the system approach cycle is the **Evaluation** phase, the continuous learning process. The aim of this step is to determine the influence (direct and indirect) of the proposed adaptation and mitigation measures at other elements (stocks and layers) of the city, positive and negative effects.

Within a cyclic approach this is not the last step, but again the first where the new state is analyzed.

PERCEPTIONS OF USABILITY OF THE UCF

In order to examine the usability of the UCF, interviews were conducted with several municipal services (OBR, DCMR, dS+V, Gemeentewerken), a housing corporation and Rotterdam city council. Beforehand the interviewees received an explanation to the UCF and the table including information about Rotterdam. During the interviews the focus was on the usability of the UCF as a practical method for the specific services, and improvement of the content and structure of the supporting table.

It is recognized by all municipal services that there is a need for a wider perspective of urban development and the integration of climate issues. The scale of traditional approaches (e.g., environmental impact assessments) is often too narrow to make well considered decisions; the scope should be broader, not only in terms of space but also in terms of sectors and disciplines. The UCF is seen as an opportunity to bring more disciplines together, and act as a discussion method for people from different backgrounds. There is a particular interest in connecting the physical and social component in urban development and climate change issues. In some cases, social developments are barely considered in physical investments, while there are a lot of potential win-win solutions. This is also recognized by the municipal Urbanism and Housing of Rotterdam (dS+V) department for whom 'social-physical connection' is one of its central spatial planning themes. Combining physical and social (non)climate issues in the UCF approach could help to make this connection.

Lack of communication is indicated as one of the main problems in finding effective integral solutions for climate and non-climate urban issues. Furthermore, services are looking for new cooperation strategies in order to jointly provide and combine relevant information on climate and environmental aspects in development projects. The interviewees indicated the suitability of UCF to support this process by enabling knowledge to be shared in a structured way. The use of the UCF table in a multi-disciplinary interactive process is seen as a powerful method for sharing knowledge and raising awareness for coherence between different domains. Much ignorance still prevails, both on this coherence and on the (indirect) effects of and responses to climate change.

For the practical use of the framework in interactive processes, people indicated that the table offers a good base for organizing information. However, a single session will probably not be enough to consider the whole framework on a city scale. The framework should be comprehensible for people unfamiliar with this kind of information and system approach. Including more facts and figures like maps and working with flowcharts can contribute to this. Furthermore, a clear explanation about the context and purpose of the framework before use is necessary. The interviewees also noticed that the information in the table can be quite subjective. A clear distinction between objective and subjective data in relation to how the UCF table is used (knowledge base or discussion method) is important.

The climate is not on the priority list of the housing association. Only when the effects of climate change start to have an influence on the interests of the association, like building costs and materials, will it become an issue. Either that or the association must take a moral stance to improve their business profile and show social commitment. The housing association, as in other sectors, cannot oversee the climate issue. The issue is complex in its components and time-frame, therefore it is easier to keep out of the complexity and keep to one's own turf. That is why integration between the sectors, necessary to get a grip on the climate issue, has failed to transpire despite all the efforts. The fear of complexity and of the unknown puts up an enormous barrier; the UCF is an analysis method that could help sectors look beyond their borders and gain some insight into the complexity free of fear.

From the political viewpoint the UCF could be useful to make the necessary connection between the macro and the micro levels; some climate measures are required at micro level (in the urban system) in order to have impact on the macro level (of climate change). This connection is crucial but often difficult to make clear to the stakeholders that micro-level participants. It is especially difficult to create support amongst city residents to make changes to improve the macro climate since they are more connected and concerned with the micro climate in their direct surroundings and their culture. By relating climate issues to the cultural aspects of the city (e.g., in the layer users), the UCF can contribute to the better incorporation of climate measures at micro level.

In a city like Rotterdam, where the harbor and its industry are the main contributors to the city's CO2 emissions, macro-level mitigation measures are much more effective than micro-level measures and even though the latter up to effects on a macro level, it could be more effective to tackle the industrial sector than the private sector. The UFC makes it possible to weigh up the measures and effects on the different levels.

The interviews show that all public services in Rotterdam are struggling with the complexity of incorporating climate issues into urban development, and that there is need for methods or instruments to support this process. The systematic, integral approach of our framework has been very positively received and further development of the UCF as a practical method is being strongly encouraged.

WORKSHOPS TESTING THE UCF

The theoretical model is tested by master students in urbanism at the University of Technology in Delft. They applied the UCF and their experiences were used in the construction of the framework. A draft version of the framework was applied to the city of Rotterdam. The functioning and potential use of the model was discussed with the different domains in Rotterdam as described above. Then the final model was tested in two workshops: the Climate Campus workshop and for the 7-UP study by a group of urbanists and architects.

The first workshop was held at the opening event of the Delft Research Initiative Environment of the Technical University in Delft. In the interdisciplinary research initiative the campus is used as a pilot project to test technology that is developed in the faculties in real time. The workshop used the UCF as a brainstorm method to get grip on the problems in the public space and the domains of the measures to solve those. This was an interesting way of using the framework because the issues in the other layers were not prominent in the discussion but can easily be connected to the outcome of this workshop. The results were, see Figure 4, that the most urgent problems in the public space of the campus are water and wind. With heavy rain storms the water can not be processed by the artificial water system that is constructed there. The high buildings take the wind down and this is very uncomfortable, especially since the new design of the campus is all about hanging around on the new lawns.

The use of the layers by the UCF made clear that a lot of solutions for the problems in one layer are situated in the other layers.

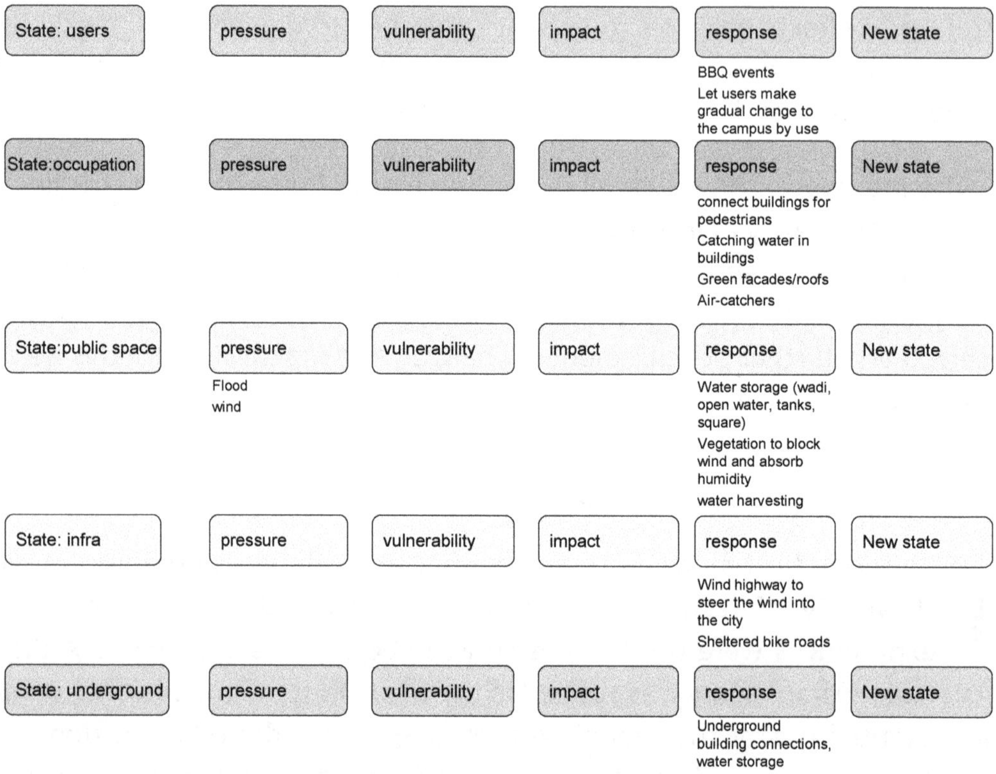

Figure 4 *Scheme of UCF Applied to the Climate Campus.*

The second workshop was based on the research done into a district in Rotterdam, Zevenkamp, by a group of spatial designers. This is an area in Rotterdam built in the 1970s. They focused on the generic aspects of the neighborhood to develop more insight into what to do with the 1970s legacy that is in the line for restructuring. Their analysis concerns almost all the physical layers of the city and making the neighborhood climate proof (there is no question of urban heat island and flooding).

As a test the group put all the information of the Zevenkamp into the framework and conclusions were drawn. Firstly, the layers contained information from the study only and, secondly, this was done by one of the members of the project team so were not representative of the professional domain of the different layers. The framework was experienced in a positive way because it was able to incorporate all the very different information (social, physical, economic) and the activity of filling in of the table made the workshop participants think about and discuss their analyses again. Filling in the framework is time-consuming but it is necessary to indicate per layer on what scale the information is situated. For example, energy systems are on a larger scale than green structures.

The results of the workshop are worked out for three layers (as a sample), see Figure 5. The connection between the different layers is made visible. The impacts on the layers *housing* and *occupation* have a secondary impact on the layer *users* and the other way around. Also the responses in the different layers have effects on the new state of the other ones. In this example the explicit climate tasks were not focussed on but in these layers also the effects and measures of the climate tasks can easily been taken in. For example the causalities of heat could be prevented by the cooling of the city in the layer of public space that contains trees for shadow and cooling water surface.

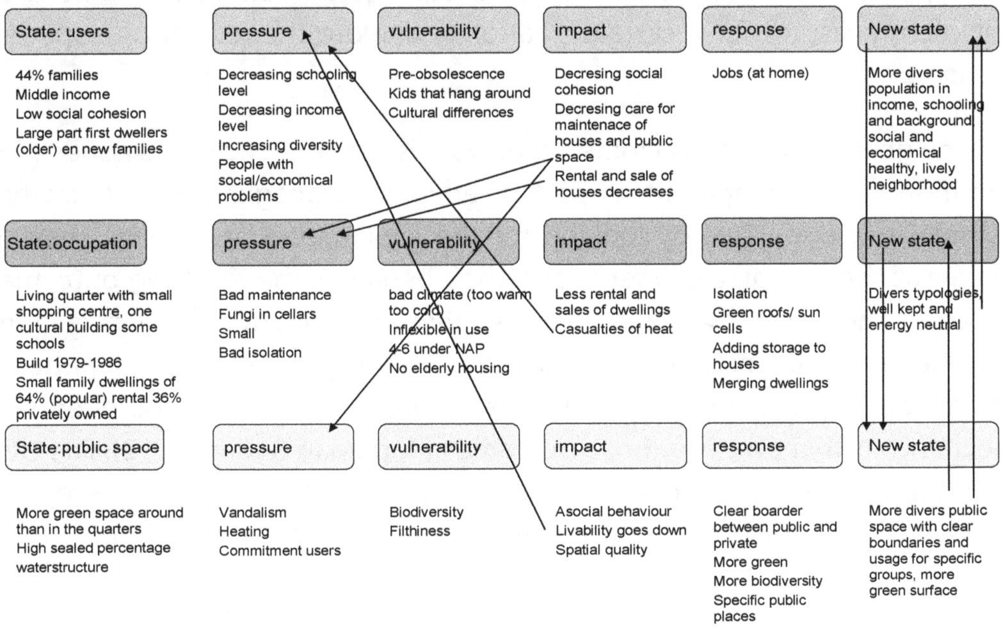

Figure 5 *Scheme UCF for Zevenkamp.*

The workshop confirmed that the framework produces a combination of information that leads, as expected, to discussion, with the subjectivity of input from people playing a significant role. The use of the UCF is easier when well defined area is examined since the information is more concise and precise. Filling in the table for the whole of Rotterdam was evidence of this. That means that the use of UCF is optimized when the location and the goal of the system analyses are very distinct because the results are dependent on the actors in the session. They define the relevant elements of the system and decide what type of guidance is necessary on evaluating the multitude of extremely difficult empirical questions they have.

To use the framework as an interactive instrument it would be interesting to learn more about dynamic system modelling, which would make the session conclusions interdependent, and also possibly showing this in a time-related model.

CONCLUSIONS

The urban climate framework is not a method that can be used for long-term predictions or to describe scenarios. It is a transdisciplinary approach in bringing together agents and their knowledge fields and giving insight in the "performance landscape" to be able to find a balance between measures for a certain goal or to explore the results of a specific scenario.

The goal of the urban climate framework is to connect climate issues with common city development as part of adaptive management. It takes the complexity perspective by looking for the explanation of decision making mainly at the interplay of agents and their behavior, the resulting outcomes and how the interactions within the system are influenced by interactions in other system. Complexity perspective requires analyzing the patterns of interaction, the resulting punctuated equilibrium and the feedback mechanisms that produce those outcomes (Koppejan & Klijn, 2012). The UCF does this by offering a framework where this takes place in order to integrate the climate issues in day to day city development: adaptive management. The capacity of a society to *anticipate* uncertain future developments means that the developments will be such that adjustments can be effected to provide safeguards from climate changes in the coming centuries. Measures that are within this range of the adaptive capacity are 'no-regret' actions that make sense or are worthwhile regardless of the additional or exacerbated impact of climate change. 'Win-win' actions that provide adaptation benefits while meeting other social, environmental or economic objectives, include climate change mitigation and 'low-regret actions', measures with relatively low costs for which benefits under climate change scenarios are high. The measures taken should not imply the solution of an end situation (for a certain budget) but must always be open, adaptive and future oriented to safeguard society from climate change.

ACKNOWLEDGEMENTS

With thanks to the interviewees Rick Grashoff (alderman Rotterdam), Peter van Haasteren (OBR), Patrick Kaashoek (Woonstad Rotterdam), Edwin Koopmanschap and Boukje van der Lecq (DCMR), Lissy Nijhuis (Gemeentewerken Rotterdam), Nico Tillie (dS+V Rotterdam), Eric-Jan Wesemann (policy services Rotterdam).

REFERENCES

Armitage, D., F. Berkes, and N. Doubleday (2007). Adaptive co-management. Collaboration, learning and multilevel governance. University of British Columbia Press, Vancouver, British Columbia, Canada.

Bai, X. (2003). The process and mechanism of urban environmental change: An evolutionary view. International Journal of Environment and Pollution, 19(5): 528-541.

Benioff, R. and J. Warren (eds.) (1996). Steps in Preparing Climate Change Action Plans: A Handbook. Washington, DC: U.S. Country Studies Program.

Benton-Short, L., Short, J.R. (2008). Cities and Nature. Critical Introductions to Urbanism and the City. Edited by M. Miles and J. R. Short, Oxon: Routledge.

Brooks, N. (2003). Vulnerability, risk and adaptation: A conceptual framework. Tyndall Centre for Climate Change Research, working paper 38.

Döpp, S.P., Bosch, P., Deelen, K. (2009) Climate Change in the Netherlands: Challenges for a safe and attractive urban environment. Fifth Urban Research Symposium 2009, Marseille.

Flood, R. L. (2001). "Local Systemic Intervention," European Journal of Operational Research, 128: 245-257.

Folke, C., Hahn, T., Olsson, P., Norberg, J., 2005. Adaptive governance of socio-ecological systems. Annual Review of Environment and Resources, 30: 441-473.

Graaf, R.E. de (2009). Innovations in urban water management to reduce the vulnerability of cities: Feasibility, case studies and governance. Delft: University of Technology.

Grosskurt, J., Rotmans, J. (2005) The Scene Model: Getting A Grip On Sustainable Development In Policy Making. Environment, Development and Sustainability, 7: 135-151.

Heeling, J.; Meyer, V.J. and Westrik, J. (2002). De Kern van de Stedebouw in het perspectief van de 21ste eeuw. Dl. 1. Het ontwerp van de stadsplattegrond. Amsterdam, SUN.

Heylighen, F. (2000). "Referencing pages in Principia Cybernetica Web", in: F. Heylighen, C. Joslyn and V. Turchin (editors): Principia Cybernetica Web (Principia Cybernetica, Brussels.

Holling, C.S. (ed.) (1978). Adaptive Environmental Assessment and Management. International Institute for Applied Systems Analysis, and Wiley, Toronto.

Holling, C. S., L. H. Gunderson and G. D. Peterson (2002). Sustainability and Panarchies, Chapter in Panarchy: Understanding Transformations in Human and Natural Systems, L. H. Gunderson and C. S. Holling (ed.). Washington, DC, USA: Island Press.

Hoog, M. de, Sijmons, D, and Verschuren, S. (1998). 'Herontwerp van het Laagland' In: Het Metropolitane Debat, Bussum, THOTH.

IPCC (2001). Impacts, Adaptation, and Vulnerability for Climate change, Third Assessment Report of the IPCC. Cambridge University Press, UK.

Koppenjan, Joop and Erik-Hans Klijn (2011). Complexity, which complexity? Three perspectives on complexity applied to decision-making on revitalizing the Center of Rotterdam South. Paper for ASPA (American Society of Public Administration) conference "Challenges of making public administration and complexity theory work" Rotterdam , Erasmus University Department of Public Administration, June 23-25.

Klein, R.J.T., S. Huq, F. Denton, T.E. Downing, R.G. Richels, J.B. Robinson, F.L. Toth (2007). Inter-relationships between adaptation and mitigation. Climate Change 2007: Impacts, Adaptation and Vulnerability. Contribution of Working Group II to the Fourth Assessment Report of the Intergovernmental Panel on Climate Change, M.L. Parry, O.F. Canziani, J.P. Palutikof, P.J. van der Linden and C.E. Hanson, Eds., Cambridge University Press, Cambridge, UK, pp. 745-777.

Kuypers, V.H.M. en E.A. de Vries (2008). Groen voor klimaat, Alterra.

Laukkonen, J., Blanco, P.K., Lenhart J., Keiner, M., Cavrice, B., Kinuthia-Njenga, C. (2009). Combining climate change adaptation and mitigation measures at the local level. Habitat International 33: 287–292

Lemos, M.C., Agrawal, A. (2006). Environmental governance. Annual Review of Environment and Resources, 31: 297-325.

Maguire, S., B. McKelvey (1999). Complexity and Management: moving from fad to firm foundations, Emergence, 1(2): 19-61.

Rhodes, M.L. (2008). Complexity and emergence in Public Management: the case of urban regeneration in Ireland, Public Management Review, 10(3): 361-380.

Rotmans, J. & Dowlatabadi, H. (1998). Integrated assessment of climate change: Evaluation of methods and strategies, in S. Rayner & E. Malone, eds, 'Human Choice and Climate Change: An International Social Science Assessment', Batelle Press.

Scholz, J. T. and B. Stiftel (2005). Adaptive Governance and Water Conflict: New Institutions for Collaborative Planning. Resources for the Future: Washington, D.C. USA.

Schaick, J. van & Klaasen, I.T. (2007). Dynamics of urban networks as basis for the re-development of layer approaches. In: Conference Proceedings International Seminar on Urbanism and Urbanization. TUDelft, Faculteit Bouwkunde.

Sheppard, S., & Shaw, A. (2007). Future visioning of local climate change scenarios: Connecting the dots and painting pictures to aid earth system governance. Vancouver: Department of Forest Resources Management/Landscape Architecture Program, Forest Sciences Center.

Turner B.L., Kasperson R. E., Matsone P. A., McCarthy J.J., Corell R. W., Christensene L., Eckley N., Kasperson J. X., Luers A., Martellog M. L., Polsky C., Pulsipher A. and Schiller A. (2003). A framework for vulnerability analysis in sustainability science, PNAS, 100: 8074-8079.

UNEP/RIVM (1994). An Overview of Environmental Indicators: State of the art and perspectives. UNEP/EATR.94-01, RIVM/402001001, Nairobi.

VROM (2008). Structuurvisie Randstad 2040, naar een duurzame en concurrerende topregio. Ministerie van Volkshuisvesting Ruimtelijke Ordening en Milieubeheer. Den Haag. [Randstad 2040 Structural Vision, in Dutch, summary available in English: http://doemee.vrom.nl/randstad2040/publicaties/structuurvisie-randstad-2040/randstad-2040-summary-of-the-structural-vision].

Whitty, S.J. and H. Maylor (2009). And then came in complex project management. In: *International Journal of Project Management*, 27: 304-310.

Sonja Döpp received a Masters Degree in Earth Sciences at the Vrije Universiteit in Amsterdam in 2007. From 2007-2011 she worked as a researcher for TNO, the Netherlands Organization for Applied Scientific Research. Currently, she is project coordinator at the Foundation Knowledge for Climate, the Dutch national research program on climate adaptation. Her projects have been in areas of climate change adaptation, sustainable urban development and corporate social responsibility. Her specific interest is the integration of knowledge from different disciplines that are relevant for sustainable development, and the translation thereof into effective strategies for policymakers.

Fransje Hooimeijer studied Art and Culture Science and has worked as an independent researcher in the fields of architecture, urban design and landscape architecture since 1997. Besides various publications and exhibits she has done research for government and corporate clients. She has been working part-time towards a Ph.D. from 2003 till 2009 at the Faculty of Architecture at Delft University of Technology, in the department of Urbanism. This research, under the title The tradition of Making: Polder cities, was focused on the relationship between the laws of water, the technology of water control and urban design. In this topic she was mentor to graduate students and wrote many publications. From 2009 on she works as a researcher at the TU Delft and at TNO. In both positions she works on the technology of urban development in the light of climate change and the energy transition. In her spare time she finished her Ph.D. research and defended it in 2011.

Nienke Maas has studied civil engineering, transport and logistics in Delft. Right after her graduation she joined TNO in 1997. By working in several domains, like mobility and infrastructure, building process innovation, spatial development, she gained a broad experience in research and consultancy in the built environment. Important recent projects are risk management for municipalities, climate change area development. Her specific interest is the strategic policies for cities to deal with future challenges and the follow up with implementation of measurements.

Chapter 12

ADVANCING PUBLIC SERVICE GOVERNANCE THEORY USING A PERFORMANCE LANDSCAPE MODEL

Mary Lee Rhodes (Trinity College Dublin, IRL)

This paper assesses the fit of empirical data relating to the Irish housing system to a complex adaptive system framework with the aim of contributing to the development of governance theory in relation to public services. The 'performance landscape' (PL) model (Kauffman, 1993; Siggelkow & Levinthal, 2003) was selected as the most appropriate complex adaptive systems framework to facilitate the development of governance theory of public services. The PL model is mapped against data on the Irish housing system to assess the fit of the model to the data and to determine what, if any, public administration behavior could not be addressed based on this model. Implications of this mapping for theory and practice are described in the conclusion and summarized in an appendix to this chapter.

INTRODUCTION

For over a decade there have been calls for more research into the applicability of complex adaptive systems (CAS) to organizational theory in general, and public management theory and practice in particular (Anderson, 1999; Boston, 2000; Weber, 2005). Numerous insights have been drawn from such exercises and several special issues of relevant journals have been dedicated to this effort[1]. However, there is still significant work to be

1. *Emergence: Complexity and Organization* (2005, 7.1); *Public Administration Quarterly* (2005, 29.3); *Public Management Review* (2008, 10.3); *The Leadership Quarterly* (2007, 18).

done to tighten up the relationship between what we know about managing organizational systems (theory), what complex systems models can / do incorporate, and what empirical evidence tells us about how human beings, organizations and environment(s) interact over time to produce patterns and outcomes. This paper attempts to advance that work, drawing on research conducted over nine years in the Irish housing system, which was undertaken with the objective of contributing to the development of public management (governance) theory (Rhodes, 1997; Lynn *et al.*, 2000; Kjaer, 2004) based on a complex adaptive systems framework.

There are two main questions addressed in this paper. The first is whether and how a complex policy domain (housing in Ireland) can be mapped onto a complex adaptive systems model chosen for its apparent alignment with existing theory in public administration and management. The second question is what implications do the result of this mapping have for theory development and practice. The paper is organized into four sections. The first section identifies the proposed elements of a performance landscape model of organizational systems in general—which, it has been argued, is a basis for public service systems models. The second section gives a brief summary of the research undertaken to 'test' the model against a real public service system—housing in Dublin, Ireland between 1998 and 2006. The third section describes in detail the elements of the housing performance landscape and highlights some of the difficulties in applying the model to empirical data. In the discussion and conclusion section, we summarize the findings and highlight the key insights and issues in relation to public management theory and practice arising from this research.

A PERFORMANCE LANDSCAPE MODEL FOR PUBLIC MANAGEMENT

In Rhodes (2008) and Rhodes & Donnelly-Cox (2008), the case was made for using performance landscape models as the computational modeling approach of choice for public service / non-profit organizational systems. The case will not be made again here, but instead we will simply define the main elements of this approach and suggest that the reader wishing to delve into this in more detail consult those articles as well as the articles by David Levinthal, Nicolaj Siggelkow and their collaborators[2] in which many of the

2. Levinthal (1997), Levinthal & Warglein (1999), Siggelkow & Levinthal (2003), Rivkin & Levinthal (2003), Siggelkow & Rivkin (2005).

relevant details of the computational approach may be found. Their models are based on Stuart Kauffmann's (1993,1995) NK fitness landscapes, which have been referenced and applied by numerous organizational CAS researchers.

The elements of the performance landscape (PL) model based on Kauffman's NK model, applied by Siggelkow and Levinthal (2003) and amended by Rhodes (2008) consist of the following elements:

1. Independent, heterogeneous, *organizational agents* with individual performance objectives;
2. The *performance landscape* made up of:
    a. The range of strategic decisions that agents can make,
    b. The range of environmental factors that can affect decisions,
    c. Interdependencies among strategic decisions and factors,
    d. An assigned level of performance for each decision / factor, and
    e. Rules regarding the movement of agents on the landscape;
3. *Individual agent outcomes* derived from the agent's location on the landscape and *systemic outcomes* derived from aggregate outcomes and patterns in agent locations over time.

Agents may enter the landscape at any random location and can move from that location depending upon their performance objective(s) and the performance improvements that are possible given the geography of the landscape and the rules governing agent movement. The geography of the landscape is defined by the number of different decisions that are possible, the relevant environmental factors affecting outcomes and decisions, the interdependency among decisions / factors and the performance outcomes associated with each decision. The rules governing movement apply at both the landscape and the agent level, i.e., there are 'global' rules that all agents must follow and also agent-specific rules that depend upon characteristics of the agent. Agent outcomes are simply the performance level achieved by a given agent at a point in time, while systemic outcomes are exogenously defined measures of interest to the systems modeler (such as the number of agents on various performance peaks, the average performance level achieved, etc.) and do not normally affect the behavior of the agents themselves. Figure 1 provides an overview of the dynamics of agents, outcomes and the performance landscape.

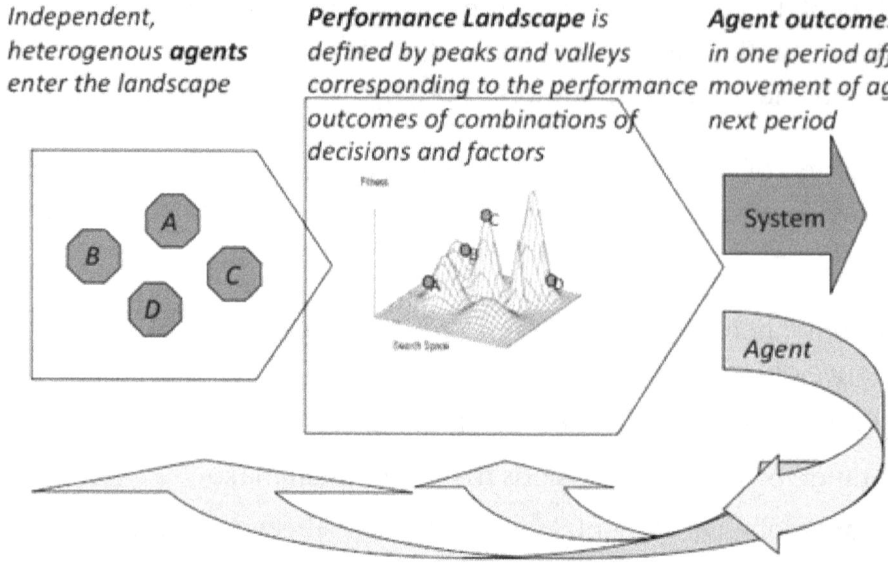

Figure 1 *High Level Dynamics of Performance Landscape Models*

Note that in the development of the performance landscape model above, it was found that the basic structure of the model could not accommodate a range of interactions among agents identified in organizational literature (e.g., Pfeffer & Salancik, 1987; Ulrich & Barney, 1984; Granovetter, 1985) without significantly changing the specifications of the model. Essentially, the gap arises from the fact that the NK model focuses on interdependencies among the elements (Ns) that define the performance landscape, e.g., the genetic 'options' that an organism has in the evolutionary process, or the strategic choices that a firm has over the period of its existence. K is a measure of the interdependencies among choices in determining performance. It does not measure the influence that *agents* have over one another's choices. There is no capacity in the NK model to address the dependency that one organism might have on another organism, which might influence its 'choices'. Furthermore, the concept of social networks and how the ties between people / firms might influence choices of organizational agents cannot be directly modeled using NK. Cellular automata or neural networks directly model the influence of one agent's behavior over those that are 'connected' in some way to the agent, e.g., through proximity or through a social/communication tie. The exclusion of these elements from the PL model has implications for the development of public service governance theory based on such a model and these are discussed in the conclusion.

'TESTING' THE MODEL AGAINST EMPIRICAL DATA

In order to test the validity of the proposed PL model as a framework for developing complex models for public services, we examined the housing system in Ireland to assess the fit between the model and 'reality'. The research conducted combined case study research techniques (Yin, 1994; Eisenhardt, 1989; Barzelay *et al.*, 2003) with a soft systems analysis (Checkland, 1981) to explore the characteristics of the Irish housing system over the period 1998-2006) and to progress towards a theory of governance in housing. The research followed the eight steps required for theory building using cases as described by Eisenhardt (1989), but also incorporated key elements of the 'systems thinking' steps in Checkland's (1981) soft systems methodology, notably the development of a conceptual model of the phenomenon as a particular type of system (the 'root metaphor') that was generally recognized and accepted by system participants.

The data collected included semi-structured interviews with managers across a range of housing activities, secondary documentation about the organizations included in the interview process, housing statistics, policy documents, relevant literature (as reference material against which to compare emerging categories/hypotheses) and a survey of factors considered by managers in their strategic decision-making. Interviews covered organizational history, key decisions made in the recent past or under consideration, and environmental factors affecting decision-making and numbered 63 in all, although only 48 of these were used in the in the analysis of the production landscape. The organizations / individuals to be interviewed were selected based on their role (main activity) in the housing system—following the exploratory phase (see below)—and using a 'snowball' sampling approach. Interviews were added until no new information about the decision processes of agents from a particular role type was discovered. The survey of decision factors followed a random sampling approach of agents in the housing system and generated a 30% usable response rate (including 119 responses).

The entire research project spanned nine years (albeit with a 2 year hiatus) and comprised three distinct phases (see Figure 2 below). Phase I took two years and was an 'exploratory' case study of housing in Ireland in which it was concluded that (1) the range of interviewees (actors) needed to be extended beyond that which was originally thought, (2) that the range of actions, environmental features, and outcomes relevant to public management governance theory was extremely broad, and (3) that many participants and observers perceived housing as a system of some kind. Phase II extended the case study to encompass a wider range of agents, while at the same time introducing a more structured analytic framework through the application of systems thinking and stakeholder review and participation to define the 'root metaphor' applicable to housing as a system. It was in this phase that a particular type of complex adaptive system model—the 'performance landscape'—was identified as being appropriate to building governance theory for public services.

Phase III focused on teasing out the specific details of housing as a performance landscape through a comparative case analysis of decision-making by 48 of the 63 actors included in the original study. These actors were identified as 'production' agents, for which sufficient primary and secondary data were available to pursue detailed specification of the performance landscape. Features of the landscape (decisions / factors) were identified and the case study concluded with a statistical analysis to establish interdependencies among elements in the performance landscape.

Overall, the combination of a two-phased case study approach—which allowed for a rich contextual analysis of the phenomenon studied—with a soft systems analysis in between to support the explicit and detailed specification of a human system based upon the system participants' own understanding of the nature of that system, proved to be a solid basis on which to develop a conceptual model of a particular public service that was both internally consistent and reflected the complex characteristics of the phenomenon as perceived by participants. The initial exploratory case and literature review supported the development of a high-level conceptual framework that was then explored through the soft systems analysis to further specify the nature and key features of the system to be analyzed. The categories and relationships inherent in the systems model identified in the second phase could then be further refined through examining multiple mini-cases focusing in on the decision-making of agents and seeing how well or how poorly the model features corresponded with reality.

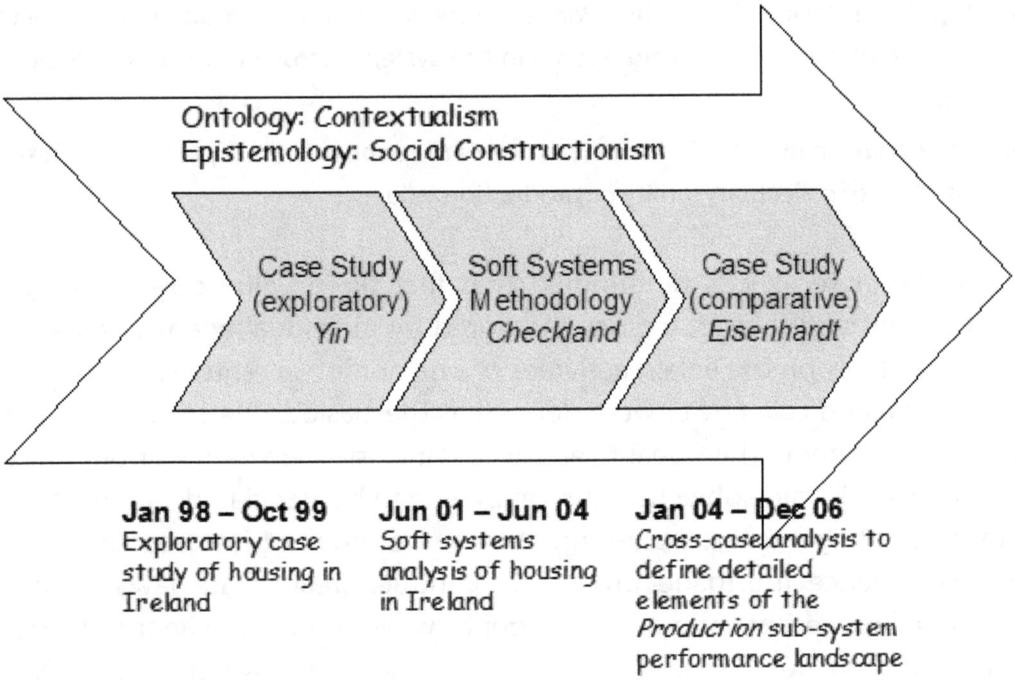

Figure 2 *Research Process*

FINDINGS FROM EMPIRICAL MAPPING OF HOUSING SYSTEM AGAINST PL MODEL ELEMENTS

The first finding in relation to the performance landscape for housing in Ireland arose from the detailed analysis of agents. This was that the landscape appeared to consist of three different landscapes, which interact to generate the outcomes of the system as a whole, as well as the rules that govern organizational behavior. This was 'discovered' principally through the analysis of different agents' objectives in respect of value creation, an exercise that Anderson (1999) suggests is one of the key challenges of developing organizational theory using complex systems. The value creation objectives of agents in the housing system were found to be:

1. **Utility**: an increase in *individual satisfaction* arising from the acquisition or use of goods and services;
2. **Production**: an increase in the *number of goods/services* produced through the transformation of inputs into outputs;
3. **Collectivity**: an increase in the *degree to which participants agree on the rules* governing organizing activity;
4. **Wealth**: an increase in an individual's *ownership of goods and resources*;

5. **Equity**: a rebalancing of the *relative distribution* of utility, production or wealth value created by organizing agents in the system across individuals in society, and;
6. **Inclusion**: an increase in the *proportion of the population that participates* in the creation of collectivity, utility or production value

Furthermore, it was determined that first three value objectives may be achieved and their outcomes measured by individual agents, while the latter three rely on the linked activities of one or more agents. Hence, utility, production and collectivity were defined as the 'basic' value objectives in an organizing system, while wealth, equity and inclusion are derived from these. The three basic value objectives also appeared to be associated with different sub-types of agents, i.e., utility agents may be grouped into search (estate agents), finance mortgage providers), contract (real estate lawyers) and exchange (brokers of various types) agents, while production agents include developers, builders, landlords, materials and service providers. Collectivity agents also display a tendency towards functional specialization in the areas of issue identification (policy consultants), development of alternative actions/rules (consultants and civil servants), concentrating power (unions, trade associations, political parties), and negotiating/agreeing solutions (legislatures, social partnership talks, various more transitory fora).

In addition to giving rise to different functional specialization, value objective was of central importance in determining the fitness function used to evaluate outcomes, and the agent's 'schema'; i.e., their perception of the actions available to them as well as the relevant features of the environment. In fact, the differences in agent behavior associated with value objective were so great as to suggest a major modification of the performance landscape model—that there are three interacting performance landscapes as shown in Figure 3 below.

These landscapes interact through information that is generated and used by agents on one landscape, but that also may be used by agents on another landscape in decision-making. The information in question consists of rules/norms that govern decision-making and performance outcomes achievable and is generated by collectivity agents (institutional rules), the interactions of utility agents (market information), and the activity of production agents (capacities / output). These initial findings in respect of agent characteristics

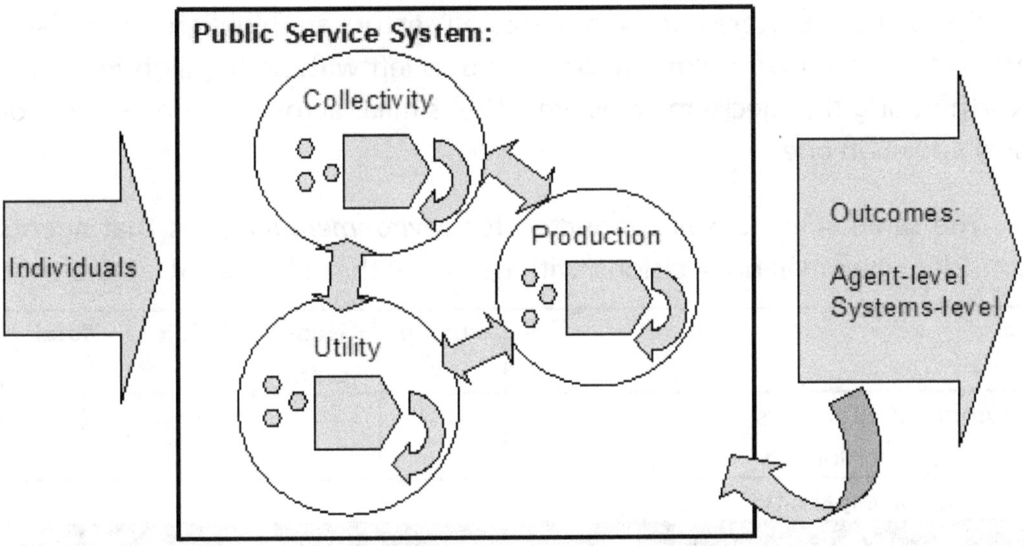

Figure 3 *Interacting Performance Landscapes in the Housing System*

and the nature of interlinking landscapes set the stage for the detailed analysis of the performance landscape as it related to the production activities in housing (phase III of the research).

Mapping the Housing Production Performance Landscape

As discussed in Section 2, the main source of data for this phase of the research were 48 out of the 63 interviews conducted with decision-makers across the private, non-profit and public sectors over the period studied. Note that, while legal status of agents was not initially assumed to be of importance in defining the features of the landscape, it did turn out that legal status (and it relationship to value creation objectives) was a vital differentiator for many of the elements.

Beginning with the types of strategic decisions that agents make, 15 types of decisions (not including innovation as it was only mentioned by one interviewee) were identified and grouped into five categories: 1) production capability, 2) production outputs, 3) relationship management, 4) target beneficiaries, and 5) value proposition. The table below (Figure 4) provides a summary of the types of decision and number of times a given decision type was reported by interviewees in organizations by legal status (public, non-profit, private).

Decisions that were common across all sectors are highlighted in black; across two sectors or unique to one sector is left with no highlighting. It is worth noting that decision areas are either similar across all three sectors or only appear in one.

The same analysis was conducted for *environmental factors* that agents consider when making decisions, with the result that 11 types of factors were

		Private	**Non-profit**	**Public**	**Total**
Number of interviews		(14)	(17)	(17)	(48)
Strategy Decisions					
Production capability:					
1	Resource access	7	12	7	26
2	Change organizational mode	7	6	3	18
3	Develop info processing	1	1	4	6
Production outputs:					
4	Grow	7	13	6	26
5	Diversify	1	6	4	11
Relationship Management:					
6	Build/maintain relationships	7	13	8	28
7	Compete / co-operate (explicit)	2	3	1	6
Target beneficiaries:					
8	Location	5	4	0(a)	9
9	Market Type	4	2	0	6
10	Segment	1	11	0	12
Value Proposition:					
11	Default to value objective	3	6	7	16
12	Customer focus	3	7	4	14
13	Cost / value for money	4	3	3	10
14	Quality	4	0	0	4
15	Liaising / information sharing	0	0	4	4
16	Innovation	0	1	0	1
	Total number of decisions	56	88	53	
	No. decisions per interviewee	4	5.2	3.1	

(a) While interviewees from the public sector did not report location as a strategic decision, location decisions were reported as playing a key role in the successful development and management of local authority housing estates and so location has been classified as a decision that is common across all sectors.

Figure 4 *Production Agent Decisions by Legal Status of Agent*

identified and grouped into three categories: 1) internal, 2) external and 3) emergent (see figure 5). Internal factors relate to the agent itself, while external factors are perceived to be outside the control or influence of the agent and/or the sector. Emergent factors are those that are directly affected by the actions and interactions of the agents within the system and are an integral part of the system, but not necessarily of the wider environment. The table below shows how many times a particular factor was mentioned by interviewees in each sector. Factors common across all sectors are highlighted in black, factors that appear in only two sectors are in grey and factors unique to one sector are not shaded.

We can see that internal factors are ahead of the other categories is terms of the number of times they are mentioned. Also, agents from all sectors appear to be affected by largely the same internal and external factors—apart from social factors, which only affect non-profit and public sector agents and the planning system which is a concern only to the private sector. But the

	Private	**Non-profit**	**Public**	**Total**
# interviewees **Factors**	**(14)**	**(17)**	**(17)**	**(48)**
Internal:				
1. Resources	7	16	17	40
2. Relationships	11	14	12	37
3. Org. Capacity	4	9	11	24
External:				
4. Political	8	10	9	27
5. Social	1	10	9	20
6. Economic	5	5	7	17
7. Planning system	6	0	2	8
Emergent:				
8. Innovation/learning	5	8	6	19
9. Sector identity	3	9	1	13
10. Sector assoc.	5	6	0	11
11. Coordination	0	4	6	10
Total factors	55	93	79	227
No. factors per agent	3.9	5.5	4.6	4.7

Figure 5 *Environmental Factors Affecting Outcomes and Strategic Choices by Sector*

social factors identified are effectively 'demand' factors for the public and non-profit sectors—similar to economic factors for private sector agents—in that these are drivers of demand for their outputs. The planning system in Ireland applies to all agents equally, so it may be that the private sector is simply more aggravated by these rules than are agents in the other sectors. With this in mind it may be said that, at least in regards to internal and external factors, agents operate on similar landscapes.

However, this is not the case for the *emergent* factors. These factors display significant variation among sectors as, aside from innovation/learning, the other four emergent factors appear in only one or two sectors, but not in all three. This is an important finding, particularly in comparison to the commonality of factors across sectors for internal and external factors. We return to this point in the discussion section.

As the next step in the mapping, an analysis of interdependencies among decisions and factors (Ns in the NK model used) was undertaken using standard statistical methods (correlations / chi-squared) to estimate 'K'. Agents of different legal statuses exhibited quite different interdependencies, so much so that few statistical correlations appearing among decisions and factors across the full range of agents. Nevertheless, by analyzing agents in sub-groups of different legal status, there appeared to be some potential for significant interdependencies among decisions/factors, however additional observations are required to provide statistical support for these. Hence, the analysis of interdependencies in the Irish case provided inconclusive evidence relating to the interdependencies (K) among decisions / factors on the landscape.

Determining the relevant performance outcome measures to link to specific decisions and/or environmental factors at first appeared to be simply a case of finding empirical data that related to the six value objectives identified, and performing statistical analysis over time to understand how different decisions / factors were correlated with particular 'levels' of performance in these measures. However, when it came down to looking at the Irish system for evidence of useful measures, it turned out that the only relevant reported measures were profitability (of private sector agents) and housing output and sales. These can be mapped to the Wealth and Production objectives, but there were no published measures of relevance to Utility, Equity, Collectivity and Inclusion. Of greater concern was the dearth of proposals in policy or in literature for establishing measures of this kind, although there were some possibilities in recent reports from the OECD to measure well-being and some interesting work by the New Economics Foundation (NEF) in the same area[3] that could be adapted for the housing system. Given the lack of measures for performance outcomes, it was an impossible task to attempt to determine the link between particular decisions and/or factors to levels of performance. In industry or business systems modeling based on NK assumptions, the single performance outcome measure is profit, making the landscape modeling exercise—at least for this element—rather uncomplicated. It is clear, however, that this is not the case for a public service system and significant research - not to mention a measurement overall in the public sector - is required.

The final component of the performance landscape model to be mapped to the Irish housing system were the rules for system dynamics—i.e., the movement of agents and changes in the performance environment. Compared to the other elements of the landscape model, the movement of agents appeared to be relatively easy to specify from the empirical data and, without going into the detailed analysis undertaken to identify these, they are: 1) agents will be formed when there is a perceived opportunity for value creation, 2) agents will exit the landscape according to their evaluation of outcomes and constraints as described in the fitness function, 3) agents have decision blind spots based on their legal status and therefore may not be able to move to certain locations on the landscape, 4) on average, agents appear to make four decisions in a given period, but this may vary based on their legal status (see Figure 4 above), and 5) the level of environmental uncertainty will affect agent movement.

The last of these requires a bit of explanation, however, before concluding this section. Uncertainty about changes in political rules or economic (price) factors were cited by several interviewees as keeping them from making decisions, i.e., freezing them in place on the landscape. While it may be argued that not making a decision is, in fact, a decision, the interviewees did not express it as such, using phrases such as "We were unable to move..." and "the policy was unclear...", and "the uncertainty in the market made it difficult to..." The nature and level of uncertainty among agents and its impact on the likelihood of moving on the landscape would need to be incorporated into a model of production agents in the housing system. Measures such as the

3. See OECD 'Society at a Glance' reports and NEF 'National Accounts of Well-Being'.

volatility of house prices (from the Utility landscape) and politically defined rules and association-defined norms (from the Collectivity landscape) are likely candidates for inputting into the decision-making of agents on the production landscape.

Taken as a whole, the mapping exercise suggests that it is conceivable that a 'real' public service system (as described in the Irish housing data) could be modeled as a performance landscape—at least in so far as production is concerned. This provides some comfort that the model had 'empirical warrant'—defined by Frederickson & Smith (2003) as the potential for "gaining empirical confirmation for the hypotheses and probabilistic assessments generated by the theory (p. 230)". However, this conclusion ignores the fact that the management theory and the empirical data suggests that relationships among agents are central to decision-making and the perception of the environment in which agents operate. As highlighted in the beginning, this feature is not part of the performance landscape model and there is no obvious way to incorporate same. The Irish case data suggests that this is a fairly big problem as discussed in the conclusion.

DISCUSSION OF FINDINGS AND IMPLICATIONS FOR THEORY AND PRACTICE

The purpose of this paper was to report on a research effort that attempted to map empirical data to a model of public service governance based on an NK-based performance landscape framework—which is a type of complex adaptive systems model. In so far as the overall contribution is concerned, the key findings (highlighted in italics below and summarized in the Appendix) help to progress the development of governance theory based on a complex adaptive systems framework. The use of a complex systems framework for building such a theory follows calls by public administration theorists (Boston, 2000; Weber, 2005) to draw upon the vast body of research and model-building activity being conducted by complexity theorists working in a wide range of disciplines to reinvigorate as well as reintegrate the "drifting and largely ignored" field of public administration theory. Furthermore, the grounding of the CAS-based theory development in an empirical case study addressed the oft-repeated criticisms of the use of complex systems frameworks, that they are too abstract and/or are only suitable as 'metaphors' for reality, instead of models (Rosenhead, 1998; Burnes, 2005).

In terms of the specific findings presented here, these may be summarized as relating to: 1) the key characteristics of **agents** in a public service governance system, 2) the features of the **performance landscape**(s) on which agents move in pursuit of their objectives and 3) the need for **CAS models** that integrate the features of NK models with neural network or social network models. We take each of these areas in turn and discuss how the findings inform theory and may have policy / practice implications.

Characteristics of Agents in Public Service Settings

Agents in public service systems were identified as organizing initiatives with particular value creation objectives that drove their formation and movement on a performance landscape. There were six types of value objective identified, of which three: Production, Utility and Collectivity were observed to result in profound differences in agent behavior and functionalization. So much so that *different performance landscapes were proposed as necessary to capture the full range of relevant actions and interactions among agents in a public service governance system.*

The six value objectives identified (production, utility, collectivity, wealth, equity and inclusion) *provide useful guidance to policy-makers and practitioners in that they provide a kind of 'checklist' of policy/management objectives against which the specific goals of a particular policy or individual agency may be compared.* Policy-makers in particular would do well to ensure that a given policy considers the impact of a proposed intervention on each of the six outcomes in order to avoid unanticipated and unwelcome results in one, through a blinkered focus on another. Irish housing policy during the 20th century is littered with examples of just this kind of pathological outcome. Managers of organizations engaged in public service provision would also benefit from a clear understanding of the relative importance of each of the six outcomes to the mission of their organization to facilitate better prioritization of strategic options and to improve communication to staff and stakeholders.

Furthermore, *it was found that the legal status of agents made a significant difference in the prioritization of value objectives, and also on how the agent perceived the range of options and how they moved on the landscape(s).* Figure 4 showed that decisions are either common to all sectors or unique to one. There appear to be slightly more decisions in common than unique to a particular sector, which suggests that a common performance landscape—at least in terms of decisions—could be used for modeling production agents across sectors. However, there are clearly 'blind spots' in each sector, which will need to be reflected in agent schema—i.e., agents of a certain legal status type may be unable to perceive decision options that are perceived by other agents. Not only does this imply that some decisions are 'unavailable' to some agents, but also that *a source of variation and change in the system may be "boundary-spanning" interactions between organizations in different sectors in which new strategic decisions are observed.*

Furthermore, there appear to be differences in the number of decisions that agents of a given legal status will make in a given period, as seen from the figures in Figure 4 for the number of decisions reported divided by the number of interviewees in each sector. It would not be wise to make too much of this finding given that the interview protocol was not designed to capture the total number of decisions made, but rather was focused on distinguishing types of decisions. Nevertheless, as there was no difference in the protocol between the sectors, this observation raises the question as to why differences in the number of decisions described arose.

Features of the Performance Landscape

Aside from the lack of performance outcome measures, the findings in relation to the performance landscape were encouraging and, in some cases, revelatory. The data suggested that *public service systems are really three interconnected landscapes*, each with a core value objective, with the connections between landscapes consisting of rules and other kinds of information (see Figure 3). Specifically, rules governing the utility and production landscapes are created by agents operating on the collectivity landscape, and may also emerge from the movements of agents on their own landscape. This is well-trodden ground in institutional theory (North, 1990; Scott, 1995). However, it was also observed that other types of information such as house prices, product standards, product availability, technological innovations and the like had a similar provenance to rules in that this type of information emerged over time from the actions and interactions of agents. Furthermore, some agents were observed to play a role in institutionalizing this type of information, such as the banks and brokers that published house price indices and the sector associations that circulated information about new products and new technologies.

The decisions and factors considered by production agents were more alike than different—supporting the proposition that these agents operate on a common landscape. The differences among agents appeared to arise from their different legal statuses. This pattern of common and unique 'N's on the performance landscape, suggested the existence of overlapping landscapes based on legal status as indicated in Figure 7 below. There were fifteen different **decisions** that were observed and these were classified into five separate categories following established strategy literature. Eleven different **factors** were observed and these could also be classified into three different categories, with two being well established in strategy literature and one category, that of 'emergent' factors, being a relatively new category relating to how features of the environment emerge and change over time. The observation that *public, private and non-profit agents in Irish housing face broadly similar decision landscapes, albeit with blind spots and often with different priorities guiding their choices, is another contribution to governance theory in public services.*

Much of the literature in public administration and strategy suggests a rather different picture of the decision landscape in that the different sectors are seen to operate in largely separate spheres of activity, driven by their values, priorities and target populations. While this perspective may be accommodated in the proposed model, it is also possible to conceive of a public service production landscape in which public, private and non-profit

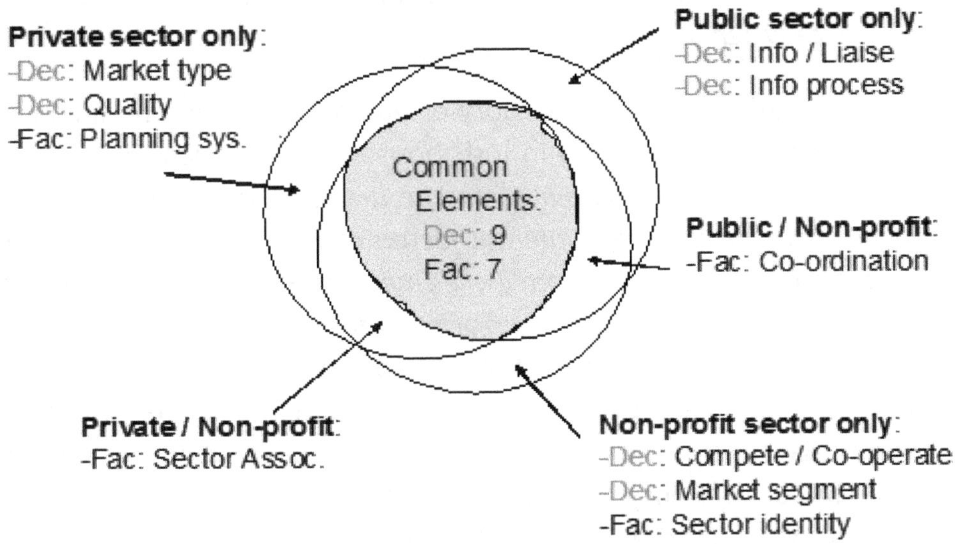

Figure 7 *Overlapping Performance Landscapes Based on Agent Legal Status*

agents operate under the same basic assumptions and rules, with as much potential for competition or cooperation with agents from different sectors as with those from their own. This was an unanticipated finding arising from the research that will require follow-up testing to determine if the observations in the Irish housing case are robust across different services and different societies.

It also appears that, while organizing agents across all sectors have much in common with respect to their external environments and how they must configure themselves internally in order to achieve their strategic objectives, *over time the interactions among agents of a particular legal status create emergent features of their environment which are unique—or unusual—within the wider system.* In the Irish system, non-profit agents interact to establish an identity for the sector, which then creates both opportunities and issues for agents in subsequent periods. Coordination across agents becomes crucial to achieving individual agent objectives as agents proliferate in the public and non-profit sectors. In the private sector, the response to proliferation of agents is, of course, competition—at least that is what economic and strategy literatures would lead us to conclude—the level of which will affect future behavior. Sector associations have relevance to private and non-profit agents as they are supported (or not) by constituent agents and acquire influence (or not) over agent behavior or other collective decision-making organizations. *As a potential source of diversity and unanticipated outcomes in a public service system these emergent factors, and the interactions/relationships from which they emerge, may be central to understanding the complex cause and effect processes in public service systems.*

The identification of a new category of 'emergent' environmental factors affecting agent decision-making in Irish housing is also a lesson for policy-makers. *To the extent that policy interventions are based on assumptions about behavior driven by 'external' and 'internal' factors alone, they are likely to result in unanticipated outcomes arising from these emergent factors, i.e., sector identity, levels of competition/coordination, rates of innovation, and the role and influence of sector associations.* Incorporating these factors into policy models, however, is no easy job as there are few available indicators that measure these factors and, even if there were, it is not clear exactly how these factors mitigate or amplify the impact of policy interventions. Nevertheless, simply being aware of these additional factors and their nature may improve the sophistication of policy-making and the reliability of the anticipated policy outcomes.

Uncertainty regarding these rules and/or the market/product information was incorporated into the proposed framework as being generated by agents operating on the three landscapes, and affecting agent behavior both on the landscape on which it is generated as well as on other landscapes. The inclusion of uncertainty as a key feature of the organizational environment affecting agent behavior is broadly supported in organizational literature going back several decades (Thompson, 1967; Perrow, 1970; Galbraith, 1977). *In the Irish case, relatively high levels of uncertainty appeared to have a kind of 'freezing' effect on the movement of agents in the production landscape, causing them to limit the range of changes considered and/or acted upon.* Uncertainty understood this way represents an expansion of the conceptualization of 'institutions' in public services, the definition of which is generally limited to the rules or 'structures' that shape the behavior and outcomes of organizational actors. While uncertainty is not the same as a 'rule' in that it is not created by a law, or group norms, it nevertheless appears to have a similar role in organizational behavior, i.e., that it arises from the interactions of many agents, becomes a part of the cognitive environment of decision-makers and hence influences agent behavior. The observations around the impact of uncertainty suggests that *public policy-makers need to take into account not only the direct effect of the rules they create, but also the indirect effect that changing the rules may have on agent behavior.* For example, over the period of the research, the government changed the rules governing tax incentives, standards and legal status relating to rental property every couple of years with the result that the rental market became 'frozen' in the words of one interviewee, with investors and property managers choosing to hold off on major decisions until the rules became 'clearer'. The lesson here is that change comes with a price in terms of the uncertainty that it creates, and policy-makers need to incorporate this realization into their models and deliberations around the probable impact of policy interventions.

The last significant policy/practice lesson arising out of the analysis of the housing system as a performance landscape was that the unique strategic activity of the public sector was not related to different or broader value objectives as suggested in strategy literature (Johnson & Scholes, 2001; Whittington, 2001). Instead, *it appears that public sector agents, at least in Ireland, are uniquely concerned with informational and coordination activities across all agents in the system and consider these to be part of their strategic mission.* To the extent that these activities are undertaken in the private and non-profit sectors, they appear as elements of implementation efforts, and are rarely undertaken for the benefit of multiple agents, never mind the system as a whole. If information sharing and coordination of efforts across agents is a condition of success in a public service system (as much of the current public administration literature suggests it is), then this finding has two clear implications for policy/practice: 1) that policy guidance for public sector agents should attend to the coordination and information management role of government agencies more explicitly, and 2) that measuring the performance of public sector agencies will fail to account for a significant component of their activity, since it is only indirectly related to the six performance outcomes identified. This is not to suggest that additional performance outcomes should be introduced into the model, but rather that some of the strategic decisions of public sector agencies contribute to (or may be a barrier to) the ability of other agents to achieve their performance objectives.

CAS Model Limitations

The major limitation of the model was also revealed in the findings, i.e., that the inability to incorporate relationships among agents directly into the NK model did appear to assume a way a significant set of features present in the empirical data. Specifically, it was not obvious how to account for resource dependencies among agents, social network ties and/or direct information exchanges. The analysis of the empirical data suggested this was a problem—in the main due to the following findings:

1. **Importance of relationships in decision-making**: The first finding was *the identification of relationships among agents as core decision and factor elements on the landscape.* The building/maintaining of relationships was the decision type observed in the interviews most often, while relationships of one kind or another featured as the second highest factor (to resources) impacting on decisions. Clearly relationships figure hugely in the Irish housing system, where the old adage, "it's not what you know, it's who you know that's important" appears to be well and truly embedded.
2. **Relationships as drivers of emergent factors**: There is quite a lot of literature in public administration that purports to deal with the nature and effects of relationships among individuals and organizations, however little of this clearly defines the nature of relationships, nor are there empirical studies of how particular relationships affect decision-making.

The second key finding is that *relationships among agents in public service systems in that they appear to result in 'emergent' environmental factors that affect decision-making, such as coordination across agents, learning in a sector/industry, the strength and focus of sector associations and the degree to which agents compete or cooperate with one another.* More on the nature and dynamics of emergence overall may be found in a range of complexity literature and in this volume in particular. The emergence of these factors from agent interactions and relationships is clearly an important dynamic of the system that must be incorporated not only into the proposed model but also into future research into governance of public services.

3. **The information/coordination role of the public sector**: The third relevant finding arising from the empirical data mapping exercise is the particular role of the public sector around information management and coordination across agents. This was an interesting finding in that *it suggests that a unique and critical function of private sector production agents is the facilitation, through information sharing, of production activities of multiple agents in the system.* This finding supports the case for focusing on the 'governance' role of public agencies (Ingraham & Lynn, 2004) and, more specifically, on the information management and coordination activities in which these agents engage. Further research is required to determine whether or not there is a relationship between achievement in this particular value activity and improved outcomes across multiple agents, and, if confirmed, this would be a significant feature of public service governance theory.

These findings highlight the fact that the exclusion of relationships decreases the 'descriptive capacity' (Frederickson & Smith, 2003) of any theory built upon the conceptual model proposed. Therefore, some accommodation of the nature, dynamics and influence of relationships among agents needs to be incorporated into the performance landscape model—and, in truth, into any complex adaptive systems model—for it to be able to support theory building in the governance of public services. Relationships are targets of decision-making as well as factors that influence decision-making, and also appear to play a role in managing uncertainty, hence they are pervasive features of the system that must be addressed. In particular, three enhancements appear to be required. The first is the incorporation of the structure and dynamics of social networks among agents into the model, probably along the lines of

Granovetter's (1985) ideas on organizational 'embeddedness', which would account for the nature and impact of organizational ties on both agent decisions and performance outcomes. The second enhancement required is a direct modelling of resource dependence among agents (again both in terms of the patterns of dependence and the dynamics of change) along the lines of Pfeffer & Salancik (1978) and Ulrich & Barney (1984). Finally, an enhancement is required to address the dynamics of information creation and dissemination among agents which can be linked into some relating to information complexity. These enhancements represent a rather ambitious research and model development agenda, but certainly not an impossible one.

REFERENCES

Anderson, P.W. (1999), 'Complexity theory and organizational science', *Organization Science*, 10(3): 216-232.

Barzelay, M., F. Gaetani, J.C. Cortázar Velarde, and G. Cejudo (2003) 'Research on Public Management Policy Change in the Latin America region: a conceptual framework and methodological guide', *International Public Management Review*, 4(1): 20-42.

Boston, J. (2000) *The Challenge of Evaluating Systemic Change: The Case of Public Management Reform*, paper presented at the IPMN Conference, 'Learning from Experience with New Public Management', Macquarie Graduate School of Management, Sydney, March 4-6.

Burnes, B. (2005), 'Complexity theories and organizational change', *International Journal of Management Reviews*, 7(2): 73-90.

Checkland, P. (1981), *Systems Thinking, Systems Practice,* New York, NY: John Wiley & Sons.

Eisenhardt, K. (1989) 'Building Theories from Case Study Research', *Academy of Management Review*, 14(4): 532-550.

Frederickson, H. G. and Smith, K.B. (2003), *The Public Administration Theory Primer,* Boulder, CO: Westview Press.

Galbraith, J.R. (1977), *Organizational Design,* Reading, MA: Addison-Wesley.

Granovetter, M. (1985), "Economic Action and Social Structure: The problem of embeddedness", *American Journal of Sociology*, 91(3): 481-510.

Johnson, G. and Scholes, K. (eds.) (2001) *Exploring Public Sector Strategy*, Essex, UK: Pearson Education Ltd.

Kauffman, S.A. (1995), *At Home in the Universe: The Search for Laws of Self-Organization and Complexity,* New York, NY: Oxford University Press.

Kauffman, S.A. (1993), *The Origins of Order: Self-Organization and Selection in Evolution,* New York, NY: Oxford University Press.

Kjaer, A.M. (2004), *Governance,* Cambridge, UK: Polity Press.

Levinthal, D. (1997), "Adaptation on Rugged Landscapes", *Management Science*, ISSN 0025-1909, 43(7): 934-950.

Levinthal, D & Warglein, M. (1999) 'Landscape Design: Designing for Local Action in Complex Worlds', *Organization Science*, ISSN 1047-7039, 10(3): 342-357.

Lynn, L. Jr., Heinrich, C. & Hill, C. (2001), *Improving Governance: A New Logic for Empirical Research,* Washington, D.C.: Georgetown Univ. Press.

Lynn, L. Jr., Heinrich, C. & Hill, C. (2000), 'Studying Governance and Public management: Why? How?' in *Governance and Performance: New Perspectives*, Washington, D.C.: Georgetown Univ. Press.

North, D.C. (1990), *Institutions, Institutional Change and Economic Performance,* Cambridge, UK: Cambridge University Press.

Pfeffer, J. and Salancik, G. (1978), *The External Control of Organizations: A Resource Dependence Perspective,* New York, NY: Harper & Row.

Perrow, C. (1970) *Organizational Analysis: A Sociological Review,* Belmont, CA: Wadsworth.

Rhodes, M.L. (2008), "Complexity and Emergence in Public Management: The case of urban regeneration in Ireland", *Public Management Review*, ISSN 1471-9037, 10(3): 361-379.

Rhodes, M.L. and Gemma Donnelly-Cox (2008), "Social Entrepreneurship as a Performance Landscape: The case of Front Line", *E:CO*, 10(3): 35-50.

Rhodes, R.A.W. (1997), *Understanding Governance: Policy Networks, Governance, Reflexivity and Accountability,* Maidenhead, Berkshire: Open University Press.

Rivkin, J.W., Siggelkow, N. (2003), "Balancing Search & Stability: Interdependencies among elements of organizational design", *Mgmt Science*, ISSN 0025-1909, 49(3): 290-311.

Rosenhead, J. (1998), *Complexity Theory and Management Practice,* http://www.human-nature.com/science-as-culture/rosenhead.html.

Scott, W.R. (1995), *Institutions and Organizations,* Thousand Oaks, CA: Sage Publications.

Siggelkow, N. & Levinthal, D. (2003), "Temporarily Divide to Conquer: Centralized, Decentralized, and Reintegrated Organizational Approaches to Exploration and Adaptation", *Organizational Science*, ISSN 1526-5455, 14(6): 650-669.

Siggelkow, N. & Rivkin, J. (2005), 'Speed and Search: Designing Organizations for Turbulence and Complexity', *Organization Science*, ISSN 1526-5455, 16(2): 101-122.

Thompson, J.D. (1967), *Organizations in Action,* New York, NY: McGraw-Hill.

Ulrich, D. & Barney, J.B. (1984), 'Perspectives in organizations: Resource dependence, efficiency, and population', *Academy of Management Review*, 9(3): 471-481.

Weber, J. (2005), 'Introduction to Chaos, Complexity, Uncertainty and Public Administration: A Symposium' *Public Administration Quarterly*, 29(3): 262-267.

Whittington, R. (2001) *What is Strategy—and Does it Matter?* (2nd ed), London, UK: Thompson Learning.

Yin, R.K. (1994) *Case Study Research: Design and Methods (2nd Ed.),* Beverly Hills, CA: Sage Publishing.

Mary Lee Rhodes (B.A., M.Sc., MBA, Ph.D.) is an assistant professor in Business at Trinity College, Dublin. She has an MBA from Wharton, an MSc from the London School of Economics (Systems Analysis & Design) and completed her PhD in Public Management at Trinity College. Dr. Rhodes has experience in managing banking operations and technology with JP Morgan, and as an economic analyst with the US Department of Agriculture. She is—or has been—a member of the Board of Directors in several government and non-profit organisations in Ireland. Her research activities span a range of public, non-profit and private management topics with an integrating emphasis on the use of complex systems theory and models to explore organisational and institutional phenomena. Her most recent book is entitled "Public Management and Complexity Theory: Richer Decision-Making in Public Services" published by Routledge in 2011. Her current research is in performance management in hybrid organizations and social impact analysis in the public service domain.

APPENDIX: KEY FINDINGS FROM MAPPING EXERCISE

Finding	Theory	Practice
1. Different, but interconnecting performance landscapes (production, utility and collectivity) are necessary to capture the full range of relevant actions and interactions among agents in a public service governance system	X	
2. The six value objectives identified (production, utility, collectivity, wealth, equity and inclusion) provide useful guidance to policy-makers and practitioners in that they provide a kind of 'checklist' of policy/management objectives against which the specific goals of a particular policy or individual agency may be compared		X
3. The legal status of agents made a significant difference in the prioritization of value objectives, and also on how the agent perceived the range of options available and how they moved on the landscape(s).	X	X
4. A source of variation and change in the system may be "boundary-spanning" interactions between organizations in different sectors in which new strategic decisions are observed.	X	
5. Some agents were observed to play a role in institutionalizing information shared between landscapes, such as the banks and brokers that published house price indices and the sector associations that circulated information about new products and new technologies	X	
6. Public, private and non-profit agents in Irish housing face broadly similar decision landscapes, albeit with blind spots and often with different priorities, guiding their choices.	X	
7. Over time the interactions among agents of a particular legal status create emergent features of their environment that are unique—or unusual—within the wider system. As a potential source of diversity and unanticipated outcomes in a public service system these emergent factors, and the interactions/relationships from which they emerge, may be central to understanding the complex cause and effect processes in public service systems.	X	
8. To the extent that policy interventions are based on assumptions about behavior driven by 'external' and 'internal' factors alone, they are likely to result in unanticipated outcomes arising from these emergent factors, i.e., sector identity, levels of competition/coordination, rates of innovation, and the role and influence of sector associations.	X	X
9. High levels of uncertainty appeared to have a kind of 'freezing' effect on the movement of agents in the production landscape, causing them to limit the range of changes considered and/or acted upon. high levels of uncertainty appeared to have a kind of 'freezing' effect on the movement of agents in the production landscape, causing them to limit the range of changes considered and/or acted upon.	X	X
10. Public sector agents, at least in Ireland, are uniquely concerned with informational and coordination activities across all agents in the system and consider these to be part of their strategic mission. It suggests that a unique and critical function of private sector production agents is the facilitation, through information sharing, of production activities of multiple agents in the system.	X	X
11. The identification of relationships between agents as core decision and factor elements on the landscape, which suggests that the inability of the PL model to incorporate these is a major gap.	X	
12. Relationships among agents in public service systems result in 'emergent' environmental factors that affect decision-making, such as coordination across agents, learning in a sector/industry, the strength and focus of sector associations and the degree to which agents compete or cooperate with one another.	X	

Chapter 13

BALANCE BETWEEN ORDER AND CHAOS AS EMERGING VALUE IN COMPLEXITY LEADERSHIP

Geert Teisman (Erasmus University Rotterdam, NLD)
Sibout Nooteboom (Erasmus University Rotterdam, NLD)
Ytsen Deelstra (Royal Haskoning DHV, NLD)

Leaders in modern societies are confronted with a paradox: simplicity in terms of goals and results are desired; they have to be achieved in a reality that is complex. We elaborate on the question how three 'professionals in charge' reflect on the impact of their leadership actions on interaction patterns and on the change that results from actions and interactions. The hypothesis that they are balancing between the 'self' and 'network' drivers and between 'self-referential' (conservative) and 'dissipative' (adaptive) communication seems to fit their reality. Are they balancing between the order of self and the chaos in interactions? Combining self interest and network needs and as well as self-referential knowledge and external knowledge seems to be the most effective style, but impossible to achieve for a longer period. Complexity leadership combined daily home based action and temporary uphill interactions that can create innovations.

THE LEADERSHIP PARADOX

In democracies, a voter's choice is based on what he sees: a leader's communicative abilities and stated ambitions. Leaders need to show that they are in charge and control. That creates attractiveness. At the same time, the existing complexity of their action field causes implementation gaps, and unforeseen and undesired side-effects. Complexity leadership theory suggests that leadership in terms of governing the results emerges in networks, in which the actions of a variety of leaders generate leadership interaction patterns from which the results emerge (e.g., Uhl-Bien *et al.*, 2007). This generates a paradox. Whilst voters tend to keep individual leaders accountable for the results they are confronted with, leader's effectiveness depends on the interactions patterns that emerge in the networks *they* are confronted with.

We elaborate how leaders perceive the relationship between their own action and leadership networks. We assume that individual actions, network interactions, and the—more structural—interaction patterns within the network (which we define as Leadership Interaction Patterns, LIPs) reinforce or attenuate each other. Policy outcomes aiming to contribute to societal development result from this often unknown interaction. Synchronization of leadership actions and style toward actions and interaction patterns in the network, with the aim of achieving personal and joint network goals seems to become an important element of complexity leadership. If synchronization is part of a leader's actions, the structure of these actions can be identified as a Complexity Accepting Leadership Style (CALS).

We expect to find complexity leadership 'on the edge of chaos and order': simple goals and actions make leaders understandable, but what looks simple in terms of aims and ambitions often will become complex in action in networks. Are leaders reflection on the possibilities to apply a complexity-accepting strategy, that focuses on the chain between action, interactions in networks and achieving system changes?

We assume that a first element of complexity-acceptance is dealing with one's knowledge partiality. Leaders embracing 'clear solutions' tend to overestimate the known and desired impacts of the solution and underestimate the unknown and undesired impacts. Leaders accepting complexity try to adapt as soon as effects of preliminary actions emerge, already during the process of getting to and implementing a decision. They are open to new knowledge that comes to them from known near sources as well as from unexpected sources elsewhere in the network.

In search for indications of what complexity leadership can look like we elaborate a set of three leader's subjective views. Do these leaders reflect on the leadership paradox? Do they develop expectations about the impact of their actions on interaction patterns, and how these patterns contribute to a desired development in their action field? Do they adapt to the evolving patterns in their action field? Do they combine personal and network motives? Do they use knowledge about the action field that is new to existing leadership networks?

THEORIES OF COMPLEXITY LEADERSHIP

Interaction between a leader and his network is conceptualized in Complexity Leadership Theory (CLT) (Uhl-Bien *et al.*, 2007; Uhl-Bien & Marion, 2009). CLT is grounded in complexity theory, and focuses on leadership in social Complex Adaptive Systems (CAS). CAS, which above we have termed 'action field', have an internal structure that allows leaders to be more influential than other agents. CAS are assumed to be composed of a large number of people, the most elementary agents, each of which can be member of several organizations. Each organization acts as a whole agent, having an internal structure. People have different roles: they act in organizations where they work, as citizens they influence politics, and as consumers they influence markets. Large CAS encompass these three types of subsystems (organizations, politics and markets), and rarely have a single leader at the top, controlling all its resources. They are not clearly bounded.

CLT distinguishes three types of leadership: administrative leadership, adaptive leadership and enabling leadership. Administrative and adaptive leadership cover the basic dynamics of a CAS. Administrative leadership concerns the ability to make formal decisions within the boundaries of organizations. Adaptive leadership concerns the ability of CAS to deal with complexity and unforeseen dynamics in such a way that it generates results that satisfy many of its participating agents for a longer period of time and hence can be characterized as resilient. Organizational goals, produced in a complex process of goal setting, tend to be rigid, but may need to become flexible and constructive. Networked agents taking initiatives (i.e., adapt their personal actions) to help a CAS adapt, show adaptive leadership. However, they are only in control of limited resources. Resources for their actions are offered to them within the very goal structure they want to change by these same adaptations. Their adaptive actions therefore need to set in motion a coevolution with their structure. It starts with the autonomy of the adaptive leaders: they bring in means not under control by administrative leaders and coordinate their actions to increase their influence. They invite others, having complementary discretionary resources. In complexity theory, this is identified as self-organizing capacity. One important resource is their time (e.g., time to explore new ideas), another is their influence on administrative leaders in their own organization. The CAS can only change if administrative leadership at some point adopts proposals from adaptive leadership. This can be done by an adaptation of official goals, which sometimes coincides with a replacement of leadership itself. In that sense, adaptive leaders are 'unfaithful' to the official goals, policies and habits within their own organizations. At the same time, adaptive leaders loose position in their own organization if they are not able to take sufficient care of the aims of its current administrative leadership. Enabling leadership entails specific behavior of administrative leaders. Enabling leaders give adaptive leadership the resources it needs, initially just by giving hope, before or behind the scenes, that administrative leadership can be influenced. They have some discretionary power to add new goals without a complex process to ask permission from their supporters. Since, in case of complexity, effective adaptive leadership has to emerge across organizational boundaries, so should enabling leadership. This implies that there must be administrative leaders who are aware of the need for adaptive leadership and to coordinate this in some way between administrative leaders across organizations. In other words, there need to be networks of enabling leaders in different organizations, synchronizing their support of adaptive leadership networks that cross their same organizations.

CLT shifts the focus to the boundary spanning ability of leaders. Adaptive and enabling leadership depend on the ability to 'see' the potential of synchronized emergence in several organizations from a common rationality. It then influences networks of administrative leadership, where it helps individual rationality coincide more with common rationality. Informal (adaptive and enabling) networks indicate the health of a social system. These networks occur behind the scenes of administrative leadership networks, whether or not these have formalized structures for network interactions (Folke *et al.*, 2005). Complexity leadership depends not only on the classic formal powers, but also on the ability to build up reciprocal relationships outside the given relations of power in a formal system (e.g., Huber, 2009; Kahane, 2010). Without this capacity, the effectiveness of administrative leadership becomes a matter of luck, since these networks interact with struggle and negotiation (where distrust is rewarded), which is difficult to combine with synchronized adaptation from a common rationality (where trust is rewarded). The quality of the whole of relationships in a CAS depends on the number of relationships across boundaries in the CAS, but perhaps even more on the quality of these relationships. In a trusting relationship, agents rely on others, even if their joint effects, as these emerge behind the scenes, may never be defined, pinpointed and officially rewarded. Investing in this trust requires seeing the quality of the network as a self-standing prerequisite for timely adaptation of a CAS. Otherwise, the selfish drive dominates an administrative leadership network and their official goals may coevolve contrary to the joint interest[1].

Administrative leadership therefore require a variety of skills. Leaders need to communicate in simple terms with their supporters, and at the same time need to rely on information provided by informal networks behind the scenes. To assess reliability of these networks, they need skills such as systems thinking, listening to new views, leaving options open, and trust building. Then, administrative leadership may entangle with adaptive and enabling leadership. Official decisions become inspired by wider knowledge. Theories about autopoïetic and dissipative capacities of self-organization suggest that the default origin of knowledge is the knowledge that has also produced the existing structure (e.g., Prigogine & Stengers, 1984). Complexity requires that leaders intuitively understand that, and accept complexity, so they can lead beyond the default.

If the assumptions of CLT hold, complexity-accepting leadership styles (CALS) are crucially important for the future of societies. We are therefore interested in ways of 'assessing' their application, especially by those having relatively more influence, the administrative leaders themselves. As indicated, theoretically, individual complexity leaders seek two kinds of balance for three leadership network types to emerge. The first balance relates to their focus: investing in a trusting network versus investing for personal position. Such balance could be defined as 'acting from the heart'. The second balance relates to accepting knowledge from outside the subsystem on which their power depends, whilst also leading that subsystem with simple, understandable communication, for which they expect to be rewarded enough to maintain their power base. Such balance could be defined as 'acting from the head'[2].

1. Conflict then escalates more easily, and even negative-sum games may be the result. There are many regions and nations where leaders play these games with passion, generating economic, social and physical decline of their system. 'Power play' claims all attention of 'leaders' aiming to survive themselves.

2. Complexity leaders may hope that their CALS contribute to a healthy CAS by influencing the leadership interaction patterns (LIPs): the emergence of networks of adaptive and enabling leadership. This should lead to observable changes of patterns in administrative

When we map out both balances in a 3 by 3 matrix, a model of Leadership Interaction Patterns (LIPs) emerges. Leaders may then be classified as belonging to one of the cells of this matrix by asking administrative leaders to reflect on their actions in the past, the reasons for these actions and the expected results. In a semi-open interview we then may assess if they are able to explain their own efforts in terms of both balances (i.e., they show their CALS) as contributions to balance in the administrative leadership network in the CAS as a whole (i.e., LIPs). We don't expect to find an even distribution of leaders over the cells. Research suggests that networks with selfish LIPs may be so consumed with their power play that leaders have no mental capacity to be open to knowledge from new, less trusted, sources, nor to invest in trust (Chabris & Simons, 2011).

Leaders are practitioners, not complexity theorists. They will not usually express their personal views about their own leadership in the terms of CLT. Rather, it is to be expected they will use metaphors and archetypes. These also illustrate the cells of the following matrix.

RESEARCH QUESTION AND RESEARCH METHOD

We address the following question:

How do leaders believe that their own conduct (CALS) influences leadership interaction patterns (LIPs) and, indirectly, generate desired results in their action field (CAS)?

We asked leaders in open interviews how they define their CAS and its challenges (1), the relevant LIP and its internal dynamics (2), and the link with their own conduct and the conduct of their peers (3). We also asked to reflect on motives of their peers and by whom they were inspired and how they acquired new knowledge. We asked them to document their case themselves, or with the help of a ghost-writer.

We accept that the interviewees may overestimate their personal impact on LIPs and the CAS. We aim to improve our understanding of their sense making of leaders in action, without having to reconstruct the case in which they are involved as a social system itself.

---

leadership networks: less conflict about positions and debate about 'facts', more cooperation and dialogue about 'facts'.

Dealing with driver structural simplicity and complexity	Dealing with simplicity and complexity in communication: *Adaptive/dissipative behaviour based on external knowledge*	*Assemblage of internal and external knowledge*	Conservative / self-referential behaviour based on internal knowledge
Focus on self	**1 Selfishness in action:** *aiming to be the winner and take it all.* Leaders are not concerned with helping other leaders. They are looking for new knowledge that helps them in their struggle for power. Political processes are polarized.	**2 Selfish in combination:** *aiming to apply sticks and carrots to generate success for 'self'.* Leaders look for knowledge about others to invent effective incentives that might work on them. They are not open for debate about the purpose of these instruments. Interaction is mainly negotiation about rules and the spending of funds.	**3 Selfishness in isolation**: *me against the hostile outside world.* Leaders stick to ideas emerging from their own organization. They try to convince others through a reiteration of arguments. Such networks may become paralyzed and the leadership may loose its legitimacy
Combining self and network	**4 Mutual adjustment** *to changing conditions, without keeping course.* Leaders understand each other's immediate needs. They take measures they think that would help all, inspired by different experts. They cannot identify first steps that move their CAS as a whole in the right direction.	**5 Creating clarity** *about what self needs, accommodating a variety of goals and resources, enabling coevolution and cocreation.* Leaders enable the flow of knowledge and look for joint images of a desirable future. They give influence to new groups and express doubts and intentions. They polarize on issues for the short term, but stay within limits of what they jointly think would contribute to a desirable future.	**6 Having a good time together** *in an 'old boy's network' without much change.* Leaders have a healthy empathy for their opponents as people, but they think in circles. One's network is likely to help one's career. They discuss the wider developments that touch them all. Yet, they make little effort to design joint perspectives of change
Focus on network	**7 Network lurking:** *focusing on vested network interests, avoiding painful intervention.,.* Leaders protect their network as it is, not accepting that new friends replace old friends. Leaders are worried that people will not accept new steps, e.g., since tragedy only can be prevented by measures that are painful on the short term.	**8 Plans getting stuck:** *joint network actions generating attractive proposals without enough selfishness to enable implantation.* Leaders suggest to society that they address long term issues, but actually do not see acceptable solutions, although they activate networks to develop new options based on internal and external knowledge.	**9 Much fuss about little change:** *adapting to external desires, without changing 'self'.* Leaders defend the position of the network as a whole. They think their own knowledge sources provide them with information that comfortably fits the existing status quo, in which little struggle for power is present.

Table 1 *Dealing With The Need For Simplicity And The Reality Of Complexity*

The interviewed leaders were randomly chosen from a group of leaders whom we expected to accept complexity. We expected to find patterns of sense making that can help to indicate the balance act between looking for simplicity and accounting for complexity, along both theoretical dimensions

THE EMPIRICAL REALITY OF LEADERSHIP GIVING INTERPRETATIONS OF COMPLEXITY

Leader A

For decades, governments had unsuccessfully applied investment programs to address neighborhood degradation. Fragmented national solutions were handed over to local authorities for implementation. Leaders of national and local government blamed each other for failures.

'A' became director of the national Neighborhoods programme in 2007. Her new approach was: all stakeholders in a neighborhood invited to join, with clear improvement targets and professional organization, putting residents at the heart of these targets (not ministries), facilitating mutual understanding, innovative and learning-oriented, and solidly based in the involved organizations. Professional moderators and independent knowledge organizations were invited to secure these principles. Neighborhoods were seen as laboratories.

Dozens of leaders invested time in relation-building between ministries responsible for ethnic integration, urban development and building, social security, health and education and between national level and municipalities. They focused on specific neighborhoods and the housing corporations and citizens within these neighborhoods. Measurable improvements occurred in the neighborhoods.

'A' opened dialogue with local government in terms of mutual added value, joint policy making and joint implementation. She offered help in specific neighborhoods, also in fields in which other ministries were competent. To that end, she developed relations with these other ministries as well. She invested in dozens of local gentrification projects, in which a variety of stakeholders and knowledge organizations participated. The different sectoral ministers enabled this process by making resources available—without certainty that these would lead to visible change that would be attributable to them personally. Her own consecutive ministers (responsible for neighborhoods) visited a neighborhood every week for years. She was able to make three consecutive ministers enthusiastic for her approach by explicitly discussing the dynamics in this network with them. She invited scientists to do an external review, keen on encouraging her partners to clarify how their actions contribute to improvements in terms of ethnic cohesion, quality of life and poverty issues. She perceived this as an exercise in communication about network dynamics.

The wider national political networks had not been willing to participate. When Cabinet demanded the recently privatized housing corporations to invest financially in the quality of neighborhoods, A did not avoid the issue, but managed to separate these tensions from the learning process. The cooperating networks were tested by hostile media and politicians who wanted to discredit Cabinet by indicating their policies were expensive and failing. This was countered by the goodwill that A had built up in her networks. Sufficient politicians remained willing to defend the programme despite its unclear contribution to large scale breakthroughs. A believes the results became widely visible at neighborhood level, whilst this remained contested in the public debate. The causal mechanisms only could be understood by those willing to reflect on interaction patterns rather than only on simplistic unilateral cause-effect mechanisms. Summary:

CAS	Multilevel: Improve a dozens of neighbourhoods by a better collaboration of local and national actors.
Network	Several ministries, municipalities, government agencies, scientists, consultancies, and housing corporations.
Earlier LIP	Ministers, expected to have clear stories about their own effectiveness, blamed each other and blamed municipalities and housing corporations. No focus on synergy.
CALS	A facilitated personal relations with several ministries and between national and local networks, and ultimately with citizens. She did that before reasoning toward any added value of specific organizations. She involved professional moderators and scientists to share problem analyses and ideas about positive impulses different government agencies could give in synergy to neighborhoods. She helped a complexity-accepting discourse, giving other ministries good publicity, with reduced credits for her own programme.
New LIP	Previously inactive networks became learning communities. Active mutual support between political and administrative layers. Organizations synchronized interventions. In wider politics, the programme became 'tolerated'.
Drivers	Ministries, municipalities, housing corporations invested more in the network without immediate benefit for themselves.
Communication and knowledge	Interaction was organized throughout the CAS. New informal networks were set up. Contributions by knowledge organizations, mediators, etc. became meaningful to partners.

Leader B

B was a prominent representative of small and medium enterprises (SMEs). He saw a bad fit between school education and skills required by the SMEs. Traditional politics had been shifting between a socio-democratic solution (better schools) and a liberal solution (more competition between schools, but how?). When B stated on television that the gap between education and SME's needed closing, his peers challenged him to act, and he started an enterprise called 'craftsman schools'. It develops courses which schools apply under franchise, and free of payment. In return, the enterprise supplies students with internship positions in the industries, for which the SMEs pay a small fee. Schools and the industries both hand over some of their autonomy.

B used his media exposure to create momentum, so that the required cooperation between schools and industries could grow. He bridged the gap between national policy and the local practices to which these policies relate by inviting the parties to share knowledge and to develop viable perspectives. Despite having no formal powers, he thought he was widely trusted as authentic leader acting for a larger good. When other leaders backed out of the deal without good reasons, he sometimes threatened them with bad publicity.

B deliberately wanted to be perceived as authentic, which gave him informal power. This is how he could set things in motion. In 2011, 'craftsman schools' has made franchise agreements with dozens of schools, and thousands of employees were trained. Summary:

CAS	Multilevel: Improving the fit between SMEs and the education system by way of a intervention by national government
Network	Management of enterprises and schools, the education ministry and socio-democratic and liberal politicians
Earlier LIP	Little interaction between the education system and the SMEs. Mainly accusations of a failing education system.
CALS	Drawing attention to new possibilities via the media and calling frontrunners together Erecting a unique enterprise to connect both worlds Persuading schools (and politicians responsible for education) and SMEs to work together via this enterprise. Exposing politicians backing out from commitments Building image as socially engaged liberal (bridging a political gap)
New LIP	Schools and SMEs better understand each other's needs, and started seeing new perspectives. Socio-democrats and liberals accept more there is a third way.
Drivers	B's authentic drive was contagious; other leaders accepted his authority despite his lack of formal power.
Communication and knowledge	Interaction was established between the education system and SMEs, and between the self-organizing activities on a local level and the interventions on a national level.

Leader C

C, director-general for health, started in 2001, with his state secretary, an attempt of health sector reform. Several ministers before had failed in similar attempts.

C listened to the stakeholders outside government, insurance companies and the health sector at large. This broader network gave him the opportunity to explore accessible roads. The involvement of private sector insurance companies was crucial. Emerging ideas did not match traditional ideas in the ministry. C 'de-idealized' political discourse by frequently involving his state secretary who tested ideas in the political arena, even if no part of political programs.

A discourse gradually emerged that could withstand most criticism and overarched the responsibilities of several ministries, which at first felt little ownership. C made the change acceptable for civil servants who were clinging to old discourses. In 2006, a new semi-private health insurance system was implemented. Summary:

CAS	Health sector, the health insurance companies, responsible ministries.
Network	Management of health insurance organizations, health ministry, political parties.
Earlier LIP	Highly politicized. Health ministry was ideology driven.
CALS	C developed a personal network in the field arena, his state secretary in doing the same in the political arena, and both seduced players to 'de-idealize' positions. This knowledge 'beat' ideologists in the ministry and made change acceptable.
New LIP	More driven by shared knowledge and less by taken positions. A learning interaction emerged between the political, field and bureaucratic arenas. People felt safe to endorse radical ideas.
Drivers	Relations in and between three arenas were improved. The process in each arena became less voluntary.
Knowledge	Scientists and experts were invited to the arenas; motto: 'understand what you talk about'.

Emerging Patterns of Leadership—A Case Comparison

All three leaders consciously tried to balance drivers as well as communication and knowledge in their networks. Their LIPs started off as polarized ones and with little communication and knowledge sharing. Their analysis is congruent with a conscious—and in these cases in their view also successful—CALS to help move their LIP from dominant 'selfishness in isolation' toward 'creating clarity'. We see four recurring patterns in their strategies to achieve balance between the need for order and simplicity and the reality of chaos and complexity in their action fields:

1. They invested in relationships with the other actors in networks in order to achieve ongoing interactions, aware that actors have the 'natural' tendency to stay within the boundaries of their own system (often an organization, a task or a position). They looked for the 'self-interest' that is at stake for these actors and how they can link this with the need for a joint effort. Finding a combination of self-interest and joint action seemed to be their most important trust-building strategy.
2. They looked for missing links in the network, inviting new agents and experts. Complicating existing networks in order to achieve results seems to be a strategy not only considered but also applied.
3. They accepted that development often takes place in a messy process, where success is not even assured at the final stage of the process. They all are open for failure and at the same time open for changes coming along. These insights generated the energy needed to act in a network much broader than needed for ensuring their own organization, task and position to survive.

4. They attracted participants by making joint opportunities clearly visible to them. It is more than a formal goal. They added a new attractor to the existing field of action. This ability helped to generate a robust result in terms of healthier neighborhoods, effective schools labour market relationships, and transparent and robust health insurance system.

CONCLUSION AND REFLECTION

Complex systems put conflicting demands on those who take responsibility for joint action in these systems. We presented an ideal typical model of dealing with complexity (Table 1) and asked leaders to reflect on their challenge. What kind of style do they adopt, what kind of interaction do they devote their limited time to? We assume that each leader, who is facing complexity in terms of content, actors and processes, will tend to simplify these elements of complexity in some way. Simplicity is an important attractor to them. At the same time, there is evidence that many combined simple solutions and simple approaches do not automatically add-up to generate the results desired.

Paraphrasing Einstein, anyone can know that the complexity of a leadership approach has to fit with the complexity of the issue at stake. And if our modern Western world is mainly dealing with complex problems, a more complex leadership approach is needed. We are in search of leaders who can combine the need for simplicity with the need for complexity.

In this contribution we portray three leaders, telling their own stories. We try to find patterns in these stories from our frame of reference based on complexity leadership theory. Our three respondents explicitly reflected on the interdependency between leadership style, interaction patterns and CAS. They implicitly used the driver and communication dimensions for balance between order and chaos from our frame of reference.

The three studies stem from a broader project in search for complexity leadership. In a book recently published on this topic, we suggests that leaders share elements of a complexity accepting style and do in fact develop archetypical LIPs (Nooteboom *et al.,* 2011). Several of the CALS and archetypes derived from CLT we presented here are observed by others in the recent past. For example, leader B's strategy is described by Andreoni *et al.* (2003) and Avolio *et al.* (2005). It is worthwhile to look for CLT archetypes in the management literature, and which of them act as strange attractors; values meaningful to—at least some—leaders.

We have tested the usefulness of the matrix with nine archetypical leadership interaction patterns. In interaction with the three cases presented we have found some confirmation that LIPs in the central cell are capable of dealing with complexity, but at the same time rather unreachable. It acts on the edge of chaos and order, creating a maximum of innovations chances without collapse. Our interviewed leaders are aware of that. At the same time they are aware that the balancing act on both dimensions requires so much extra time and resources, that it cannot be applied as a daily leadership strategy. It is like climbing the Mount Everest. You can only do it with the help of basic camps and support and only for a short period. There is a natural tendency of LIPs to move back to the corner cells.

Complexity leadership therefore looks like an uphill battle for a common future, where most time is spent in the lowlands, searching for an uphill route. Sometimes a CAS reaches the top, but staying there is difficult. And the landscape changes perpetually. To help complexity leaders, CLT should look for methods of observing these processes in more detail, clarifying interdependencies dynamics, self organization and appropriate boundary judgement.

REFERENCES

Andreoni, J., W. Harbaugh, L. Vesterlund (2003). The Carrot or the Stick: Rewards, Punishments, and Cooperation, The American Economic Review, 93(3): 893-902.

Avolio, B.J.,T. William, L. Gardner (2005). Authentic leadership development: Getting to the root of positive forms of leadership, The Leadership Quarterly, 16: 315-338.

Chabris, C., D. Simons (2011). The invisible gorilla and other ways our intuition deceives us. Harper, London.

Flood, R.L. (1999). Rethinking the fifth discipline: learning within the unknowable. London, Routledge.

Folke, C., Hahn, T., Olsson, P., and Norberg, J. (2005). Adaptive governance of social-ecological systems. Annual Review of Environment and Resources, 30(1): 441-473.

Huber, F. (2009). *Social Capital of Economic Clusters: Towards a Network-based Conception of Social Resources*. Journal of Economic and Social Geography (TESG), 100(2): 160-170.

Kahane, A. (2010). Power and love. A theory and practice of social change. BK, San Francisco.

Mintzberg, H. (1979). The structuring of Organizations, Prentice-Hall.

Mintzberg, H. (1983). Structure in fives: Designing Effective Organizations, Prentice-Hall.

Nooteboom, S. Y. Deelstra, G. Teisman, J. Van den Berg, W, Kesseler (2011). Quartermakers of the future. Mastercircle, Deventer.

Nooteboom, S., Deelstra, Y., Kesseler, W, van den Berg, J. (2011). Kwartiermakers van de toekomst. Mastercircle, Deventer. www.kwartiermakersvandetoekomst.nl (in Dutch).

Pressmann J. & A. Wildavsky (1973). *Implementation*, Berkeley.

Prigogine I. & Stengers I. (1984). Order out of chaos. Man's new dialogue with nature. Toronto: Bantam books.

Teisman G.R., L. Gerrits, A van Buuren (2009). Managing Complex Governance Systems, Routledge.

Uhl-Bien M., R. Marion (2009). Complexity Leadership in Bureaucratic Forms of Organizing: A Meso Model. The Leadership Quarterly, 20(4): 631-650.

Uhl-Bien M., R. Marion, B. McKelvey (2007). Complexity Leadership Theory: Shifting Leadership from the industrial age to the knowledge era. *The leadership quarterly*, 18(4): 298-318.

Dr. ing. **Geert R. Teisman** is professor in Public Administration at the Erasmus University Rotterdam, focusing on Complex Decision Making and Process Management. He practices several functions at the edge of scientific research and practical applications. On a regular basis he gives advice to governments and private organizations on the topics like complex decision making, strategic planning, public-private partnerships, process management, inter-governmental cooperation in metropolitan areas and policy evaluation.

Sibout Nooteboom has worked as consultant at DHV in environmental, transportation, spatial and water policies and governance issues from 1990 until 2010, in The Netherlands, Europe and in developing countries. Since 2010, he had worked as management and organization consultant inside the Dutch government. In 2006, he received his Ph.D. degree in public management at Erasmus University Rotterdam, where he has taught on a part time basis since.

Ytsen Deelstra MPA, is a consultant at Royal HaskoningDHV and a Ph.D. student in Public Management at the Erasmus University Rotterdam. He works in the field of water policy and water governance and is associated to the Dutch Water Governance Centre in the Hague. The PhD thesis of Ytsen, due 2013, analyses leadership interactions within the Delta Program (a large scale long term policy program of the Dutch Government aiming at flood protection and fresh water supply).

SECTION III
HAPPY MARRIAGE?

Chapter 14

THE MICRO-MACRO PROBLEM AS THE CORE PROBLEM OF COMPLEXITY THEORY OF POLICY PROCESSES

Göktuğ Morçöl (The Pennsylvania State University at Harrisburg, USA)

The case is made for a complexity theory of public policy. It is argued that the micro–macro problem should be the core of this theory. The three sub-problem areas of the micro–macro problem—assumptions about the nature of individual actors, how macro structures and processes emerge from the interactions of these actors, and how macro structures and processes affect individual actors—are discussed. The implications of complexity theory in these three areas are compared with those of the institutional analysis and development framework. These comparisons show that complexity theorists still need to conceptually separate their theory from others and further refine and empirically verify their conceptualizations.

WHY A COMPLEXITY THEORY FOR PUBLIC POLICY?

Public policy processes are complex, as most policy theorists have recognized since the beginnings of policy studies. There has been a tension between this recognition and the tendency of policy analysts and theorist to simplify. Sabatier (2007) points out that this tendency to simplify is inevitable, because theorization and analysis require simplifications. These simplifications do not diminish the complexities in policy processes, however; they lead to further complications, according to Sharkansky (2002). Then, can we develop an understanding of policy studies, a framework that would factor in the complexities of policy processes systematically?

Complexity theory has the conceptual tools that can be used to develop such a framework. Complexity theory is not a universal, unified, cohesive, and well-developed theory (see Mitchell, 2009), but some of the conceptual tools that are associated with it—nonlinearity, self-organization, and emergence—can be used to develop a complexity theory of policy processes.

Because of its epistemological implications, complexity theory is not suitable for developing a common framework for all realms of study, as I discussed elsewhere (Morçöl, 2012). Briefly, complexity theory suggests a contextual understanding of the phenomena under study, which means that concepts that are borrowed from other realms should at least be modified to make them applicable in the target realm. For instance, Prigogine and Stengers's (1984) insight that thermodynamic systems have the tendency to move toward far-from-equilibrium conditions, rather than toward equilibrium conditions, can be helpful in understanding the perpetual dynamism of socials systems as well, but the specific mechanism of the respective dynamisms of thermodynamic and social systems are not the same. Social systems are composed of information-processing and self-conscious individuals, but the atoms of thermodynamic systems do not have these capabilities.

When complexity theorists apply concepts that were developed for physical or biological phenomena in understanding social phenomena, they should borrow suitable concepts from the social sciences as well. I propose that to develop a framework for understanding policy processes complexity theorists should adopt the *micro–macro problem* of sociologists and political economists as the core theoretical problem. In doing so, complexity theorists should also contend with the established theories policy processes, particularly those theories in which the micro–macro problem is the core problem, such as the institutional analysis and development (IAD) framework (Ostrom, 1990, 2005) and the advocacy coalition framework (ACF) (Sabatier and Jenkins-Smith 1993).

In this paper, I will outline my argument and compare the implications of complexity theory with the IAD conceptualization of policy processes briefly. Both IAD and ACF are elaborate and sophisticated theories, and thorough comparisons of complexity theory with them remain to be the tasks for future studies.

EXISTING CONCEPTUALIZATIONS OF PUBLIC POLICY AND COMPLEXITY

Most policy theorists and practitioners understand that textbook definitions of public policy, such as Dye's (1992) (it is "whatever governments do or not do" (2)) and C. A. Simon's (2007) ("what governments *ought or ought not do*, and *does* or *does not do*" (1)) are simplifications. The underlying assumptions in these definitions are that a *singular actor* (government) decides and acts to achieve pre-determined goals and that there is a *causal relationship* between governmental decisions and "policy outcomes." These assumptions are misleading, because first multiple individual and organizational actors influence policy processes at various points. "Governmental decisions," such as laws, ordinances, and resolutions are not acts of singular actors, but products of the actions of multiple actors. The literature on governance networks attests to this observation (e.g., Koliba, Meek & Zia, 2011; Koppenjan & Klijn, 2004). Second, most scholars and practitioners would also agree that it is difficult, if not impossible, to establish direct causal links between the abstract principles and rules that are spelled out in legislative documents and "policy outcomes." A lot happens in the implementation process, as Pressman and Wildavsky (1973) and others demonstrated with many examples.

Many policy theorists recognize that policy processes are complex and some make complexity part of their conceptualizations of policy processes (e.g., Roe, 1998; Dunn, 2008: 75-76). According to Dunn, for example, dynamic interactions among multiple interdependent policy actors and interdependencies among multiple policy domains (energy, economy, etc.) make policy processes complex.

Such recognitions and conceptualizations of the complexity of policy processes laid the groundwork for a complexity theory of public policy and several theorists and researchers have applied complexity theory concepts to public policy and administration (e.g., Kiel, 1994; Morçöl, 2002; Teisman, van Buurenand & Gerrits, 2009; Rhodes, Murphy, Muir & Murray, 2011). These applications were critiqued, most pointedly by Pollitt (2009) and Klijn (2011). These critics ask: Do these applications demonstrate that complexity theory provides a better understanding of policy processes than the established theories of the field? I begin answering the challenge by outlining a framework in this paper. My main argument is that a complexity theory of public policy should be built around the micro–macro problem.

THE MICRO-MACRO PROBLEM AND COMPLEXITY THEORY

Policies are systems of collective action. Even in the simplified textbook definitions of public policy I cited earlier, the "government" is viewed as a collective actor that acts on behalf of people, or some segments of people. In more sophisticated conceptualizations, such as IAD and ACF, policy processes are viewed as actions by multiple actors that lead to collective (public, social, macro) outcomes. How exactly individual actions turn into collective outcomes is a theoretical and empirical problem, also known as the micro–macro problem, the "agency–structure problem" (Georg Simmel's sociology), the "transformation problem" (European sociology), or the "collective action problem" (rational choice theories) (Coleman, 1986). In Coleman's words, the micro–macro process is:

> *...the process through which individual preferences become collective choices; the process through which dissatisfaction becomes revolution; through which simultaneous fear in members of a crowd turns into a mass panic; through which preferences, holdings of private goods, and the possibility of exchange create market prices and a redistribution of goods; through which individuals' task performance in an organization creates a social product; through which the reduction of usefulness of children to parents leads families to disintegrate; through which interest cleavages lead (or fail to lead) to overt social conflict.* (1321)

As this description suggests, the two levels of analysis in the micro–macro problem are beliefs, motivations, and actions of individual actors (the micro level) and collective processes and outcomes (the macro level). One might argue that social processes are more complex than this dichotomous conceptualization: There should be multiple levels of analysis, not two. Theorists who use micro–macro conceptualizations do recognize that policy processes should be analyzed at multiple levels. In Ostrom's (2005) IAD framework, for example, there are operational, collective-choice, constitutional, and metaconstituional levels (58-62). Complexity theorist Axelrod (1997) specifically studied micro–macro transformations at multiple levels. He demonstrated how collective political actors emerge from the actions of individuals and then they become actors in larger social processes. Ostrom's and Axelrod's multi-level conceptualizations do not invalidate the micro–macro problem, but they underscore the fact that it is a framework that needs to be supported with more refined descriptions that are specific to the situations under study.

In its most generic form, there are three subsets of problems/questions within the micro–macro problem:

1. What is/are the nature(s) of micro units (e.g., individual actors)? Are they intelligent, rational, etc.? Or, how, to what extent, are intelligent, rational etc.? I will call answers given this set of questions: *assumptions about micro units.*
2. How do actions of and relationships among micro units (e.g., individual actors) lead to the creation/aggregation/emergence of macro processes and structures? I will call answers given this question: assumptions about *micro to macro relations.*
3. Once they are created, do (or how do) these macro structures and processes affect the beliefs, motivations, and actions of individual actors? I will call answers given this question: assumptions about *macro to micro relations.*

In the following sections, I will summarize and discuss the answers given by the IAD and complexity theorists to these questions, to demonstrate the similarities and differences between the two.

Institutional Analysis and Development Framework

To understand how the IAD framework deals with the micro–macro problem, one has to go back to its philosophical roots in rational choice theory. Rational choice is a major theoretical framework that has been applied in the social sciences in recent decades. As Lichbach (2003) and MacDonald (2003) observe, rational choice is more than a theory. It is a paradigm, in the Kuhnian sense of the term: a set of beliefs among a group of scientists who apply a set of commonly accepted instruments of investigation. Several well-known theories operate within this paradigm: public choice, principal-agent theory, new public management, polycenticism, and institutional rational choice (the IAD framework).

The IAD framework deserves the particular attention of complexity theorists, not only because it is arguably the most advanced and elaborate version of rational choice theory, but also, and more importantly, Elinor Ostrom, its leading theorist, is keenly aware of the implications of complexity theory and incorporates some of its concepts, such as nonlinearity, self-organization, and feedback loops, into her own work (see for example Dietz, Ostrom & Stern 2003; Ostrom 2005).

To begin understanding how IAD theorists conceptualize the micro–macro problem, one should go back to the foundational assumptions of rational choice. MacDonald (2003) sums up these assumptions succinctly. In the rational choice paradigm individual actors are rational. MacDonald describes the rationality assumption of rational choice as follows:

> *The rationality assumption consists of three components: purposive action, consistent preferences, and utility maximization. Purposive action posits that most social outcomes can be explained by goal-oriented action on the part of the actors in the theory, as opposed to being motivated by habit, tradition, or social appropriateness. Consistent preferences refers to preferences that are ranked, are transitive, and do not depend on the presence or absence of essentially independent alternatives. Utility maximization posits that actors will select the behavior that provides them with the most subjective expected utility from a set of possible behaviors.* (552)

MacDonald further states that in rational choice theory:

> *[M]acrosocial outcomes are the sum of discrete, intentional acts by preconstituted actors.... [It] is the purposive, intentional, self-propelled behavior of individuals that aggregate into outcomes; structures neither constitute this behavior nor constitute the actors.* (558)

Assumptions about Micro Units

There are answers to the three subsets of questions I asked earlier in the above description by MacDonald. The answer to the first question (question about the nature of micro units) is that individual actors are rational. Different versions of rational choice theory agree with this assumption at varying degrees, as MacDonald points out, but the most orthodox version of the theory assumes that all individuals are purposive, consistent in their preferences, and utility maximizing in all kinds of different settings. In other words, the rationality assumption applies universally.

The rationality assumption has two components: that individuals are self-interested and that they have the cognitive capability of maximizing their utilities and they actually use these capabilities in their real life decisions. Both components have been criticized from various angles. As MacDonald (2003) notes, several social scientists showed that not all forms of human behavior are motivated by self-interest. Behavioral and cognitive psychologists demonstrated that the utility maximization assumption ignores the behavioral and cognitive biases and predispositions in the human cognitive processes (Edwards & Tversky, 1967; Hogarth & Reder, 1986; Friedland & Roberston, 1990). The most well-known of these critics is Herbert Simon. In his theory of "intended and bounded rationality" Simon (1979) posits that individuals are indeed purposive, but they cannot rank-order their decision choices consistently. Nor can they predict the outcomes of their decisions, as the utility maximization assumption requires. Individuals have "ineradicable" cognitive limitations that prevent them from rank-ordering their preferences and predicting the outcomes of their behaviors. Rationality, in other words, is inherently bounded.

Elinor Ostrom, the most prominent theorist of the IAD framework, adopts Simon's notion of bounded rationality, but she is not ready to give up on the possibility of full rationality yet. This tension is evident in her writings. On the one hand, she stresses that the bounded rationality assumption should be substituted for the "extreme assumptions such as unlimited computational capability and full maximization of net benefits" (1999: 45) and that the results of her and her colleagues' research are consistent with the assumption of bounded rationality (2005: 118). On the other hand, she thinks that the rationality assumption is still the best assumption under some conditions (2005: 99-133) and a "useful starting point" for her analyses (2005: 103).

Assumptions about Micro to Macro Relations

How do actions of and relationships among micro units (e.g., individual actors) lead to the creation/aggregation/emergence of macro processes and structures? This is a central question for rational choice theorists. It is commonly known as the "aggregation problem" (e.g., Buchanan & Tullock, 1971; Ostrom, 2005). The term aggregation is used quite vaguely in the literature. Does it signify a simple sum of individual acts? In other words, is it additive, in mathematical terms? Or, do the micro–macro transformations take on more complex forms?

MacDonald (2003) observes that most rational choice theorist subscribe to the view that aggregation is additive: "macrosocial outcomes are the sum of discrete, intentional acts by preconstituted actors.... [It] is the purposive, intentional, self-propelled behavior of individuals that aggregate into outcomes..." (558). Cochran and Malone's (1995) view of public interest also reflects this additive view: "'[P]ublic interest' may be understood as the entirety of...individual preferences expressed as choices" (5). In this way of thinking, when individuals are allowed to choose freely in economic (political or organizational) markets, their behaviors will add up to a state that is beneficial for all.

Simon and his associates (1992: 49) raise doubts about this additive view of aggregation. They point out that there is actually so much unknown about how individual behaviors aggregate into collective outcomes and that this should be a major area of research and theorization for future scholars. Ostrom (1999, 2005) also thinks that this additive approach is not helpful in understanding how actually interests aggregate in a society. She does not offer a specific alternative, however.

Assumptions about Macro to Micro Relations

Do (or how do) macro structures and processes affect the beliefs, motivations, and actions of individual actors? In other words, to what extent do individuals' pictures of the world "contain representations of the emergent macropatterns" (Sawyer, 2005: 26), or to what extent are macropatterns internalized by individuals? This is what sociologists call *social causation*, or *downward causation*.

In the orthodox version of rational choice theory, individual actors are preconstituted, as MacDonald (2003) puts it. They are uniformly utility-maximizing rational actors. They do not have any histories or cultural affiliations that could create variations among them in the ways they make their decisions. In other words, macro structures and processes do not have any bearing on individual beliefs, motivations, and actions.

Ostrom (2005) offers a more sophisticated and complex view of actors. She acknowledges that because of the different mental models individuals use, various feedbacks they receive from the world, their shared culture, and the emotional states they are in, their behaviors are more complex and context-dependent than the rationality assumption suggests (103, 105, 112, 119). Ostrom's recognition of contexts of individual actions is important and her conceptualization deserves a close look.

In Ostrom's IAD framework, the "attributes of a particular community" (i.e., culture) and institutional structures constitute the context in which complex individuals make their decisions and act (Ostrom, 1999: 57-58, 2005, passim).

Among those institutional structures, "action situations" are the most directly relevant to actors' actions. An action situation is the "immediate structure affecting a process of interest for the purpose of explaining regularities in human actions and results, and potentially to reform them" (Ostrom, 2007: 29). More specifically, she points out, these situations are the rules of the game, such rules about who are allowed to play the game and the level of control participants have over their choices.

It is important to understand that in Ostrom's conceptualization these action situations *frame* the actions of actors *externally*. In other words, Ostrom acknowledges that there is some sort of social causation, but in only the form of *external inducements* for action (Ostrom & Parks, 1999: 292). In other words, they do not permeate individual beliefs, preferences, or values; they are not internalized by individual actors.

In Ostrom's framework, "attributes of community" frame the actions of actors externally as well. This is important to stress, because what she means by the "attributes of community" is "culture" (2007: 43). A culture is supposed to be composed of beliefs, values, and norms that are shared and internalized by the members of a community, but in her framework it is an exogenous variable. In other words, the cultural traits—she cites language and shared beliefs about the foundations of social order as examples (43)—are external to individual actors. The beliefs of these individuals about the foundations of social order are somehow external to them!

It is not my intention to criticize Ostrom for the inconsistencies, or conceptual gaps, in her framework, but to stress that in her view macro structures are separated from, external to, the properties of micro units: individual actors. This view is common among rational choice theorists; as MacDonald (2003) notes, in their view: "structures neither constitute this behavior nor constitute the actors" (558).

The relations between social structures and the behaviors of individual actors is an important issue for complexity theorists as well. I will come back to it in the next section.

OUTLINE OF A COMPLEXITY THEORY OF POLICY PROCESSES

What is complexity theorists' answer to the micro–macro problem? How does it differ from that of the IAD framework? There is no coherent and universally agreed-upon conceptualization of the micro–macro problem among complexity theorists. Some complexity theorist may not even agree that the micro–macro problem is a problem at all. What I discuss next is my proposal for other complexity theorists to consider.

Assumptions about Micro Units

There is very limited theoretical discussion among complexity theorists on the nature of actors. The assumptions about the nature of actors can be elicited mostly from the applications of agent-based simulations, primary methodological tools of many complexity researchers.

Miller and Page (2007) point out that there are two conceptual problems regarding the nature of agents in agent-based simulations: the problem of "generalized attributes" and the problem of "level of sophistication" (238-239). The first is the problem of to what extent the actors are modeled homogenously in a simulation. Put in other words, are all actors alike in their characteristics, such as their beliefs, preferences, and abilities to act? Are they uniformly self-interested and utility-maximizing rational actors, for example? Remember that the orthodox version of rational choice theory would answer this question affirmatively. In complexity theory, this is a conceptual problem.

There is no common and clearly stated position among complexity theorists on the issue of agent homogeneity. In agent-based simulations agents are programmed at different degrees of homogeneity. In some simulations agents are more homogenous then others.

Miller and Page observe that most social scientists have the tendency to homogenize the behaviors of individuals in their analyses by averaging out differences among individuals and point out the potential dangers in doing so: "the introduction of just a bit of heterogeneity into homogenous world may substantially alter the system's behavior" (239). Those researchers who ignore this may miss out a lot in understanding the systems they study, Miller and Page remind us. Indeed, as complexity theorists know very well, small changes in the behaviors of individual actors can lead to big changes in the system's macro behavior. The implication is that complexity researchers who use agent-based simulations should acknowledge agent heterogeneity and build this assumption into their simulations as much as possible.

The second problem Miller and Page point out is the "level of sophistication" of agents. The question is: To what extent agents should be modeled as rational, informed, and able to act in the way they desire to act in agent-based simulations (239)? As I mentioned earlier, orthodox rational choice theorists assume that all agents are utility maximizers. Simon modifies this assumption by pointing out the cognitive limitations of actors: They cannot maximize their utilities because of these limitations; they can only satisfice. Ostrom and many others use bounded rationality as one of their core assumptions.

Some complexity theorists and researchers adopted the bounded rationality assumption as an umbrella concept (e.g., Holland, 1995; Axelrod, 1997; Cohen, Riolo & Axelrod, 2001; Epstein, 2006; Miller & Page, 2007; Bednar & Page, 2007), but there is a dichotomous conceptualization actors that characterize the assumptions of agents by complexity theorists better. This is the "cognitive agent" versus "reactive agent" dichotomy. Cognitive agents are intentional and deliberative agents who have beliefs about their environments, their actions, and their impacts, whereas reactive agents do not have any internal representation of the world and they simply react to external impulses (Sawyer, 2005).

I argued elsewhere that there are problems with the dichotomous conceptualization of cognitive versus reactive agents (Morçöl, 2012). It is better to ask the question: To what extent do agents have the knowledge of the structural properties of the systems that emerge from their interactions? Furthermore, the issue is not only the degree of knowledge of structural properties; human beings have different types of internal representations of their worlds. To better represent human societies in agent-based simulations these gradations and variations of knowledge should be modeled.

This is a particularly important issue in understanding policy processes because if all policy actors/agents had the full knowledge of the consequences of their and others' actions and the interactions among them, then presumably they collectively would be able to solve collective action problems. Of course this is not a reasonable assumption. Then the question is: To what extent agents have the knowledge of their worlds? This is a question than can be answered only empirically for specific situations.

Miler and Page (2007) recognize the diversity among individual agents and suggest that agent-based simulators should factor this into their simulators by using bottom-up approaches. Instead of using generalized assumptions about agent behaviors and capabilities, such as that all agents are utility maximizers, complexity researchers should build variations among agents into their simulations and let agent behaviors evolve, they suggest. Epstein (2006: 346) agrees that simulators should model "psychologically richer and coevolving" agents, but, he points out, there are immense technical challenges in modeling the infinite complexity of human beliefs, motivations, and behaviors in simulations.

Assumptions about Micro to Macro Relations

I mentioned earlier that rational choice theorists either offer additive answers to the aggregation problem or avoid offering specific answers to it. Now the question is: Does complexity theory have a different and better answer?

Complexity theorists do not have a common or coherent answer, but two concepts of complexity theory are particularly relevant for developing an answer: self-organization and nonlinearity.

Actors of a social system (e.g., a policy system) are self-organizational. Macro structures and processes emerge from their interactions, without a central authority or control. Rational choice theorists share the notion of self-organization. They tend to favor self-organizational political and economic mechanisms, such as decentralized organizational forms, democratic participation, and free markets, for ensuring that the interests of the public are served. Ostrom (2005) raises some doubts about this blanket assertion. She argues that self-organization does not always happen, nor does it lead to successful results. In particular she identifies a set of conditions that would make self-organizational processes successful in the governance of common-pool resources: a common understanding of issues among actors, availability of good indicators to measure success, etc. (pp. 244-245).

Some complexity theorists agree, at least in general terms, with the rational choice assertions that decentralized organizational forms, democratic participation, and free markets work better than their alternatives (e.g., Kauffman, 1995). Others point to empirical findings that self-organizational mechanisms may actually lead to the emergence of hierarchies and skewed resource distributions in societies (Kellert, 1995; Epstein & Axtell, 1996).

I propose that self-organization should be viewed as an open-ended process with uncertain outcomes: Both positive and negative collective outcomes may emerge self-organizationally. Complexity theorists should investigate self-organizational processes and their outcomes empirically.

There are some complexity theory concepts that have been used and can be used in investigating self-organizational processes. All complexity theorists agree that the relations among the units of a system are more often nonlinear than linear. Nonlinearity is mathematical concept that connotes unevenness in relationships and disproportionality in outcomes of relationships. It can be posited, for example, that policy interventions can have disproportionate outcomes. Small-scale interventions by governmental actors may have large-scale outcomes, and vice versa. To solve a massive crime problem in a city, the police department does not have to utilize proportionately massive resources. Instead, as the "broken windows theory" suggests, preventing petty crimes in neighborhood with the application of small, but targeted, resources can disproportionately contribute preventing large-scale crimes.

Nonlinearity is a core concept of complexity theory understanding of the world, but there are more specific concepts complexity theorists developed to understand how micro actions aggregate to macro structures. Holland (1998) and Barabási and Alberts (1999) identified two particular mechanisms of how hierarchical organizational structures emerge: *differential persistence* and *preferential attachment*. Holland's basic observation is that both in biological and social systems those elements that were more successful in the past because of the ways they acted will be more successful in shaping the future patterns of the macro behaviors of the systems (227-229). Barabási and Alberts observe that the hierarchical structures of "scale-free networks," such as the World Wide Web and airline "hub" systems, emerge through a process in which new elements attach to the ones that are already well-connected (509). Future studies and conceptualizations by complexity theorists can help us understand these mechanisms better and bring forth other specific mechanisms of self-organization.

Assumptions about Macro to Micro Relations

Rational choice theorists conceptualize rational actors as decontextualized "preconstituted actors," who do not have any histories or cultural affiliations that could affect the ways they make decisions and act, as I mentioned earlier. Ostrom recognizes the importance of context in understanding actors' behaviors. In Ostrom's view, social structures affect individual actions as external inducements, not as internalized values. So she sidesteps the issue of social causation.

There is no unified conceptualization of social causation among complexity theorists, but they are aware of it and its implications. In most general terms the issue is conceptualized as a dynamic interaction between external stimuli and actors' internal models of the world (Holland, 1995; Mitchell, 2009). In the "cognitive actor" conceptualization I mentioned earlier, actors have of internal models.

The problem is how to build agent-based models that would include representations of social structures in individuals' internal cognitive models. Epstein (2006) notes that this can be done through "feedback between macrostructures and microstructures, as where newborn agents are conditioned by social norms or institutions that have taken shape endogenously through earlier agent interactions" (pp. 51-52). Some complexity researchers have taken on this challenge in their simulations (e.g., Cohen, Riolo & Axelrod, 2001; Bednar & Page, 2007) at varying degrees of success. Technical problems remain in such applications, which can be addressed in future studies.

CONCLUSIONS

Complexity theory is a work in progress. It may not be possible to develop a universal theory of complexity, but it is possible to develop a complexity theory of policy processes. To accomplish that, I argued in this paper, complexity theorists should borrow concepts from other theories and engage the existing theories of policy processes.

In this paper I proposed the micro–macro problem of sociologists and political economists as the core problem of a complexity theory of policy processes. Complexity theorists will have to address three sets of theoretical problems/questions related to this core problem: the nature of micro units (individual actors) in policy processes, how macro structures emerge from these micro units, and how emergent social structures influence the behaviors of micro units.

I proposed that complexity theorists should also engage theories of policy processes that are also built around the micro–macro problem, those theories whose conceptualizations of policy processes include interactions among multiple actors and the collective (macro, public) outcomes of those interactions. I cited the institutional analysis and development (IAD) framework and the advocacy coalition framework as the best examples of such theories and discussed the former specifically in this paper. I compared the foundational assumptions of rational choice theory and the modifications of them by the IAD theorists with the assumptions of complexity theory.

These comparisons are both humbling and promising for the prospects for a complexity theory of policy processes. They are humbling, because they show that brilliant theorists like Ostrom have already considered some of the complexity theory concepts, such as self-organization, and incorporated them into their own frameworks. Complexity theorists face a serious challenge in demonstrating that theirs is a distinct and better theory.

The comparisons are also promising, because they show that complexity theorists have made serious advances in key areas like conceptualizing and modelling the internal models and behaviors of individual agents and the effects of social causation on these internal models and behaviors. It is also humbling to observe that despite the advances, many complexity theorists still rely on generic conceptualizations like boundedly rational agents.

Much more work remains to be done by complexity theorists in refining their conceptualizations and empirically verifying them. I propose the outline I presented in this paper as a starting point of future conceptual and empirical works.

REFERENCES

Anderson, P. (1999). "Complexity Theory and Organization Science," *Organization Science,* ISSN 1526-5455, 10: 216-232.

Axelrod, R. (1997). *The complexity of cooperation: Agent-based models of competition and collaboration*, ISBN 978069101567.

Barabási, A-L. and Albert, R. (1999). "Emergence of Scaling in Random Networks." *Science,* ISSN 1095-9203, 286: 509-512.

Bednar, J. and Page, S. (2007). "The Emergence of Cultural Behavior Within Multiple Games," *Rationality and Society*, ISSN 1043-4631, 19: 65-97.

Buchanan, J.M., and Tullaock, G. (1971). *The Calculus of Consent: Logical Foundations of Constitutional Democracy,* ISBN 0472061003.

Cochran, C.L. and Malone, E.F. (1995). *Public policy: Perspectives and Choices*, ISBN 0070115281.

Cohen, M.D., Riolo, R.L., and Axelrod, R. (2001). "The Role of Social Structure in the Maintenance of Cooperative Regimes" *Rationality and Society*, ISSN 1043-4631, 13: 5-32.

Coleman, J.S. (1986). "Social Theory, Social Research, and a Theory of Action," *The American Journal of Sociology*, ISSN 0002-9602, 91(6): 1309-1335.

Dietz, T., Ostrom, E., and Stern, P.C. (2003). "The Struggle to Govern the Commons," *Science,* ISSN 0036-8075, 302: 1907-1912.

Dunn, W.N. (2008). *Public Policy Analysis: An Introduction* (4th Ed.), ISBN 0205677924.

Dye, T.R. (2004). *Understanding Public Policy* (11th Ed.), ISBN 0205757421.

Edwards, W., and Tversky, A. (eds.) (1967). *Decision Making*, Middlesex, UK: Penguin.

Epstein, J.M. (2006). *Generative Social Science: Studies in Agent-Based Computational Modeling*, ISBN 0691125473.

Epstein, J.M. and Axtell, R. (1996). *Growing Artificial Societies: Social Science from the Bottom Up*, ISBN 0262550253.

Friedland, R., and Robertson, A.F. (eds.). (1990). *Beyond the Marketplace: Rethinking Economy and Society*, ISBN 0202303705.

Hogarth, R.M., and Reder, M.W. (Eds.) (1986). *Rational Choice: The Contrast between Economics and Psychology,* ISBN 0226348571.

Holland, J.H. (1995). *Hidden Order: How Adaptation Builds Complexity*, ISBN 0201442302.

Holland, J.H. (1998). *Emergence: From Chaos to Order,* ISBN 0738201421.

Kauffman, S. (1995). *At Home in the Universe: The Search for Laws of Self-organization and Complexity,* ISBN 0195111303.

Kellert, S.H. (1993). *In the Wake of Chaos: Unpredictable Order in Dynamical Systems,* ISBN 0226429741.

Kiel, L.D. (1994). *Managing Chaos and Complexity in Government: A New Paradigm for Managing Change, Innovation, and Organizational Renewal,* ISBN 0787900230.

Klijn, E-H. (2011). *Complexity Theory and Public Administration: A Flourishing Relation?* Keynote Speech at the Conference on Challenges of Making Public Administration and Complexity Theory Work (COMPACT), Rotterdam, the Netherlands, June 2011.

Koliba, C., Meek, J.W., and Zia, A. (2011). *Governance Networks in Public Administration and Policy,* ISBN 9781420071269.

Koppenjan, J. and Klijn, E-H. (2004). Managing Uncertainties in Networks.

Lichbach, M.A. (2003). *Is Rational Choice Theory All of Social Science?* ISBN 9780472068197.

MacDonald, P.K. (2003). "Useful Fiction or Miracle Maker: The Competing Epistemological Foundations of Rational Choice Theory," *American Political Science Review,* ISSN 0003-0554, 97(4): 551-565.

Miller, J.H., and Page, S.E. (2007). *Complex Adaptive Systems: An Introduction to Computational Models of Social Life,* ISBN 0691130965.

Mitchell, M. (2009). *Complexity: A Guided Tour,* ISBN 9780195124415.

Morçöl, G. (2002). *A New Mind for Policy Analysis: Toward A Post-Newtonian and Postpositivist Epistemology and Methodology,* ISBN 0275970124.

Morçöl, G. (2012). *A Complexity Theory for Public Policy,* 9780415518277.

Ostrom, E. (1990). *Governing the Commons: The Evolution of Institutions for Collective Action,* ISBN 9780521405997.

Ostrom, E. (2005). *Understanding Institutional Diversity,* ISBN 0691122075.

Ostrom, E. (2007). "Institutional Rational Choice: An Assessment of the Institutional Analysis and Development Framework." In P. A. Sabatier (Ed.), *Theories of the Policy Process,* ISBN 0813343593, pp. 21-64.

Ostrom, E., and Parks, R.B. (1999). "Neither Gargantuan nor the Land of Lilliputs: Conjectures on Mixed Systems of Metropolitan Organization." In M. D. McGinnis (Ed.), *Polycentricity and Local Public Economies: Readings from the Workshop in Political Theory and Policy Analysis,* ISBN 047211038, pp. 284-305.

Pollitt, C. (2009). "Complexity Theory and Evolutionary Public Administration: A Skeptical Afterword," in G. R. Teisman, A. van Buuren, and L.Gerrits (eds.), *Managing Complex Governance Systems: Dynamics, Self-Organization and Coevolution in Public Investments,* ISBN 0415459737, pp. 213-230.

Pressman, J.L. and Wildavsky, A.B. (1984). *Implementation: How Great Expectations in Washington are Dashed in Oakland; or, Why It's Amazing That Federal Programs Work at All*, ISBN 0520053311.

Prigogine, I., and Stengers, I. (1984). *Order Out of Chaos: Man's New Dialogue with Nature*, ISBN 9780553340822.

Rhodes, M.L., Murphy, J., Muir, J., and Murray, J.A. (2011). *Public Management and Complexity Theory: Richer Decision-Making in Public Services*, ISBN 041545753X.

Sabatier, P.A. (2007). "The Need for Better Theories," in P. Sabatier (Ed.), *Theories of the Policy Process*, ISBN 9780813343594, pp. 3-17.

Sabatier, P.A. and Jenkins-Smith, H. (1993). *Policy Change and Learning: An Advocacy Coalition Approach*, ISBN 0813316499.

Sawyer, R.K. (2005). *Social Emergence: Societies as Complex Systems*, ISBN 9780521606370.

Sharkansky, I. (2002). *Politics and Policymaking: In Search of Simplicity*, ISBN 1588260844.

Simon, C.A. (2007). *Public Policy: Preferences and Outcomes*, ISBN 0205744826.

Simon, H.A. (1979). "Rational Decision Making and Business Organizations," *The American Economic Review*, ISSN 0002-8282, 69: 493-513.

Simon H.A. and Associates. (1992). Decision Making and Problem Solving. In M. Zey (Ed.), *Decision Making: Alternatives to Rational Choice Models*, ISBN 0803947518, pp. 32-53.

Teisman, G.R., van Buuren, A., and Gerrits, L. (eds.) (2009). *Managing Complex Governance Systems: Dynamics, Self-Organization and Coevolution in Public Investments*, ISBN 0415459737.

Göktuğ Morçöl is an associate professor of public policy and administration at the Pennsylvania State University, Harrisburg. The areas of his research interests are metropolitan governance, with a particular focus on business improvement districts, complexity theory applications in public policy and policy analysis, and research methods. He has authored, edited, and co-edited the following books: *A Complexity Theory for Public Policy* (Routledge, 2012); *Business Improvement Districts: Research Theory and Controversies* (CRC Press, 2008); *Complexity and Policy Analysis: Tools and Methods for Designing Robust Policies in a Complex World* (ISCE Publishing, 2008); *Handbook of Decision Making* (CRC Press, 2007); *A New Mind for Policy Analysis: Toward a Post-Newtonian and Postpositivist Epistemology and Methodology* (Praeger, 2002); and *New Sciences for Public Administration and Policy: Connections and Reflections* (Chatelaine Press, 2000). His articles have appeared in *Public Administration Review, Administrative Theory and Praxis, Policy Sciences, Public Administration Quarterly, Politics and Policy, International Journal of Public Administration, Emergence*, and others. He served as the chair of the Section for Complexity and Systems Studies of the American Society for Public Administration.

Chapter 15

WHY TAKING COMPLEXITY SERIOUSLY IMPLIES A PARADIGM SHIFT FOR POLICY STUDIES

Derk Loorbach (Erasmus University Rotterdam, NLD)
Niki Frantzeskaki (Erasmus University Rotterdam, NLD)

This paper argues that Public Administration should draw more fundamental lessons from complex system sciences than it has so far. Taking societal complexity seriously in the context of persistent societal challenges like the transitions in energy, food production, health care and the economy requires a new way of thinking about the role of policy in society. We argue that in complex, uncertain and ambiguous contexts, policy is one of many factors influencing long-term change. So far, policy studies have mainly used complexity perspective to improve the existing policy paradigm and processes. We explore how a complexity perspective could help to develop a new policy perspective as well as new policy practices to influence the speed and direction of transitions.

INTRODUCTION

In this paper we explore how a complexity perspective can contribute to Public Administration and policy research. As the world is increasingly faced with interlinked and highly uncertain disruptions, the question how to effectively deal with these becomes increasingly urgent. This is visible not only in the financial crises, but also in the fundamental changes in modern societies in areas such as provision of food, energy, materials and 'climate-robust' living environments. Our basic starting point is that these wicked societal challenges combined with the increased power of business and (civil) society requires a fundamental redefinition of the role of policy and public administration. We argue that policy effectiveness should not be the sole focus, but that efforts need to be made to create space to collectively deal with dynamics emerging at the intersection of the public sector, civil society and business in a novel way.

We witness a growing attention in Public Administration research for this question, illustrated by the research into networks, multi-level and complexity governance. Our argument is that so far, these perspectives and concepts have been mainly used within the frame of policymaking to explore possible limits to policy-making (processes) and how to deal with these. We argue that the perspective of complex adaptive systems offers a fruitful basis for producing theoretical innovations as well as practical governance innovations (Van Buuren & Loorbach, 2009); when we view policy as an embedded element within a broader context of societal governance processes. We argue that this broader complexity approach could offer new insights and a source of inspiration towards redefining the role of policy and governance in light of broader long-term societal challenges.

As in many areas (e.g., energy, food, mobility, water management, social cohesion, health-care) more fundamental changes are looming (Grin *et al.*, 2010), this also questions the fundamental characteristics and basis of the policy-making regimes that have evolved over the last decades. The challenges related to so-called persistent problems are inherently long-term, highly uncertain and requiring systemic changes in institutional structures and engrained paradigms. The nature of the policy-making process has over time become dominated by short-term cycles, a focus on minimizing risks and uncertainties and on 'manageability' issues. The challenge of how developed societies can escape an unsustainable 'lock-in' and tip towards sustainability (Westely *et al.*, 2011), is therefore also a question of how the inherent complexities, ambiguities and uncertainties related to this can be effectively addressed by new forms of governance.

In other words: how to facilitate societal dynamics that in the end might threaten the legitimacy, control and position of the policy-making process and institutions themselves? Such societal dynamics are evinced by for example the emergence of social movements (Nichols, 2006), new environmental movements, and the self-organizing initiatives of communities and citizens to directly address problems and cluster around interests mediated by the communication technologies and networks (Johnston, 2010). In this continuous changing world, a question of how the inherent complexities, ambiguities and uncertainties related to this can be effectively addressed by new forms of governance requires investigation. We argue that complexity theory can comprise a ground for governance practices and new governance paradigms.

In this paper we first give an overview of how complexity theory is now used in public administration and policy research, before arguing for a broader conceptual scope that could offer starting points for forming complexity-based governance.

ON COMPLEXITY

Societies have evolved into complex constructs with interrelated and interdependent subsystems and their respective functions. The complexity of modern societal systems is not a new property discovered when a complex systems' perspective is adopted, but a characteristic inherent to them. Because of the highly non-linear behavior, large uncertainties, diffuse causal relationships and numerous feed-back mechanisms, the predictability and thus manageability of large scale societal change is by definition limited. The development however of new approaches such as complexity theory and complex adaptive systems with their new analytical lenses and concepts help us understand that societal systems are too complex to be analyzed and/or managed in a straightforward manner.

Complexity Theory

Complexity theory, otherwise known as complex systems theory, has its roots in the general systems theory (Von Bertalanffy, 1968) that originated in the 1930s. Systems theory is an interdisciplinary field of science that studies the nature of complex systems in society, nature, science and technology (Midgley, 2003). It provides a framework by which a group of interrelated components that influence each other can be analyzed. That can be a sector, branch, city, organism, or even a society. Systems theory evolved over the last century. The first generation, roughly from the beginning till the sixties of the last century, focused on general systems theory and was somewhat deterministic, arising from cybernetics and control engineering. In the 1970s and 1980s integral systems theory became an important field, focusing on the integration of social, economic and ecological processes (Nelson & Winter, 1982; Varela, Maturana & Uribe, 1974).

During the 1970s and 1980s soft systems theory emerged, taking a qualitative approach rather than a quantitative approach, mostly applied to companies and organizations (Senge, 1990). In the 1990s complex systems theory was introduced, focusing on the co-evolutionary development of systems. The establishment of the Santa Fé institute in New Mexico in the US in the early 1980s functioned as incubator for a new research movement, which laid the basis for complex systems theory (Holland, 1995; Kauffman, 1995) and a new approach from within: Complex Adaptive Systems' Approach.

Complex Adaptive Systems

Complex adaptive systems have four distinct properties according to Holland (1995): (a) aggregation, (b) non-linearity, (c) diversity, and (d) flows. Aggregation refers to the multi-level and multi-scalar character of systems that manifests their interconnectedness. Non-linearity refers to the behavior of complex adaptive systems and specifically refers to the fact that "nonlinear interactions almost always make the behavior of the aggregate more complicated than would be predicted by summing or averaging" (Holland, 1995: 23). Non-linearity refers to the fact that a relatively small stimulus can bring in return large changes and impacts to the system. At the same time, non-linearity manifests the complexity of processes and structures. Diversity refers to the differentiated responses and constellations of complex adaptive systems. Diversity of complex adaptive systems "is the product of progressive adaptations. Each new adaptation opens the possibility for further interactions and new niches." (Holland, 1995: 29). Flows of energy and resources are present in complex adaptive systems and are characterized by two distinctive properties: the multiplier effect and the recycle effect. The multiplier effect refers to the impact and complexity networks within the complex adaptive systems and "it typically jeopardizes long-range predictions based on simple trend" (Holland, 1995: 25). The recycling effect refers to cycles or feedback loops that take place in networks within the complex adaptive system. According to Holland (1995: 26) "recycling can increase output (...), but the overall effect in a network with many cycles can be striking."

Complex adaptive systems are a special type of complex systems that can adapt to changes and learn from their experiences (see also the review of complex adaptive systems by Rotmans & Loorbach, 2009). Complex adaptive systems co-evolve, self-organize and exert emergent behaviors. Co-evolution refers to the mutual irreversible changes between the system and its environment or between two systems or subsystems. Self-organization refers to the imminent capability of complex adaptive systems to organize itself without external involvement. Self-organization is inherent and distinctive to complex adaptive systems (Prigogine & Stengers, 1984). Emergence describes the presence of processes at a high level of abstraction that are not related linearly to lower level decisions or structures. In agreement with this view, Hedstrom (2005: 74) explains social emergence in a coherent way: "From an epistemological point of view, social emergence refers to social properties that cannot, in practice, be predicted by knowing everything there is to know about the pre-emergent properties of the parts." Complex adaptive systems are also anticipatory, meaning that they have a memory that refers to past lessons and experiences. A decision and experience of the past is taken into account by complex adaptive systems not due to their adaptive capacity but due to their system memory that assigns them with anticipatory responses.

Although the theory of complex adaptive systems is far from mature, it has attracted a great deal of attention and has many applications in diverse research fields: in biology (Kauffman, 1995), economics (Arthur, Durlauf & Lane, 1997), public administration (Teisman & Klijn, 2008), and policy analysis (Morçöl, 2010). In the relatively young field of 'transition research' (Grin *et al.*, 2010), transition management is developed as a prescriptive, complexity-based form of governance. It takes a starting point not to adopt a complexity perspective within an established discipline, but explores in practice as well as theory what type of (meta-) governance could be developed that guides and accelerates long-term non-linear change or transitions towards sustainability.

COMPLEXITY AS AN ANALYTIC BASIS

In this section we further discuss two different ways in which a complexity perspective might inform policy- and governance studies. One in which the focus in on a complex governance system and the other in which the focus is on complex societal systems of which a governance system is a subset. We argue that these two perspectives are not contradictory but complementary, in that they should and could inform one-another. In practice however, these two perspectives seem hard to merge and efforts to this end are so far without much result. The danger in pursuing either one of the two perspectives lies in a suboptimal framing and in turn, missing out on important insights. For example, a focus on the governance system solely might obscure alternative routes and meta-governance mechanisms that are influencing societal change in different ways. Alternatively, a focus on societal systems alone might lead to a too shallow understanding of the nature and impact of governance and policy processes, and lead to unrealistic ideas on how to shape society. In this section we will present the insights and contributions that the two different approaches bring forward.

Analyzing governance systems using complexity theory to inform policy and governance practices	**Analyzing societal systems using complexity theory to inform policy and governance practices**
• Subject of research is the governance system and its relations to actors and to infrasystems • The societal system and its dynamics are perceived /conceptualized as context.	• Subject of research is the societal system and its processes, complexity theory is used as either an analytical framework or an ontological framework for theory building, • The focus is on societal system and the relation of societal dynamics to actors' practices and infrasystems. • The governance system in this conceptualization is considered as an entity that has to respond and adapt to the societal complexity, and is seen analytically as part of the context.
in an image: 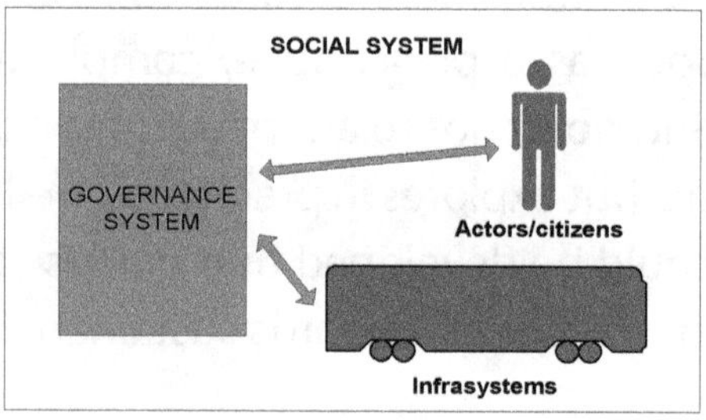 With the light grey highlighting the focus of the analytics when the subject of research is the governance system	in an image: 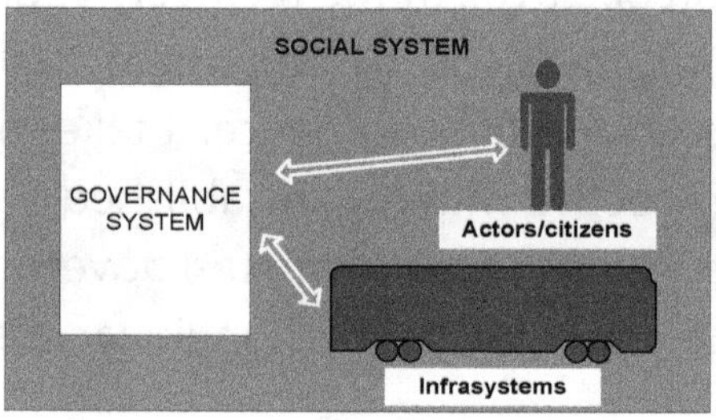 With the light grey highlighting the focus of the analytics when the subject of research is the societal system

Table 1 *An Overview of the Conceptualizations of the Two Approaches that Use Complexity Theory as an Analytical Framework*

	SELECTED JOURNAL PUBLICATIONS & BOOKS	REFERENCES FROM ASPA - COMPACT 2011 Conference	Comments related to complexity
Complexity as an analytic framework			
Analyzing governance processes using complexity	**Subject of research: Governance processes & systems** • On dynamics: Klijn, 2008; Van Buuren & Gerrits, 2008 • On co-evolution: Klijn, 2008; Van Buuren & Gerrits, 2008 • On emergence and non-linearity: Rhodes, 2008; White, 2001 • On self-organization: Klijn, 2008; Butler & Allen, 2008; White, 2001 • On fitness landscape: Rhodes, 2008 • Public policies as complex systems: Morcol, 2010	**Subject of research: Governance processes & systems** • On dynamics: Koppenjan & Klijn, 2011; Koliba & Zia, 2011; Morcol, 2011 • On co-evolution: Eppel, 2011 • On emergence and non-linearity:: Eppel, 2011; Koppenjan & Klijn, 2011; Rhodes, 2011; Morcol, 2011 • On self-organization: Eppel, 2011; Morcol, 2011 • On fitness landscape: Koppenjan & Klijn, 2011; Rhodes, 2011 • On shocks and cascades: Koliba & Zia, 2011	Klijn (2008): • self-organization limits governance • complexity theory to analyze wicked problems Van Buuren & Gerrits, (2008): • Policy processes are conceptualized as complex adaptive processes that co-evolve • Policy decisions are seen as temporary equilibria in a complex evolving policy process Gerrits & Meek, (2011) (Compact Conference): • Complexity theory for public administration has not reached the level of a truly grounded theory. (p. 17) • Complexity theory to be employed so as "to reach a deeper understanding of the contextual dynamics at work in public administration practice" Koppenjan & Klijn, (2011) (Compact Conference) • Complexity theory spots the fluidity of policy making (p. 20) • Complexity theory shows that the manageability of complexity is far less than the approaches assume (p. 20) • Complexity theory focuses on explaining the process and not explaining the outputs of complex processes (p. 22)

Table 2 *Complexity Theory and its Concepts for Explaining Governance Processes and Societal Systems. Recent publications from public administration scholars and societal transition scholars.*

In public administration literature, Van Buuren and Gerrits (2008) analyzed policy processes with the lens of complexity theory to better describe and understand them. In their analysis, policy decisions are seen at temporary equilibria in the continuum of policy processes (p. 397). In their analysis they concluded that: by analyzing policy dynamics with complexity theory, governance efforts have a limited impact on complex policy processes and policy decision equilibrium states. More specifically they state: "Knowing the context and content of a specific policy process in detail is helpful but not enough. Influencing public policy remains anything but a predictable job." (p. 397)

In their analysis they avoided giving any recommendations or deducing lessons for governance based on complexity-related findings. In the same vein, Teisman and Klijn (2008: 289) employed complexity theory as a tool to better understand policy and governance processes. They argued that by understanding processes with the lens of complexity, effectiveness of policy may be improved. By analyzing governance processes with complexity as an analytical framework, it is revealed that the policies employed are just a part of a complex loci hence their impact is limited (Teisman & Klijn, 2008: 291).

In his recent paper, Klijn, (2008) reviewed how existing theories of public administration have dealt with issues similar to those brought forward by complexity theory and he reflected upon the concepts of dynamics, self-organization, and emergence. Klijn (2008) focuses on the complexity of governance processes and interrelations, and he does not refer to complexity of the system. In his effort to show that there are no new insights from complexity theory since public administration scholars have been proactive in dealing with complexity, he brought forward some critical points about the limitations of public administration:

> *From a complexity theory perspective, it can be argued that* **self-organizing** *systems are hard to govern and/or influence.* (Klijn, 2008: 308)
>
> *The notion of* **co-evolution** *challenges researchers to search for more complex relations in decision making. More specifically, researchers should expand the existing notions of public administration, such as connected arenas or separate streams, even further and look for other patterns of relations between decisions and developments.* (Klijn, 2008: 311)

From policy studies, Eppel (2011: 10) notes that by accepting a complexity theory view on policy processes "implies recognizing that it is impossible to control their evolution of policy in conventional ways." Morçöl (2010) (re) conceptualizes public policy as a complex system and recognizes that public policy systems can manifest self-organization and emergence. For explaining these systemic properties and behaviors of the public policy, Morçöl (2010: 55-56) focuses on the policy agents and their interactions and practices as manifestations for self-organization and emergence of public policy. Even though the actor-centered approach in understanding complexity in public policy is not new (Scharpf, 1997), Morçöl's (2010) conceptualization of public policy as a system rather than as a standing-alone means to a decided goal is refreshing and opens up a dialogue for understanding policy as a dynamic element with inherent complexity rather than as a static reducible means. The conceptualization of public policy as a complex system may however imply that society-based governance processes are either externalized to the public policy system or conceptualized to be located at a meta-level hence being un-influenced by the public policy system and its outputs. In the same contribution, Morçöl (2010: 59) argues that the concept of emergence for public policy is not giving fruitful insights on workings and dynamics of governance and policy; despite its appeal in describing aggregate behavior and patterns.

From management science, White (2001: 247) notes that complexity offers a new perspective for understanding governance processes and that "(...) emerging governance structures appear to exhibit properties of complex systems".

Critique and Reflection

The cited literatures show that complex system thinking is predominantly used in Public Administration research to illustrate the limitations of policy-making (processes). While it is identified as an interesting and perhaps even necessary perspective for Public Administration research, scholars seem to struggle with translating complexity insights into meaningful, practical and relevant suggestions on how policy could deal with social complexity. The dominant line of thinking seems to be that social complexity either leads to uncontrollability and 'unknowability' of social issues or that it leads to overly complex analyses that do not provide any basis for choice and/or action (Teisman & Klijn, 2008).

At the same time, it seems that only little effort is put in exploring and understanding the nature of complexity of governance systems, as also pointed by Morçöl and Wachhaus (2009: 52) in: "social science concepts that were developed for studying strategic games and institutions are quite advanced and elaborate, but the nature of complexity, which is the source of substantive uncertainty, it is not understood well by social scientists." Morçöl and Wachhaus (2009: 53) suggest that interfaces and interactions between politics and science, natural and social systems can be better understood with concepts and lenses of complexity theory rather than network theory as a governance approach. Their conclusions remain on how complexity theory can comprise a new analytic for deepening the limitedly understood issues in public policy such as politics, expert information impact or the nexus of social-ecological systems; albeit recognizing the usefulness of complexity theory for public policy field.

Paradoxical Persistence!

From the above presentation of the employment of complexity theory as an analytical lens to explain governance processes, we observe that public administration scholars analyzed governance processes by adopting a complex theory perspective to conclude that policy and governance interventions are limited and/or of limited impact due to complexity. The paradox is that even after revealing the limited nature of interventions in such complexity, the majority of their conclusions summarize that interventions can still make a difference (albeit limited) without altering the underlying assumptions or foundations of the employed governance approach.

The use of complex systems thinking as a solely descriptive framework results in limited insights in non-linear behavior and tipping points of such complex systems. The complexity of societal systems is revealed when dose-response patterns proved insufficient to explain their behavior. At the same time, complex adaptive systems' approach and innovation studies (with the focus on technology diffusion and development) provided new insights and conceptualizations for the inherent complexity of societal systems and the problems' experienced by them. The reason is twofold: a different kind of dynamics is present in complex adaptive systems and those systems experience a different type of problems that persist intervention given that the problems as behaviors of the system are rooted in the deep structures of the system itself. Thus complex adaptive societal systems ask for a governance paradigm that starts from complexity itself. In other words: a governance perspective that perceives policy as one of the components in interaction with other elements and in co-evolution with other elements and context dynamics.

Analyzing Societal Complexity

In research on transition dynamics (Frantzeskaki & De Haan, 2009; de Haan & Rotmans, 2011), conceptualizations on societal complexity and societal systems have been presented using complexity theory as an analytic framework. The transition perspective starts from analyzing societal dynamics, using different conceptual frames, and asks how actors interact with and relate to these. The starting point for thinking about governance of transitions is thus a complex persistent problem, of which historically developed policy structures and practices are often a central element (Grin *et al.*, 2010). Related to persistent societal problems, such as major sustainability transitions in areas like energy, food, health-care, water management and so on, the central question for governance is how to understand and if needed change the role of existing policies that have created and become part of these problems.

Taking a transition's approach, societal complexity encompasses the complexity of the societal system and its manifestations that include complex problems and complex processes (both systemic processes and processes of change). Modern societal systems are complex, given that they involve multiple actors with diverse interests and resources and a great number of interdependencies between them. The complexity of societal systems does not only rest on the presence of multiple actors but also results from the function of institutions and from the interaction between institutions and actors (Frantzeskaki, 2011). Complex societal systems can remain in a dynamic equilibrium for a very long time but ultimately always enter a period of non-linear and highly unpredictable behavior (Rotmans & Loorbach, 2009).

In the context of sustainability challenges, the societal complexity is further driven by the inherent uncertainties and ambiguity of these processes. The uncertainty arises from the unpredictability of future developments in technology, society, economy or ecology. The future in general is unknown, and in the context of long-term systemic changes planning and predicting change is almost impossible. The ambiguity relates to how actors interpret these uncertainties and the conflict between different sets of values and expectations. On the short-term, this produces a continuous process of competition, conflict and co-evolution between discourses, interests, ambitions, values and solutions. Sustainable development from this perspective is the continuous process of variation and selection through which social structures are reinforced and reformed. Sustainability transitions are thus a conceptualization of long-term societal change as non-linear systemic dynamics produced by broad societal interaction between all actors that seek to influence, change and preserve societal structures.

The transitions perspective builds on historical cases as well as on insights from a variety of disciplines that draw on complex system theories to explore the possibilities of such shifts in the relatively near future. As much as such possibilities are hypothetical, there are many indicators suggesting such tipping points are coming closer. Certainly many broader trends that drove the post-war modernization period are changing fundamentally: rapid demographic and economic growth, abundant availability of resources and societal consensus on what progress meant. This perspective implies at least exploring the possibilities of the breaking down of existing regimes, if only for the possibilities this would offer to accelerate towards more benign and sustainable ones.

Persistent Problems

Societal complexity is expressed by a new type of wicked problems; *persistent problems*. The definition of persistent problems is introduced by Rotmans, (2005); Dirven, Rotmans & Verkaik (2002) and Loorbach (2007: 14). Persistent problems are complex, have their origins in the deep structure of the system and are difficult to manage since they involve multiple actors which operate in a multitude of sectors and governance levels and have divergent interests (Dirven, Rotmans & Verkaik, 2002). Persistent problems tend to reappear when marginal actions that focus on resolving their symptoms are only undertaken. From a transitions perspective, persistent problems are seen as symptoms of the un-sustainability of a societal system (Loorbach, 2007; Loorbach *et al.*, 2009). They relate predominantly to the inability of actors, specifically government, to adequately frame or understand, and deal with these problems at the root causes as they primarily focus on the symptoms. The recognition of the existence of new types of problems is also found in writings of multi-actor policy analysis and public administration. From policy analysis research, Brewer and deLeon (1983) address that complexity of social systems produces new types of problems that add to existing complexity.

"Because of their complexity, social systems are capable of producing problems neither expected nor results intended. Participants may perceive these surprises as occurring outside of their spheres of interest or responsibility; with increased complexity, beneficial and harmful externalities (...) seem to happen more often." (Brewer & deLeon, 1983: 93)

From public administration writings, we conclude that complexity theory can inform policy practitioners about the characteristics of wicked problems and about persistent problems. It is also seen as a suitable conceptual framework for wicked problems by Klijn, (2008: 314). Koliba and Zia (2011: 2) also recognize that "wicked problems persist because of our failure to understand their complexity." In this way, persistent problems are not only unmanageable due to their inherent complexity but also due to the limited understanding we have about the nature of their complexity.

Transitions as a Form of Complex Societal Processes

Transitions are driven by developments at different levels of scale (a broader societal landscape, incumbent regimes and innovation in niches) that produce different type of systemic behavior over time. Typically, transitions take a long time to develop speed and direction (called pre-development) before passing a tipping point after which a relatively rapid, chaotic and largely uncontrollable acceleration takes place in which a system reconfigures. In a pre-development, alternatives to the dominant regime are explored typically by pioneers, frontrunners and experts, while mainstream public, media and politics ignores, denies or even resists acknowledging the fundamental problem. Over time however, the pressure on an existing regime can increase so much that it in the end inevitable will open up and reconfigure. As much as these processes are unpredictable in terms of outcome and timing, historical studies (Geels, 2006) increasingly suggest this pattern of punctuated equilibria to be generic as well as inevitable in social systems.

The more fundamental message here is that over the last decades, society has increasingly acquired means in terms of skills, knowledge, communication tools, funds and perhaps even confidence to self-organise societal change. This increasing influence of (civil) society on social change has been incrementally taken up by policy through involving social actors in policy-making processes. The whole movement towards participation and network governance can be interpreted as an adaptive response of governance systems to the increasing power of civil society and business. This incremental improvement however seems to be too little and too slow to accommodate and adequately deal with the accelerating processes of social innovation in the mentioned areas. From this perspective the challenge of governance in sustainability transitions is thus not how to better involve societal dynamics in the policy-making process, but rather how the policy-making process can be better aligned with social innovation dynamics. In other words: we quest the shift of how to move from public participation in policy to policy participation in society.

Critique and Reflection

What transition studies reveal is that (a) governance processes are only a small piece in the puzzle of continuous co-evolving societal dynamics and (b) policy changes are seen as requirements for allowing institutions to be reflexive towards the continuous social and technological changes and advancements.

Transitions studies focus on understanding the dynamics that underlie both problems and processes as products of entangled systems' evolution and co-evolution and in this way exposing the breadth of complexity present. By doing this, untraceable and unmanageable complexity is revealed often making governance propositions appear 'too little' to deal with complexity. A response to the need for governance in view of this complexity comes with promises that societal and technological innovations that emerge in niches have the potential to create broader innovations in governance when space is created for them to get empowered. Both transition management (Loorbach, 2010; Rotmans & Loorbach, 2009) and strategic niche management (Schot *et al.*, 1994) support this view.

Even though this belief in innovation's capacity relies on research on historical transitions, the proposition that social and technological innovations will aspire or get empowered to aspire broader and deeper governance innovations is risky. At the same time, the in-depth analytics of societal complexity (de Haan & Rotmans, 2011; Frantzeskaki & de Haan, 2009; Holtz *et al.*, 2008) inform policy studies and public administration only to a limited extend. Policy makers and policy advisors are often left alone to translate governance propositions (Geels, 2010; Voss *et al.*, 2009) to operational policies that can respond to societal complexities. This is because the only way to translate the generic complexity-based governance principles such as the ones defined by transition management literature can be operationalized is in an experimental learning-by-doing manner in a specific context.

Despite the fruitful efforts towards understanding and unraveling governance complexity and societal complexity, there is a need to better integrate those insights and to create a new basis for complexity-based governance rather than complexity-informed governance.

TAKING COMPLEXITY SERIOUSLY

The recent attention of public administration scholars fits nicely in a longer historical trend in which policy and public administration scholars are increasingly focusing on the limits to policy and governance and how to remedy these. Set against a background of processes of liberalization, democratization and informatisation, society has developed a growing capacity for self-governance and can influence policy itself (Johnston, 2010). Especially since the 1990s, participation, interactive and network governance have been developed as strategies to develop support for policy through engaging stakeholders. One could say that the recent attention to the complexity perspective from public administration scholars is in part because of the obvious theoretical and practical limitations to these approaches: they tend to complicate the policy process itself and by no means provide insurance that no problems will arise in the implementation phase, as many empirical studies illustrate.

While policy-processes certainly play an important role in influencing long-term change, they are only one of the factors and themselves often influenced by other mechanisms. Public opinion, scientific debate, large scale societal trends, incidents and crises, and business strategies are examples of such influencing mechanisms that shape societal change at least as much as policy does. It is obvious that policy processes play a crucial role in codifying societal consensus in terms of long-term goals and legislation, creating space for innovation and deciding upon short-term conflicts. If we only seek to optimize the efficiency of policy-making by limiting the impact of a complex societal context, this entails a risk of decreasing the effectiveness of policy-making related to long-term change. Complexity studies suggest that attempts to control complex systems in a top-down manner might have counterproductive effects. We therefore suggest applying a complex adaptive systems perspective to the broader context of societal change within which we identify policy-making processes as one of the sub-systems.

From a transitions perspective, we provocatively frame the attempts to develop more efficient and effective policy and governance strategies as part of optimizing a subsystem (regime) in society and that this ultimately is a fruitless and perhaps even counterproductive strategy. By considering that other actors in society have acquired means for shaping societal change and influencing policy, and that society is facing more fundamental changes and increasing uncertainties, a radical reframing of the role of policy herein is made implicit.

We therefore argue that the question should not be how to increase the effectiveness of policy (processes), but what a new role of policy in a broader (meta-) governance context could be and what kind of instruments, competencies and roles relate to it. Taking complexity as an inherent characteristic of societal transformations implies a redefinition of the role of policy. This should be one of the core tasks of future research for public administration and policy studies.

In the following paragraphs, we draw propositions as suggestions for future framing of a new paradigm in public administration that is complexity-based rather than complexity-phobic. Using philosophical terms, we suggest creating a basis from complexity theory as an ontology rather than as an epistemology only.

Proposition 1: Co-production and self-organization define governance dynamics by considering interdependencies and diversity of interests and solution-spaces of wicked problems.

This proposition is triggered by the need to consider the role of society and state as two equally important and critical conditions for taking on board both societal and governance complexity. In a recent publication, Robichau (2011: 117) brought forward the dichotomy in governance theories of being either state-centric or society-centric. At the same time, there are neutral approaches that account for a relational dependency between state and society, meaning that the state creates operational relations with non-state actors for achieving governance (Bell & Hindmoor, 2009).

Governments have to seize the self-organization capacity and initiatives that citizens and communities have as evinced by NGOs and community movements (see Johnston & Hansen, 2011). This is best captured by Johnston

(2010) addressing that governance capacity can be enriched and supported by the self-organization of citizens as well as by "the thoughtful application of information technology to increase the role of the public in governing" (p. 123).

We propose a social reflexive mode of governance, where social agents are not included only for issues of legitimacy and democracy but are responsible and accountable to contribute and to uptake actions and activities. This mode of governance explicitly implies that social agents are active citizens that can uphold responsibilities and are accountable (Roberts, 2004). This implies that new methods for selecting 'super agents' (Rhodes, 2008) or for enabling agents of change or empowering frontrunners (Loorbach, 2010) are needed. The social agents that will be part of the governance 'alliance' are not necessarily or only corporate agents but social agents with accountability and societal responsiveness.

Proposition 2: Government to function as a space for stimuli for societal change and as a space of response to societal stimuli.

We argue that the shift from government to governance (experienced by modern societies) did not eliminate or diminish the role of government as a regulator and provider of services. Viewing societal dynamics as both stimuli and responses depending on the context and problem, implies a dual role of government, one which simultaneously is co-producing innovations as systematically reflecting upon its own role and seeking to remove or break-down undesirable structures and practices. While public administration certainly provides insights on how a complexity perspective helps us to understand limits to governance, it does not fundamentally reflect upon the logical conclusions, namely that the explicit and implicit norms and rules upon which policy and governance regimes are based perhaps are also playing a role in limiting policy and governance effectiveness. In other words, we argue for an approach to governance that also discusses its foundations vis-à-vis a (perceived) societal complexity of which it is part.

Proposition 3: Adaptation and anticipation mechanisms to be in place for accommodating societal and governance complexity and their dynamics.

While recent focus on adaptive capacity provides insights for policy and governance of complex systems (Edelenbos *et al.*, 2009; Gupta *et al.*, 2010) a governance mode based on complexity need to acquire both adaptive and anticipatory reflexes. Public administration and policy studies knowledge are valuable for adaptive reflexes. Anticipatory reflexes require policy actors to build upon knowledge on societal complexity and societal transitions and to recognize patterns and system's responses to stimuli.

We suggest exploring possibilities for anticipatory strategies such as anticipating the opening-up of regimes and tipping-points as windows of opportunity for systemic innovation. While such 'opening-up' is frequently explored in the context of policy windows, it is not in societal terms. Public administration and public management scholars can view these propositions as open-calls for finding new public management practices that are suitable for governance for complexity. We recognize that public administration and public management fields have valuable insights on which existing practices can be altered to aid the above suggested tenets towards a new governance practice.

DISCUSSION AND CONCLUSION

In this paper we discuss the role of public administration and the related research in light of major societal challenges. We argue that in order to develop governance strategies that can effectively deal with the major societal challenges, we need to start from their characteristics. Approaching these challenges through a complex system perspective offers in our view, a basis for such an exploration whereas it highlights the limitations of the way complexity is used in public administration research.

Complexity of both problems and societal processes challenges public administration practices and paradigms. 'Embracing complexity' is at odds with the dominant logic in public administration which is based on limiting risk and uncertainty; supporting a tendency for planning and organized in relative short time-cycles. In the literature, societal complexity is therefore often seen as a complicating factor that limits policy effectiveness and thus, needs to be anticipated by lowering ambitions, and maintaining adaptive capacity.

We tried to identify the contours of a governance approach that encounters the possibilities that a complex societal systems perspective offers. This perspective, emphasizing the possibilities and underlying dynamics of non-linear systemic change, suggests that surprisingly rapid and profound transitions are possible and could provide a way to more rapidly escape current 'unsustainable' lock-ins and move to more benign social systems. As many scholars see increasing signs of destabilization in societal systems, the quest for new governance strategies and methods to deal with long-term non-linear change becomes more urgent.

REFERENCES

Arthur, W. B., Durlauf, S. N. and Lane, D.A. (1997). The economy as an evolving complex system, Adddison-Weasly, reading MA.

Bell, S., and Hindmoor, A., (2009). Rethinking Governance: The centrality of the State in Modern society, Cambridge University Press.

Boonstra, B., (2011). Unleashing self-organising capacity in urban development? The case of business improvement districts in the UK, Paper in: COMPACT - Challenges of Making Public Administration and Complexity Theory Work, Rotterdam, The Netherlands, June 23-25.

Butler, M.J.R., and Allen, P.M., (2008). Understanding policy implementation processes as self-organizing systems, Public Management Review, 10(3): 421-440.

Edelenbos, J., Klok, P.J., and van Tatenhove, J., (2009). The institutional embedding of interactive policy making, Insights from a comparative reserach based on eight interactive projects in the Netherlands, The American Review of Public Administration, 39(2): 125-148.

Eppel, E., (2011). The application of a complexity analytical lens to understanding and explaining public policy processes, Paper in: COMPACT - Challenges of Making Public Administration and Complexity Theory Work , Rotterdam, The Netherlands, June 23-25.

Frantzeskaki, N., (2011). Dynamics of societal transitions: Driving forces and feedback loops, PhD thesis, Delft University of Technology, The Netherlands, Wohram Publishing Services.

Frantzeskaki, N. and de Haan, H. (2009). "Transitions: Two steps from theory to policy", Futures, 41(9): 593-606.

Geels, F. (2006). "The hygienic transition from cesspools to sewer systems (1840-1930): the dynamics of regime transformation", Research Policy, 35(7): 1069-1082.

Geels, F.W. (2010). Ontologies, socio-technical transitions (to sustainability), and the multi-level perspective, Research Policy, 39: 495-510.

Grin, J., Rotmans, J., and Schot, J. (2010). Transitions to sustainable development: new directions in the study of long term transformative change, Routledge: London.

Gupta, J., Termeer, C., Klostermann, J., Meijerink, S., can den Brink, M., Jong, P., Nooteboom, S., and Bergsam, E. (2010). The Adaptive Capacity Wheel: a method to assess the inherent characteristics of institutions to enable the adaptive capacity of society, Environmental Science and Policy, 13: 459-471.

Haan, H.J. de, and Rotmans, J. (2011). Patterns in transitions: Understanding the complex chains of change, Technological Forecasting and Social Change, 78: 90-102.

Holland, J.H. (1995). Hidden Order: How Adaptation Builds Complexity, Helix books / Perseus books, Cambridge, Massachusetts.

Holtz, G., Brugnach, M., and Pahl-Wostl, C. (2008). Specifying "regime" – A framework for defining and describing regimes in transition research, Technological Forecasting and Social Change, 75: 623-643.

Johnston, E. (2010). Governance Infrastructures in 2020,Public Administration Review, 70(1): 122-128.

Johnston, E.W., Desouza, K., and Hu, Q. (2011). How the application of complexity and policy informatives to public administration can change the questions we ask and the solutions we discover, Paper in: COMPACT - Challenges of Making Public Administration and Complexity Theory Work , Rotterdam, The Netherlands, June 23-25.

Johnston, E. and Hansen, D. (2011). Design Lessons for Smart Governance Infrastructures. Chapter in American Governance 3.0: Rebooting the Public Square? An edited book by the National Academy of Public Administration. Ink, D., Balutis, A., & Buss, T (editors).

Kauffman, S. (1995). At home in the universe: the search for laws of complexity, Oxford University Press, Oxford.

Klijn, E.H. (2008). Complexity theory and public administration: What's new?, Key concepts in complexity theory compared to their counterparts in public administration research, Public Management Review, 10(3): 299-317.

Koliba, C., and Zia, A. (2011). Theory testing using complex systems modeling in public administration and policy studies: Challenges and opportunities for a meta-theoretical research program, Paper in: COMPACT - Challenges of Making Public Administration and Complexity Theory Work , Rotterdam, The Netherlands, June 23-25.

Koppenjan, J., and Klijn, E.H. (2011). Complexity, which complexity? Three perspectives on complexity applied to decision-making on revitalizing the Center of Rotterdam South, Paper in: COMPACT - Challenges of Making Public Administration and Complexity Theory Work , Rotterdam, The Netherlands, June 23-25.

Loorbach, D. and Rotmans, J. (2010). The practice of transition management; examples and lessons from four distinct cases, Futures, 42: 237-246

Loorbach, D. (2010). Transition Management for Sustainable Development: A Prescriptive, Complexity-Based Governance Framework, Governance: An International Journal of Policy, Administration, and Institutions, 23(1): 161-183.

Midgley, G. (2003). "System Thinking: An Introduction and Overview" in Systems Thinking, ed. G. Midgley, Sage, Thousand Oaks, pp. xvii.

Morçöl, G. (2010). Issues in reconceptualizing public policy from the perspective of complexity theory, Emergence: Complexity and Organization, 12(1): 52-60.

Morçöl, G. (2011). Why complexity theory? What kind of theory is it?, Paper in: COMPACT - Challenges of Making Public Administration and Complexity Theory Work , Rotterdam, The Netherlands, June 23-25.

Nelson, R.R. & Winter, S.G. (1982). An evolutionary theory of economic change. Belknap Press of Harvard University Press, Cambridge, Massachusetts.

Nichols, A. (ed.) (2006). Social entrepreneurship, New models of sustainable social change, Oxford University Press.

Rhodes, M.L. (2008). Complexity and emergence in public management, The case of urban regeneration in Ireland, Public Management Review, 10(3): 361-379.

Rhodes, M.L. (2011). Complexity theory and Public Management: A report from the edge, Paper in: COMPACT - Challenges of Making Public Administration and Complexity Theory Work , Rotterdam, The Netherlands, June 23-25.

Roberts, N. (2004). Public deliberation in an age of direct citizen participation, American Review of Public Administration, 34(4): 315-353.

Robichau, R.W. (2011). The mosaic of governance: Creating a picture with definitions, theories and debates, The Policy Studies Journal, 39(1): 113-131.

Rotmans, J., and Loorbach, D. (2009). Complexity and transition management, Journal of Industrial Ecology, 13(2): 184-196.

Senge, P.M. (1990). The Fifth Discipline: The Art & Practice of The Learning Organization, Random House, London.

Scharpf, F.W. (1997). Games real actors play, Actor-centered institutionalism in policy research, Westview Press.

Schot, J., Hoogma, R., and Elzen, B. (1994). Strategies for shifting technological systems, The case of the automobile system, Futures, 24(10): 1060-1076.

Teisman, G.R., and Klijn, E.H. (2008). Complexity theory and public management, Public Management Review, 10(3): 287-297.

Van Buuren, A., and Gerrits, L. (2008). Decisions as dynamic equilibriums in erratic policy processes, Positive and negative feedback as drivers of non-linear policy dynamics, Public Management Review, 10(3): 381-399.

Van Buuren, A. & Loorbach, D. (2009). "Policy innovation in isolation. Conditions for policy-renewal by transition arenas and pilot projects.", Public Management Review, 11(3): 375-392.

Varela, F., Maturana, H. & Uribe, R. (1974). "Autopoiesis: The organization of living systems, its characterizationand a model", Biosystems, 5: 187-196.

Verbong, G. & Loorbach, D. (eds.) (2012). Governing the Energy Transition; Reality, Illusion or Necessity?, Routlegde, New York.

Von Bertalanffy, L. (1968). General system theory: foundation, development and applications, New York.

Westley, F., Olsson, P., Folke, C., Homer-Dixon, T., Vredenburg, H., Loorbach, D., Thompson, J., Nilsson, M., Lambin, E., Sendzimir, J., Banerjee, B., Galaz, V., and van der Leeuw, S. (2011). Tipping toward sustainability: Emerging pathways of transformation, Ambio, 40: 762-780.

White, L. (2001). Effective governance through complexity thinking and management science, Systems research and behavioral science, 18: 241-257.

Voss, J.P., Smith, A., and Grin, J. (2009). Designing long-term policy: rethinking transition management, Policy sciences, 42: 275-302.

Dr. **Derk Loorbach** is Associate Professor and Director of the Dutch Research Institute for Transitions (DRIFT), Erasmus University Rotterdam, The Netherlands, where he received his PhD in June 2007. The central theme in his research is the development of the transition management approach as a new governance-model based on complex systems thinking, governance theories and sociology. He is currently involved in various transition arenas, innovation programmes and envisioning practices in areas of urban development, energy, healthcare and housing. His research is seen as an example of 'sustainability science', combining fundamental with action research to contribute to sustainable development in practice.

Dr. **Niki Frantzeskaki** received a Bachelor and Master of Environmental Engineering with the Technical University of Crete, Greece, in 2003 and a Master of Science in Engineering and Policy Analysis from Delft University of Technology, The Netherlands, in 2005. She received her doctorate with the Delft University of Technology in 2011 entitled "Dynamics of Transitions: Driving Forces and Feedback Loops". Since 2010 Niki works at the Dutch Research Institute For Transitions (DRIFT) at Erasmus University Rotterdam, working on governance of sustainabillity transitions. From November 2011 until June 2012, Niki was a Visiting Scholar with the Monash Water for Liveability at Monash University, Melbourne, Australia.

Chapter 16

PROPOSITIONS FOR COMPLEXITY THEORY IN APPLICATION TO THE STUDY AND PRACTICE OF PUBLIC ADMINISTRATION

Lasse Gerrits (Erasmus University Rotterdam, NLD)
Jack W. Meek (University of La Verne, USA)

Complexity theory has potential to inform the field of public administration in both theory and practice. For almost two decades, researchers have constructed research efforts that draw upon concepts derived from complexity theory to highlight characteristics and patterns in public affairs and administration. Critics of this intellectual project note the lack of analytic clarity in conceptualization and question the epistemological assumptions of complexity theory in application to social phenomenon. These constructive criticisms are viewed in this paper as translation errors that often hinder metaphorical application and assert three propositions regarding the promise of complexity theory for the study and practice of public affairs and administration. While these propositions to do not necessarily invalidate critical comments in regard to the application of complexity theory, they do attempt to establish the basis for encouraging continued examination of how concepts derived from complexity theory. Finally, these propositions do not diminish the promise of other avenues of inquiry in the field; they are intended to establish complexity theory as one avenue of that also has promise.

INTRODUCTION

It has been nearly two decades since the "new sciences" made its entry into the domain of Public Administration and Political Sciences through the work of Kiel (1989, 1994) on chaos theory, which was by that time fused with notions from complexity theory. There are a number of published works that examine the implications of chaos and complexity theory for public administration (Overman, 1996; Morçöl & Dennard, 2000). An impressive list of publications has appeared that attempt to conceptualize and directly apply complexity theory or complexity thinking to number of public policy and administration settings, including managing change and organizational renewal (Kiel, 1994), policy analysis (Morçöl, 2002), public administration (Morçöl & Dennard, 2000; Weber, 2008; Meek, 2010), public management (Teisman & Klijn, 2008); policy networks (Kickert, Klijn & Koppenjan, 1997); policy processes (Klijn, 1996); policy tools and methods (Dennard, Richardson & Morçöl, 2008) and governance systems (Teisman, van Buuren & Gerrits, 2009). What these approaches have in common is the recognition that our social systems are dynamic and evolving. As such, research efforts should aim at considering insights and approaches public policy and administration that reflect this understanding. These approaches recognize that many of our existing tools and approaches are based on questionable assumptions and in application have limited value (cf. Gerrits & Moody, 2011). As one of our colleagues stated, "I came to realize I had incredible faith in things that don't work." These research efforts cited above are attempts to improve both our theory and practice of public administration based on a different kind of understanding about the how patterns of interaction could be better understood. This led many of the researchers cited here to explore the potential of complexity theory.

Despite growing interest and research in how complexity theory can be applied to policy and public administration, approaches to conceptualization and application have varied. Confusion over what complexity theory specifically represents has complicated rather than clarified its application to issues in public administration to such an extent that it is regarded as a minority interest (Pollitt, 2009). Rooted in domains such as physics, chemistry and biology, complexity theory came with a specific epistemology and ontology that was often discarded within the target domain of public administration (cf. Kerr, 2002). The word 'theory' implies a set of explanatory variables or mechanisms whereas a considerable number of works in this field seem to be on the level of conceptualization. Conceptual elegance can be helpful in understanding something, but not necessarily so.

The troublesome nature in applying complexity theory to public administration rests in the nexus of its promising potential in identifying patterns of interaction alongside its apparent weakness in construct specificity. The former engages both the researcher and practitioner in rich interpretation and guide for action. The latter beguiles the researcher in search of clarity and specificity of behavioral phenomenon. Those who hold up complexity theory as a valuable paradigm for Public Administration and Policy Analysis see the behavioral connections and patterns identifiable in its attractive concepts. Those who note the weakness of complexity theory in application of behavioral phenomenon stress the apparent vagueness of it foundational terms and concepts.

Given the unevenness of research orientation and application, it is useful to begin to identify some common components that characterize the approach of complexity theory in public administration so as to offer a basis for future research and establishing a basis of expectations that we can gain as rely on this approach. This paper sets out to offer three propositions about the epistemological foundation of complexity that establishes its potential for application in the study and practice of public administration.

THEORY TRANSFER AND METAPHORS

Because complexity theory has a diverse background that is somewhat alien to the social sciences, it is not surprising that some things were altered or lost in translation during theory transfer. The positivist protocols of measurement certainty common in the domain of natural sciences is a challenge in the examination of social reality where many things are not unambiguous enough to be measured through positivist methods. Furthermore, associations with previous types of systemic thinking, most importantly Parsons's functionalism (1951), meant that complexity theory was scrutinized critically in the domain of Public Administration (e.g., Klijn & Snellen, 2009; Pollitt, 2009). Some had argued that complexity theory in the social sciences is nothing more than metaphors that do not hold much explanatory power (cf. Rosenhead, 1999). Given these concerns, it is necessary to discuss the process of theory transfer in order to understand the nature of complexity theory.

The use of metaphors is helpful in many ways and some argue that human thought processes are actually largely metaphorical because it can't be certain that people see the real complexity behind events and processes (Lakoff & Johnson, 2003). It is therefore natural to think of theories as metaphors and this perspective helps explaining theory transfer (Chettiparamb, 2006). Metaphors are understood as the vehicles for transplantation from one domain to another. It is accepted that the properties of the target domain, Public Administration in this case, do not correspond to the properties of the source domain, such as physics, chemistry and biology. The consequence of the difference between source domain and target domain is that the metaphor has a dynamic meaning.

There are several roles that metaphors can have in theory transfer. The most common one is the structural metaphor. Such metaphors are used to develop analogies between the two domains and can work to enrich both, as the transformation of the metaphor during the confrontation with the target domain can help to enhance understanding in the source domain (Chettiparamb, 2006). Using a structural metaphor, it can be inspiring to think of, e.g., managerial activities as butterfly wings setting in motion all sorts of turbulence or that a public organization resembles a flock of birds, as some early accounts on complexity and (public) management do. However, if the metaphor doesn't do much more than that, initial enthusiasm may collapse into disappointment (cf. Mitleton-Kelly, 2003).

The important difference is in the way causation behind the metaphor is treated. The fact that birds are able to self-organize by very simple rules of behavior does not mean that organizations are able to self-organize by simple rules of behavior. In other words, the causation present in the source domain does not mean that the same causation is preserved in the target domain. Careless transfer of the causation behind the metaphor has led to fierce criticism on several occasions (e.g., Sokal & Bricmont, 1998). Causation in social reality always has degrees of uncertainty and is in most cases simply not given. Defining causation requires interpretation and assigning meaning to what is being observed. Thus, the process of defining causation is simultaneously a process of assigning meaning. As such, causation "is best understood as an experiential gestalt. A proper understanding of causation requires that it be viewed as a cluster of other components. But the cluster forms a gestalt—a whole that we human beings find more basic than the parts" (Lakoff & Johnson, 2003: 70-71). In other words: while the exact causation as present in the source domain is not present in the target domain, the observed emergence as a whole could be understood through the process of semiosis that originates in the target domain.

The application of complexity theory in Public Administration has been hampered because, on many occasions, the application failed to address this issue of theory transfer. This does not mean that every application requires a discussion about metaphors but it requires that the concepts that are used in a given research are made operational for that particular research and that the use of concepts and the difference between the source and target domain are carefully mapped (Chettiparamb, 2006). At the same time, there is still a certain attractiveness in the analogies that enrich the target domain, as witnessed by, e.g., the fact that books about complexity management are quite popular. We postulate that the transfer of complexity theory to Public Administration was never a single, unambiguous transfer. To scholars in Public Administration working with complexity theory, it was an attempt to get a better understanding of the complex causation of the issues in this domain and many have developed their own particular way of looking at complex causation. To practitioners in this domain, complexity theory offered an alternative vocabulary that helped expressing dynamics they may otherwise not be able to express. To those studying the nature of social sciences, complexity theory questioned some of the established ideas about causality and the ways of knowing in social reality. In other words: different niches within Public Administration have an entirely different way of understanding complexity. Confusion reigns because these differences are still often not explicated. While complexity theory and its concepts as a whole resonate with a part of Public Administration, it is not consistent across the different niches within the domain. For example, practitioners are happy to work with metaphors that enrich whereas scientists are looking for ways to improve understanding of causality. This difference is exactly the core of lost translations during theory transfer through metaphors (cf. Lakoff & Johnson, 2003).

The remainder of this paper aims to explicate the ambiguities between the different niches by developing three propositions that establish the basis for utilizing complexity theory in the practice and theory development in Public Administration. These positions to not an attempt to unify the different niches within the Public Administration domain but they do help to discern between them and to identify where translation errors occur.

Proposition 1: The Importance of Dynamical Contexts

A social scientific approach to the development of complex systems through time acknowledges that social processes have a systemic character, i.e., that a process results from multiple causes that in turn are affected by that particular process (cf. Byrne, 1998). Although this implies a notion of causality, it does not imply that this causality is fixed over time such as assumed in early manifestations of functionalist systemic thinking. Complexity theorists may use the metaphor of mechanistic systems vs. thermodynamic systems to point at the temporal character of causality. The processes that are observed at a given point in time are rooted in a particular local configuration of causes where the value of (a set of) variables depends on the particular configuration of the set. That configuration can and often will change over time, which means that a particular process can be explained as long as it is understood that such an explanation is partial (Cilliers, 2005), and local in time and place (Byrne, 2002). To understand a process or a system therefore requires understanding the local configuration with the acceptance that this understanding is temporal as time moves on and configurations alter (Verweij & Gerrits, 2011).

The development of a complex system over time is a property of the interaction between the system, in itself a set of interacting elements (components, structure and mechanisms), and the environment or context. If the context is important in explaining the development of a complex system, it should not be treated as something nebulous or ambiguous but should be researched in a systematic and systemic way (Ragin *et al.* 2003; Rihoux, 2006; Rihoux & Lobbe, 2009; Byrne, 2009). Among others Byrne (2005, 2011) and Cilliers (2001) make a distinction between generic complexity and situated complexity. Generic complexity is of the type indicated by e.g., Reynolds or Langton (cf. Smith & Stevens, 1996) and assumes that complex developments can be traced back to a limited of simple rules of behavior and a limited set of variables that are determinants for those developments. The assumption goes that the complexity of a given situation can be understood by understanding the operation of the conjunction of those variables and simple rules of behavior.

Situated complexity, on the other hand, departs from the idea that complex systems are open by definition and that the constituent variables do not define the borders of what is being observed. Therefore, complexity arises not only from the constituent elements of a system but also from the fact that this constitution is dynamic in itself, i.e., that it is constantly changing. This holds that complexity cannot be traced back to simple rules and a limited set of variables. From this perspective, the origins of complexity are discursive to the extent that it is not possible to discern the afore-mentioned rules and variables.

A third take on complexity is to treat it as something complicated. Complexity is complicated but complicated things are not necessarily complex. Complicated things are structured in such a way that they reproduce the same outcome over time, whereas complexity as discussed above generates novel or near-novel outcomes every time. Thus, in the experience of an administrator, the context in which he or she operates could be complex or complicated or both. Each environment requires a specific approach to the organization of the decision/making process and the type of decision or policy. The types of contexts mentioned here are archetypical in the sense that it will be difficult to see unambiguous appearances of each of them. Indeed, some authors have argued that all policy and administrative contexts should be considered complex and call upon new approaches to policy deliberation and decision (Morçöl, forthcoming). Furthermore, the contexts are dynamic, which means that a particular type can evolve in a different context.

The relationship between context, systems and processes has been fleshed out before, e.g., in Mintzberg (etc., etc.), Leleur (2007). The argument here is therefore not that the context matters but that the context is explanatory for a specific process or outcome. Byrne (2011) notes that this contextual and dynamic approach to policy processes and decision-making is often acknowledged by practitioners but not recognized in policy analysis and practice, where order is sought even when not available. The first proposition is therefore:

> *[P1] Using complexity theory in Public Administration means first and foremost that research objects are analyzed and understood as deeply contingent phenomenon, whose outcomes can only be explained by analyzing the nature and dynamics of the context.*

Proposition 2: Observation and Action

Public administration works in two seemingly antithetical paradigms: one of science and one of action. The world of action can be informed by the world of science. But that action shapes the world we examine, changing our observations. We are both observers and actors. Any action based on an initial logic must respect and expect unanticipated outcomes and respond. Complexity theory expects this condition, but it also has science as foundation. Thus, complexity theory is both normative and scientific. The identification of both understanding complex system behavior and what can guide action in complex systems means we have both the ability to examine and, at the same time, act in complex systems.

The confluence of theory and action has been at center stage in public administration from its origin. The challenge of separating the development of a science of Public Administration from the practice of public administration is a constant challenge because of the complexity of the arenas where action or policy implementation takes place. For Vincent Ostrom (1989), implementing bureaucratic values (such as efficiency and effectiveness) came at the expense of democratic values. For the administrator, the domain of Public Administration can be theory informed, but call upon individual judgment, or what Donald Schön referred to as "the reflective practitioner." The ability to know of a pattern within a context, and, at the same time, have the ability to act and shape the pattern is the paradox of Public Administration research. Individuals can choose between seeking administrative efficiency and effectiveness through the science metaphor (act as machines) or choose to serve members of the public (complexity metaphor), constantly seeking democratic values, by democratizing through dialogue (Fung, 2004; Yankelovich, 1991) or through the mindful facilitation of leaders (Cooper & Bryer, 2008). Either way, the influence of the individual in manipulating feedback loops shape the system in very human ways. We can research these patterns of interaction and seek scientific interpretations of their determinants. We can also come to understand these patterns and influence the direction they take at the same time. The second proposition is therefore:

[P2] Complexity theory can both allow observation (thus enabling scientific inquiry) and inform a guide to action.

Proposition 3: Heuristic Concepts

Some of the criticism of the vagueness of complex systems concepts also applies to mainstream Public Administration, which is also littered with vague terms such as 'networks,' 'governance', and 'public'. There are more such blind spots. Yet, this apparent 'vagueness' has a purpose. It is not always a functional to keep on defining a certain concept because doing that can restrict a theory in development. As long as a certain concept provokes new ideas with people it already has a function. This contradicts the scientific requirement of defining, testing and redefining.

Much of the public administration research that draws upon concepts from complexity theory typically establishes relevant concepts through definitions developed initially in the physical and biological sciences. These concepts are then utilized to inform thinking in the social sciences and are linked to public policy and public administration research. The work of Booher (2008) is illustrative of this kind of application where features (concepts) of a complex adaptive system—agents, interactions, nonlinearity, system behavior and robustness and adaptation—are utilized to make sense of the structure and dynamics of process that evolved in addressing a contentious water policy case in California. In this work, Booher stresses the character of the complex adaptive system as a network of participants demonstrating the capacity of a system incorporate an extremely large and diverse set of citizenry, task-forces, subcommittees and local and regional stakeholders. He views this system as developing a form of evolving democracy based on a new kind of structure for civic engagement. For Booher, the unit of analysis for this application of complexity theory is the network of stakeholders. The central practice or management implication of this observation is the need to improve our understanding of how we can build and maintain these diverse networks.

In another example, Gerrits (2010) examines the paradigmatic assumptions of rational decision-making in public policy that ignores the contextual complexity of the targeted decision. The central feature of this research is the necessity of adopting a complex systems perspective in public policy decision-making that takes into account "that causation is complex and changeable over time, and it should abandon the anthropocentric perspective for a co-evolutionary revision" (p. 19). Gerrits applies this complexity theory to planning decisions for port extensions and the concomitant unintended outcomes. For Gerrits, the unit of analysis for this application of complexity theory is the decision making frame of reference (assumptions) regarding a complex physical system. The central practical feature or management implication of this observation is the need to incorporate a more 'co-evolutionary' consideration, or paradigmatic assumption, in our decision-making approaches that target physical systems.

There are a number of other illustrations of the varied approaches to public policy and administration stimulated by interpretations of complexity theory. Meek (2008) examined the emergent nature of collective action in metropolitan governance as a form of complex system adaptation. This work examined self-organizing features (concepts) derived from the literature in application to an emerging collectivity. The unit of analysis in this research was the metropolitan system within which various kinds of self-organizing patterns could be discerned and where one particular case was examined in depth regarding self-organizing features. Teisman, van Buuren and Gerrits (2009) also establish a number of complexity concepts—non-linear dynamics, self-organization, co-evolution—and then illustrate their analytic value in the areas of infrastructure development, port systems, metropolitan regions, and traffic management. This variety of contexts (and units of analysis) informs the authors regarding implications for public policy making and management. One central operating foci of this work is the necessity of viewing public administration as an evolutionary phenomenon where interest is focused on the development of complex systems rather than the characteristics at a certain time and place (Teisman *et al.*, 2009: 3).

From research developed to this point, it is clear that the domain of Public Administration can be informed by the study of complexity theory. In this approach, we are learning that Public Administration, in part, is a practice that is based on epistemological assumptions regarding the context and behavior of the systems where action takes place. The domain is to be viewed within a complex system that is evolving. The implications for public management include the reframing of epistemological assumptions [that influence our decision-making parameters] from determinism, order and predictability to probabilistic, partial order, emergence and uncertainty (Geyer & Rihani, 2010). This reframing will likely be uncomfortable for those who have relied upon a more orderly positivist paradigm can bring forth criticism. Our guess is that this reframing will also be more embraced by those who practice public administration management and are familiar with the nuances and complexities of policy implementation. We are optimistic that complexity theory will offer new ways of interpreting our social, economic and political patterns of interaction as demonstrated by the variety of approaches evident in the literature.

It is clear that the study of complex systems in public administration and policy to his point has developed from a number of approaches, definitions and applications. It is also clear that research has relied upon a number of differentiated units of analysis and contexts. Complexity theory concepts enable action because of their relative heuristic value in application to behavioral phenomena. What is also offered in the research are valued case applications that can begin to assist the formulation of a broader perspective from which we can develop both a scientific (therefore testable) basis for action as well as a guide to how action best serves outcomes sought by actors within the system. The third proposition is therefore:

[P3] Complexity concepts offer a frame of reference that fosters flexibility in pattern recognition that enables action.

DISCUSSION

Complexity theory encompasses multiple, sometimes conflicting, approaches to complex systems (in Public Administration). This allows for flexibility in research, philosophy and practice, which explains its potential and the enthusiasm with which many authors have embraced certain notions. However, a theory of complexity for public administration does not seek to become a unifying theory for public administration; the heterogeneity in the field of public administration remains its central feature (Ricucci, 2010).

In this paper, we assert three propositions that are useful in guiding our thinking regarding the application of complexity theory to public administration (see Table 1). First, it is recognized that the practice of public administration is rooted in addressing social-economic-political issues that have numerous characteristics and difficult to reify or simplify. This condition makes it difficult for unique disciplinary foci, bound by assumptions regarding units of analysis and behavior, to offer comprehensive understanding of the issue to be addressed. At best, disciplinary foci can offer interpretations based on simplifying assumptions and for specific units of analysis within a specified context of actors. One can imagine a large number of contending disciplinary insights for any number of social issues and public responses. In addition, the practice of public administration is often rooted in the development and implementation of policy solutions framed by multiple stakeholders—namely citizens, citizen groups, interests groups, and competing organizations/agencies—each seeking solutions unique to their interests. Facilitating these interests offer opportunities to develop solutions that enhance a broader spectrum of interests if these solutions were rooted in

	Component in Public Administration	**Propositions of Complexity** Using complexity theory in Public Administration means that:
1	Objects	Research objects are analyzed and understood as deeply contingent phenomenon, whose outcomes can only be explained by analyzing the nature and dynamics of the context.
2	Observation & Action	Complexity theory can both allow observation (thus enabling scientific inquiry) and inform a guide to action.
3	Concepts	Concepts offer a frame of reference that fosters pattern recognition and enables action.

Table 1 *Three Propositions For Complexity Theory And Public Administration*

processes that honor differences rather than calculate interests unexamined. Often issues in addressed in public administration are multi-jurisdictional, such as environmental issues and call upon cross-jurisdictional designs and arrays of accountability to reflect constituent interests. Combined, these kinds of contextual complexities (issue complexity, stakeholder complexity, jurisdictional complexity, among others) call upon solutions that represent the kinds of processes reflected in studies of complexity sciences and offer insight into ways of democratizing public administration (Booher, 2008).

Second, we assert that complexity theory embraces the seemingly difficult proposition that humans are both influenced by and can act upon the forces that are shaping our destiny. This recognition means that a narrowly defined value-free objectivity is not possible for scientific examination as the researcher is always in some way part of the system they are examining. Of course, we do recognize that there are degrees of research objectivity. There are also a number of protocols that seek to establish or strengthen research objectivity in order to claim value-free observations. One characteristic of this proposition is that individuals can seek either self-interested or altruistic framing of complexity solutions, both of which are value-based. The work of Raab and Milward (2003) has informed our field with regard to "dark networks" in the examination of drug cartels, diamond trade and terrorist networks. Still, the complex systems approach allows one to see possibilities of stakeholder participation as a way to see emergent solutions, including forms of self-organization. Participation does matter in the development of solutions and the inclusion of stakeholders can broaden respect for the outcome.

Last, the concepts of complexity theory offer both the opportunity for refinement for scientific inquiry as well as action. Complexity theory for Public

Administration has not reached the level of a truly grounded theory. At the one hand, it is because much work focuses on the translation from the source domain to the target domain. On the other hand, it is because complexity theory is ambiguous at best. This ambiguity is necessary because Public Administration has multiple, sometimes contradictory orientations that are inherent to the different realms it is applied, i.e., science and practice. We argue that this ambiguity should be cherished and developed. We think that research into complexity theory and public administration will not lead to the accumulation of knowledge in the sense of positivist science. That is, it will never enable us to state that "y will always be caused by x" or that "y will always appear under conditions x-x^{-n}". That is the type of knowledge that can't be generated through the complexity lens. Complexity is a way of thinking so as to interpret and guide action. The fact that its many concepts resonate strongly with practitioners means that it has meaningful impact for those working with issues in Public Administration. From this perspective, it is not a matter of reaching the golden standard of positivist research to generate ever-lasting knowledge but rather to reach a deeper understanding of the contextual dynamics at work in public administration practice. We acknowledge that there is still much work to do in this, much of it of exploratory nature, but we are confident that complexity will continue to inform both researchers and practitioners of how to appreciate the vast complexity of everyday reality in public decision-making.

REFERENCES

Ashby, W.R. (1962). Principles of the self-organizing system. In: Von Foerster, H. and Zopf, G.W. (eds.). Principles of Self-Organization: Transactions of the University of Illinois Symposium. London: Pergamon Press.

Axelrod, T. (1984). The Evolution of Cooperation. Basic Books, New York.

Bass, B. M. (1990). Bass & Stogdill's Handbook of Leadership: A Survey of Theory and Research, New York, NY: Free Press.

Barnard, C. I. (1938). The Individual and Organization. In: Barnard, C.I., The Functions of the Executive. Harvard: Harvard University Press.

Bondarenko, D.M. (2007). "What is there in a word?: Heterarchy, homoarchy and difference in understanding 'complexity' in the social sciences and complexity studies," in K.A. Richardson and P. Cilliers (eds.), Explorations in Complexity Thinking: Pre-Proceedings of the 3rd International Workshop on Complexity and Philosophy.

Booher, D.E. (2008). "Civic Engagement as Collaborative Complex Adaptive Networks," in Kaifeng Yang & Erik Bergrund (eds.) Civic Engagement in a Network Society. Charlotte, NC: Information Age Publishing, pp. 111-148.

Byrne, D.S. (1998). Complexity Theory and the Social Sciences. London, New York: Routledge.

Byrne, D.S. (2002). Interpreting Quantitative Data. London: Sage Publications.

Byrne, D.S. (2005). Complexity, Configurations and Cases. Theory, Culture & Society, 22(5): 95-111.

Byrne, D.S. (2009). Complex realist and configurational approaches to cases: A radical synthesis. In D. Byrne & C.C. Ragin (Eds.), The Sage handbook of case-based methods (pp. 101-119). London: Sage.

Byrne, D.S. (2011). Applying Social Science; The role of social research in politics, policy and practice. Bristol: The Policy Press.

Chettiparamb, A. (2006). Metaphors in Complexity Theory and Planning. *Planning Theory*, 5(1): 71-91.

Cilliers, P. (2001). Boundaries, hierarchies and networks in complex systems. *International Journal of Innovation Management*, 5(2): 135-147.

Cilliers, P. (2007). What is inside and what is on top? Complex systems and hierarchies. In: Richardson, K. & Goldstein, J.A. (eds.). Classic Complexity. From the Abstract to the Concrete. Mansfield, MA: ISCE Publishing.

Cooper, T. L. and Bryer, T. A. 2007. William Robertson: Exemplar of Politics and Public Management Rightly Understood. *Public Administration Review*, 67(5): 816-824.

Dennard, L., Richardson, K.A. and Morçöl, G. (2008). *Complexity and Policy Analysis: Tools and Methods for Designing Robust Policies in a Complex World*. ISCE Publishing.

Fuchs, C. (2002). Some Implications of Anthony Giddens' Works for a Theory of Social Self-Organization. *Emergence*, 4(3): 7.

Fung, A. (2004) Empowered Particpation: Reinventing Urban Democracy. Princeton: Princeton University Press.

Gerrits, L. (2010). "Public Decision Making as Coevolution," *Emergence: Complexity & Organization*, 12(1): 19-28.

Gerrits, L. and Rebecca Moody (2011) Envisaging Futures: An Analysis of the Use of Computational Models in Complex Decision Making Processes. *Emergence: Complexity & Organization*, 13(1-2): 96-114.

Geyer, R. and S. Rihani (2010). Complexity and Public Policy: A New Approach to Twenty-First Century Politics, Policy and Society. New York: Routledge.

Giddens, Anthony (1985). A Contemporary Critique of Historical Materialism. Vol. 2: The Nation-State and Violence, Cambridge, UK: Polity Press.

Harter, N. (2006). Leadership as the promise of simplification. *Emergence: Complexity & Organization*, 8(4): 77-87.

Hazy, J. K. (2006). Measuring leadership effectiveness in complex socio-technical systems. *Emergence: Complexity & Organization*, 8(3): 58-77.

Kernick, D. P. (2005). Facilitating resource decision making in public organizations drawing upon insights from complexity theory. *Emergence: Complexity & Organization*, 7(1): 23-28.

Kerr, P. (2002). Saved from Extinction: Evolutionary Theorising, Politics and the State. *British Journal of Politics and International Relations*, 4(2): 330-358.

Kickert, J.M., Klijn, E.H, Koppenjan, J.F.M. (1997). Managing Complex Networks. Strategies for the Public Sector. London: Sage Publications.

Kiel, D.L. (1989). "Nonequilibrium theory and implications for public administration," *Public Administration Review*, ISSN 0033-3352, 49(6): 544-551.

Kiel, D.L. (1994). Managing Chaos and Complexity in Government. San Francisco: Jossey-Bass.

Klijn, E.H. (1996). "Analyzing and managing policy processes in complex networks," *Administration and Society*, 28(1): 90-119.

Klijn, E.H, & Snellen, I. (2009). Complexity Theory and Public Administration: A Critical Appraisal. In: Teisman, G.R., Van Buuren, A. & Gerrits, L. (eds.) Managing Complex Governance Systems. London, New York: Routledge.

Koliba, C., Meek, J.W., Zia, A. (2011). Governance Networks in Public Administration and Public Policy. New York: CRC Press.

Lakoff, G. & Johnson, M. (2003). Metaphors We Live By. (2nd edition). Chicago: The University of Chicago Press.

Leleur, S. (2007). Systemic planning: Dealing with complexity by a wider approach to planning. *Emergence: Complexity & Organization*, 9(1/2): 2-10.

Lipsky, M. (1980). Street Level Bureaucracy: Dilemmas of the Individual in Public Services. New York: Sage Foundation.

Lissack M, Richardson K. (2001). When Modeling Social Systems, Models ≠ the Modeled: Reacting to Wolfram's A New Kind of Science. *Emergence: Complexity & Organization*, 3(4): 95-111.

Lukes, S. (2005). Power: A Radical View (2nd edition). London: Palgrave Macmillan, 192pp.

McKelvey, B. (2003). Microstrategy from macroleadership: Distributed intelligence via new science. In A. Y. Lewin and H. Volberda (eds.), Mobilizing the Self-Renewing Organization, Thousand Oaks, CA: Sage.

Meek, J. (2008). "Partnerships and Metropolitan Governance: An Adaptive Systems Perspective," in Linda Dennart, Kurt A. Richardson & Goktug Morçül (eds.). Complexity and Policy Analysis: Tools and Methods for Designing Robust Policies in a Complex World. Goodyear, AZ.: ISCE Publishing.

Meek, J.W. (2010). "Complexity theory for public administration and policy," *Emergence, Complexity and Organization* , ISSN 1521-3250, 12(1): 1-4.

Morçöl, G. (2002). A New Mind for Policy Analysis: Toward a Post-Newtonian and Postpositivist Epistemology and Methodology , ISBN 9780275970123.

Morçöl, G. (2008a). "Complexity of public policy administration: Introduction to the special issue," *Public Administration Quarterly* , ISSN 0734-9149, 32(3): 305-313.

Morçöl, G. (2008b). "A complexity theory for policy analysis," in L. Dennard, K.A. Richardson and G. Morçöl (eds.), Complexity and Policy Analysis: Tools and Methods for Designing Robust Policies in a Complex World , ISBN 9780981703220.

Morçöl, G. (forthcoming) Complexity Theory for Public Policy. Routledge.

Morçöl, G. and Dennard, L. (eds.) (2000). New Sciences for Public Administration and Policy: Connections and Reflections , ISBN 9781574200706.

Midgley, G., and Richardson, K. A. (2007). Systems Thinking for Community Involvement in Policy Analysis. *Emergence: Complexity & Organization,* 9(1/2): 167-183.

Mitleton-Kelly, E. (2003). Ten Principles of Complexity and Enabling Infrastructures. In Complex Systems & Evolutionary Perspectives of Organisations: The Application of Complexity Theory to Organisations. Amsterdam: Elsevier.

Ostrom, V. (1989) The Intellectual Crisis in American Public Administration. University of Alabama.

Overman, E.S. (1996). "The new science of management: Chaos and quantum theory and method," *Journal of Public Administration Research and Theory,* ISSN 1053-1858, 6(1): 75-89.

Parsons, T. (2005). The Social System (reprint). London: Routledge.

Pollitt, C. (2009). Complexity Theory and Evolutionary Public Administration: A Sceptical Afterword. In: Teisman, G.R., Van Buuren, A. & Gerrits, L. (eds.) Managing Complex Governance Systems. London, New York: Routledge.

Ragin, C.C., Shulman, D., Weinberg, A. & Gran, B. (2003). Complexity, generality, and qualitative comparative analysis. *Field Methods,* 15(4): 323-340.

Raab, J., &Milward, B. (2003). Dark Networks as Problems. *Journal of Public Administration Research,* 13(4): 413-439.

Riccucci, N.M. (2010). Public Administration: Traditions of Inquiry and Philosophies of Knowledge. Washington D.C.: Georgetown University Press.

Rihoux, B. (2006). Qualitative comparative analysis (QCA) and related systematic comparative methods: Recent advances and remaining challenges for social science research. *International Sociology,* 21(5): 679-706.

Rihoux, B. & Lobe, B. (2009). The case for qualitative comparative analysis (QCA): Adding leverage for thick cross-case comparison. In D. Byrne & C.C. Ragin (Eds.), The Sage handbook of case-based methods (pp. 222-242). London: Sage.

Rosenhead, J. (1998). Complexity Theory and Management practice (Scientific No. LSEOR 98.25). London: Department of Operational Research, London School of Economics.

Schon, D. (1983) The reflective Practitioner: How Professionals Think in Action. New York: Basic Books.

Scott, R. J. (2010). The Science of Muddling Through Revisited. *Emergence, Complexity and Organization*, 12(1): 5-18.

Schreiber, C., & Carley, K. M. (2006). Leadership style as an enabler of organizational complex functioning. *Emergence: Complexity & Organization*, 8(4): 61-76.

Simon, H. (1962). The Architecture of Complexity. *Proceedings of the American Philosophical Society*, 106(6): 467-482.

Smith, T. S., & Stevens, G. T. (1996). Emergence, Self-organisation, and Social Interaction: Arousal Dependent Structure in Social Systems. *Sociological Theory*, 14(2): 131-153.

Sokal, A., & Bricmont, J. (1999). *Fashionable Nonsense; Postmodern Intellectuals' Abuse of Science*. St. Martins: Picador Press.

Teisman, G.R. and Klijn, E.-H. (2008). "Complexity theory and public management: An introduction," *Public Management Review*, 10(3): 287-297.

Teisman, G.R., van Buuren, A. and Gerrits, L. (2009). *Managing Complex Governance Systems: Dynamics, Self-Organization and Coevolution in Public Investments*, New York: Routledge.

Ulrich, W. (1994). Critical Heuristics of Social Planning: A New Approach to Practical Philosophy (2nd edition). Chichester, UK, and New York: John Wiley & Sons, 1994. 504 pp.

Verweij, S. & Gerrits, L.M. (2011). A proposal to assess projects using a complexity-informed framework. Paper presented at COMPACT Work (Rotterdam, June 23 - June 25).

Vesterby, V. (2008). Measuring Complexity: Things That Go Wrong and How to Get It Right. *Emergence: Complexity & Organization*, 10(2): 90-102.

Weber, M. (1968). Economy and Society: an outline of interpretive sociology. New York: Bedminster Press.

Weber, J.E. (2008) "Symposium: Complexity of public policy and public administration," *Public Administration Quarterly*, 32(3).

Yankelovich, D. (1991). Coming to Public Judgment. Syracuse: Syracuse University Press.

Lasse Gerrits is an associate professor of Public Administration at the. His main research focuses on the analysis of the complexity of public decision-making, especially in urban planning and metropolitan governance. He also teaches master courses on governance, complexity and system's theory. Previously, he worked as a consultant for TNO Built Environment and Geosciences.

Jack W. Meek is Professor of Public Administration at the College of Business and Public Management at the University of La Verne and Visiting Professor of Government at Claremont McKenna College. Professor Meek offers courses in research methods, policy analysis and collaborative public management. His research focuses on metropolitan governance including the emergence of administrative connections and relationships in local government, regional collaboration and partnerships, policy networks and citizen engagement.

Chapter 17

COLLECTIVE DECISION-MAKING AS AN EVOLUTIONARY PROCESS

Frank Boons (Erasmus University Rotterdam, NLD)
Lasse Gerrits (Erasmus University Rotterdam, NLD)
Peter Marks (Erasmus University Rotterdam, NLD)

Collective decision making processes can be considered as a process where populations of ideas, consisting of problem framing and solution framing, evolve over time. Actions and perceptions of actors within networks, material changes induced by decisions, and chance events drive the evolution in decision making. The creation and selection of problem ideas and solution ideas within populations are intertwined and are part of the same feedback loop. The variation, selection and retention of problem and solution ideas take place within a system of heterogeneous actors and this takes place in multilevel systems. Hence, the problem-solution space evolves. Not only are actors connected to other actors through dynamic boundaries, they are also connected to other physical systems, which is not constant and exogenous to the individual species' evolution, but influenced by it. Social and physical systems can exert selection pressures on each other, influencing the variation and connected selections in a continuous process. This continuous process is not necessarily progress, but at best a local optimum or selective adaptation. We aim to conceptualize collective decision making as an evolution of process and content and to develop a research agenda to investigate the explanatory power of this conceptualization.

INTRODUCTION AND AIMS

The work of Darwin and Lamarck has inspired researchers in disciplines other than biology to conceptualize change through time as an outcome of evolutionary processes and mechanisms. Examples of such applications can be found in, among others, the philosophy of science (cf. Toulmin, 1972), evolutionary economics (cf. Van den Bergh, 2004, 2007; Silva & Teixeira, 2009), environmental sciences (cf. Norgaard, 1994; Guan & Norgaard, 2010), socio-cultural studies (cf. Sanderson, 1990; Nelson, 2006), organizational theories (cf. Ven & Poole 1995; Volberda & Lewin, 2003; Porter, 2006; Hodgson, 2010), game theory (cf. Neumann & Morgenstern, 1944; Axelrod, 1984; Arthur *et al.*, 1997), architecture (cf. Geddes, 1915), urban planning (cf. Marshall, 2009), environmental business strategies (cf. Boons, 2009), sociology (cf. Hird, 2010), and many others. These approaches differ substantially in their applications (cf. Silva & Teixeira, 2009) but they share the goal to deploy the explanatory mechanisms of evolution to explain for the seemingly incremental and destination-less changes observed in those fields.

A small number of applications to political sciences and public administration have also been made, but full-scale applications of evolutionary frameworks have yet to gain widespread popularity, despite a number of explanatory mechanisms that could be helpful in understanding collective decision-making processes (cf. Kerr, 2002; Sementelli, 2007; Gerrits, 2011). This contribution aims to conceptualize collective decision-making as an evolution of process and content and to develop a research agenda to investigate the explanatory power of this conceptualization. In brief, our argument is that collective decision-making processes can be considered as a process where populations of ideas, consisting of problem framing and solution framing, evolve over time. This evolution is driven by the actions and perceptions of actors within networks, and by the material changes that take place as a result of decisions and as a result of the conjunction of circumstances, i.e., chance events. The next section identifies the need for an evolutionary approach to collective decision-making. We then define evolution and what makes a theory evolutionary and discuss the outlines of our conceptualization, followed bu our conclusions.

THE NEED FOR AN EVOLUTIONARY APPROACH TO COLLECTIVE DECISION-MAKING

Evolutionary approaches to collective decision-making are not unfamiliar in the domain of Public Administration, albeit not always under that particular header. Lindblom's take on decision-making as an incremental process, as first presented in 'The Science of Muddling-through' (1959), is arguably a proto-evolutionary approach (cf. Scott, 2010). There is a good reason to start with Lindblom's discussion of the root and branch models of decision-making, because he argued that there is no clear relationship between decisions made at a certain point in time and the intended end-state or final goal as a result of the steps taken in between the two points (Parsons, 1995; Scott, 2010). He states that a more realistic approach to decision-making is to understand that each next step builds from the current state of affairs and moves from there by small, incremental degrees. The 'current state of affairs' is the result of steps made through previous decisions but also the result from a particular coming together of circumstances and chance events in the current situation. Thus, the relationship between what was originally intended and where the decision maker arrives at after a certain amount of time and associated events has passed is mostly disjointed (Lindblom, 1959).

Lindblom's idea that there is no clear link between what the decision maker wants at a given stage and what is achieved at a later stage, if anything at all, is one of the core puzzles that Public Administration deals with.It resonates in the work of among others March and Olsen (1976), Cohen, March and Olsen (1978), Kingdon (1984), Hogwood (1987) and Baumgartner and Jones (1993). This relationship between intention and outcome is non-linear, for lack of better term. This does not mean that each outcome is entirely novel as systems develop path-dependency and lock-ins (Pierson, 2000) but it points at the fact that there is no direct link between what one actor wants and does and the results in the system. While there are ample conceptualizations of the disconnected nature of decisions and outcomes, there are fewer in-depth analyses of the mechanisms that explain this disconnection. Sementelli (2007) asserts that Public Administration as a discipline is ill-equipped to analyze the full complexity and non-linearity of the wicked problems policy-makers and analysts are faced with because it lacks analytical tools that are appreciative of the complexity and non-linearity of structure and process in public decision making. An evolutionary theory of decision-making could help dealing with that complexity and non-linearity, according to him.

Early attempts to use evolutionary mechanisms for explaining public decision-making can be attributed to e.g., Corning (1983), Daneke (1990) and Kaufman (1991). Daneke identifies the need for a research agenda that approaches issues in Public Administration as systemic and evolutionary. Twenty years later, some elements of this research agenda appear in the current research. Of the authors mentioned here, Kaufman's account (1991) is perhaps the most well-known. He acknowledges that actors in collective decision-making overestimate the influence they have on the outcomes of certain decisions, also known as the fundamental attribution error (Ross, 1977; Ross & Nisbett, 1991). Those outcomes, Kaufman says, are the result of a partly random conjunction of several factors, which is indeed an important property of evolutionary processes as discussed before. A disconnecting of individual choice and action and overall outcome is something also found in research into evolutionary economics (cf. Van den Bergh) and sociology (Elster, 2007; Boudon, 1986).

Although Kaufman uses the vocabulary of evolution to describe a real phenomenon, he does not claim that he deploys the concepts as explanatory mechanisms. His main point that decision-maker's intentional actions bear little relation to the outcomes is something that persists throughout his work. Since Kaufman there have been very few works within Public Administration were evolutionary mechanisms were applied to collective decision making. Therefore, it is fair to say that thinking in terms of evolutionary mechanisms has gained little ground (Kerr, 2002). However, evolutionary thinking is not entirely absent from the analysis of collective decision-making. First, certain mechanisms from evolutionary theories have been used without clear reference to those theories. Second, mechanisms from evolutionary theories have resurfaced in researchers focusing on complexity theory in Public Administration research, as outlined by Daneke's research agenda (1990). Third, Public Administration is and remains a multidisciplinary field (Pollit, 2010) and that makes it sometimes hard to assign a certain research to this or a different domain. For example, Norgaard's work on evolution and co-evolution concerns a blending of economics, environmental science and politics.

Kerr, then, traces the unpopularity of evolutionary theories in this field back to attempts to replicate the positivist, objectivist research approaches common to evolutionary theories in natural sciences and to the functionalist tendencies that came with early applications, most prominently by Parsons (1966). However, more recent rethinking led to the adoption of certain ideas from evolutionary theories to political sciences and Public Administration, a long list of which is given in Kerr (2002). These authors "have demonstrated [...] that evolutionary theory can provide us with a more versatile, sophisticated and complex explanatory framework than many of the theoretical tools which we currently have at our disposal." (2002: 355) An example of this can be found in Gerrits & Marks (2008).

One particular area where evolutionary theories have resurfaced is when researchers attempted to use complexity theory to research issues in Public Administration, including Dennard (1996), Mitleton-Kelly (2003), Gerrits (2008), Teisman, Van Buuren & Gerrits (2009), and Morçöl (2010). This fusion is no surprise because evolution and systems thinking are closely related and are often linked in reviews (cf. Daneke, 1990; Silva & Teixeira, 2009). Especially 'adaptation' is often used and analyzed as part of a systemic framework. So, while there is reason to assume that an evolutionary approach to collective decision-making could help explaining the erratic nature of such decision making, it seems that most attempts at such an approach have not resulted in a coherent framework and research agenda.

WHAT IS AN EVOLUTIONARY THEORY?

Before we present our conceptualization, it is necessary to outline what constitutes an evolutionary theory. Evolutionary approaches in the social sciences originate from using concepts and insights from evolutionary biology to understand social phenomena. If we look at this in terms of metaphors (Chettiparamb, 2006), evolutionary biology is the source domain from which insights are taken to be applied in the target domain of social phenomena. The focus of the source domain is the way in which characteristics in populations of species change over time, and may lead to the emergence of change in the species' characteristics and perhaps even new species in the long run, i.e., distinguishable groups of organisms that are bounded by the impossibility to mate with other species members. Darwin (1859) proposed a theory to understand such changes by looking at the interplay between mechanisms of variation, selection and retention of characteristics that act in or on a population. The next step occurred when Huxley formulated the Modern synthesis (1942). This term refers to the combination of Darwin's insights with Mendelian genetics.

Campbell (1969) was the first to develop a systematic framework for application of Darwinian evolutionary theory in the social sciences. One of the consequences of using a scientific theory as the source domain for a metaphor is that it may change significantly over time. As a result, the application in the target domain may change over time as new insights become available. More often, it seems that basic ideas of the initial theory are taken as a source, and later developments are not incorporated. Recently, a group of theoretical biologists joined forces to describe the state of the field 60 years after the first publication of Huxley's book. To tap into the potential of more recent insights of evolutionary biology we make use of this collection of papers (Pigliucci and Müller, 2010). There are several reasons why it makes sense to further develop an evolutionary conceptualization of collective decision-making processes:

- It naturally emphasizes the possibility of patterns of change within social systems that occur without (intentional) design and planning. This is complementary to theories that focus on decision-making as an intentional and purposeful activity with traceable consequences.
- It calls for an analysis of the content of collective decision-making (in terms of the population of ideas) as well as the process (the mechanisms through which ideas emerge, are transmitted, and selected). It therefore helps linking process and content in such way that the mutual influence becomes clearly visible.
- It explains why the best ideas in terms of problem definitions and solution definitions do not necessarily survive and are the 'best' because fitness does not equal truth, rationality, or high moral standing. It means that those ideas survive because they suit a particular conjunction of circumstances.
- It allows an analysis of the interplay between social and physical systems. This is important because many of the collective issues societies are dealing with relate both systems, often in combination, such as climate change, energy provision and resource scarcity. It helps to overcome the social-physical dichotomy.

If one develops an evolutionary theory, it is important to be explicit about the *unit of selection*. We acknowledge the difference between a replicator and the unit of selection (cf. Dawkins, 1976: 47), which is especially useful in social systems where actors (individuals as well as organizations) have a life somewhat independent of the ideas and routines they incorporate. In our view, collective decision-making as a process in the end is about ideas that concern problem definitions and problem solutions and the associated actions. More specifically, it concerns [a] a set of perspectives on the outside world, which specifies that a given situation is wanted or unwanted; this is the 'problem idea' and it combines value statements with factual observations; and [b] a set of perspectives on causality that specifies how the situation identified in [a] should be dealt with; these are the 'solution ideas'. Each set [ab] has a subset of associated actions [a1b1] that affect the real world. The information that drives the evolution of ideas in the problem-solution-space is contained in people, products, books and structures. Also knowledge, science, religion, habits, institutions and politics are relatively constant under wide range of circumstances (Van den Bergh & Gowdy, 2000).

Collective decision-making can then be defined as the evolutionary process through which ideas are selected and become connected to one or more solution ideas, akin to Cohen, March and Olsen's garbage can (1972). The garbage can model is a subset of the set of processes covered by the approach outlined here, because it singles out a specific type of evolutionary process where the selection mechanism is highly variable or weak.

Taking ideas as the unit of selection does not necessitate an extreme position as taken by those advocating the study of memes (cf. Dawkins, 1976; Blackmore, 1999). As Toulmin (1972) shows, a moderate approach is possible in which the evolution of ideas is analyzed in relation to the networks of actors that 'carry' them. In this process, evolutionary mechanisms operate that are shaped by, or constitute, the institutions that are involved in collective decision-making as well as the wider socio-physical systems and their characteristics it is embedded in. These ideas will be fleshed out by further discussing the evolutionary process and its mechanisms variation, selection and retention and the role of choice, the structure in which evolution takes place, the material responses in the shape of feedback on decisions and chance events and the matter of direction of evolution.

Variation-Transmission-Selection Combination

Central to our conceptualization are populations with problem ideas and solution ideas, both with associated actions. Evolutionary theories hold that three mechanisms occur in evolutionary processes: variation, selection and retention. The latter can be divided into transmission and inheritance. Rather than just assuming that the operation of variation, selection and retention takes place and that those mechanisms conjunct at certain points in time, we take the operation of those mechanisms as the empirical variables that explain the coming-together of collective decision making.

A simplistic notion would be to regard these three mechanisms as synoptic phases, starting with the variation of problem and solution ideas, the selection of certain problems and solution ideas and the retention, similar to the synoptic approach to collective decision making. However, just like the synoptic model of decision-making has often been discarded (cf. Braybrooke & Lindblom, 1963; Van den Donk, 1998), this would be an oversimplification of the complex relationship between the three mechanisms and the role of chance and randomness. Yet, if 'time' is taken as the dimension that structures our analysis, it will be clear that there are sequences of events and that decision-making is intricately connected with that sequence. Our approach is therefore a hybrid perspective that acknowledges that decision-making is not a synoptic process that takes place in fixed sequential steps but that, at the same time, it has a longitudinal character consisting of consecutive events. The cycle of reciprocal selection is not considered to be a sequential process but a simultaneous process that consists of continuous feedback loops.

Actors are constantly engaged in the process of variation, selection and retention. A pool of problem ideas and solution ideas is present at a given time. We call this pool the problem-and-solution space. Actors attempt at finding connections between problems and solutions and it should be emphasized that this concerns perceptions of both. It constitutes a confrontation between values (about what ought to be), perceptions of the current situation and a mental map of the causations that drive the problem and that should be dealt with. The solution idea is built on a perception of the causation that would address the issue that needs to be solved, the proverbial antidote that should be applied. Collective decision-making, then, is the survival of certain problem ideas and solution ideas from the problem-and-solution space. That is: variety occurs and ideas are selected. Transmission is not trivial here (Boons, 2009). A problem or solution idea can survive at the level of an individual actor (who then might be called stubborn or persistent), but if this actor is the sole entity that adopts this particular idea, then it may remain unimportant in the collective process. This mechanism can be elaborated further by building on research into policy diffusion (Berry & Berry, 1990).

The role of chance and randomness in variation deserves some attention in evolutionary theory, because it is often misunderstood (Sober, 1993). 'Blind' or random variation does not mean that it is produced through a non-directed process; it means that the process that leads up to variation is seemingly separate from the process of selection (Gerrits, 2011). Recent discussion in biology is about the relative importance of the selection mechanism and chance variation. If variation is both random and ubiquitous (meaning that there are many different variations that pop up regularly), then selection is the dominant mechanism. However, recent evidence shows that variation is not necessarily ubiquitous, and the order in which variations appear, i.e., when options present themselves, affect the evolutionary path (*ibid.*).

Within the variety that exists in the problem-and-solution space, some options are then selected or are being selected. The creation of variety and selecting variety is considered to be a part of the same feedback loop; variation and selection are intertwined in a complex way (Foster & Hölzl, 2004). There is no neat separation between variation that is created by actors and variation that is not and exists 'out there'. The latter type of variation cannot be externalized from the selections made by actors because, although some actions can be without consequences, it is quite possible for actors' actions to lead to change and further variation with complex causality obscuring the relationship between the two. This ostensible absence of clear causation may give the impression that, while intended selection is observable, certain variation can't be related and is, therefore, blind. We argue that variation never occurs 'just like that'. Indeed, it is the result of human action or the result of the conjunction of several conditions. In both cases, there is a causation behind it that can be traced back. Variation at a future point in time can still be triggered by current selection, even when this variation is unintended, unobserved and unexpected (Gerrits, 2008). In other words, variation and selection are part of the same feedback loop (Hrebiniak & Joyce, 1985).

The act of decision-making can be seen as selection, even if it is acknowledged that a decision doesn't necessarily lead to concrete outcomes. According to Norgaard (1984), 'selection pressure' and 'feedback' are similar concepts. Feedback is a driver for change but not every feedback loop leads to change as negative feedback loops confirm the current state of affairs whereas positive feedback loops provoke change (Norgaard, 1994). Feedback therefore becomes selection pressure as a response to an incentive leading to change. Such feedback loops consist of information for actors, upon which they act accordingly (Foster & Hölzl, 2004). That is, they compare the new information with an existing pool of information that allows them to assess the nature of the new information. There are then two basic types of responses to this new information: adoption and adaptation. Adoption occurs when information is absorbed but not acted on. This includes the rejection of information. Adaptation occurs when information is not only absorbed but also leads to changes in response to this information (Dopfer, 2005). It should be noted that these two basic options are not synonymous with the passive or active processing of information as actors can actively decide not to alter their internal structure when confronted with new information. In this processing of information, actors are limited in their cognitive capacities and restrained by their values.

The distribution of variation in a population, which comes via mutation and recombination, is the information that drives evolution. Units have some degree of durability and resilience and pass these characteristics to other units (heredity or transmission or retention). Selection pressures such as scarcity of resources and space but also public pressure through changing problem and solution ideas makes actors compete among each other in a selection environment consisting of physical as well as technological, social, market, institutional and organizational constraints. The space of problem-and-solution-ideas changes continuously but within a certain bandwidth because of this competition. Actors make choices within this space. The choice can be made to aim for certain problem and solution ideas but it should be remembered that choosing a certain causal construct is also a choice not to pursue other possibilities. This, again, means that selection and variation are connected.

It is not possible for actors to fully analyze their own situation or the environment and design optimal strategies. Actors can still direct their own actions, they are not mere victims of their environment. In that, they shape the environment and influence the course of events within certain restrictions. That is, actors design strategies that are based on successful and unsuccessful strategies from the past, to what others have been doing and with their selection environment (Sotarauta & Srinivas, 2006). Of importance here is trial and error, cultural learning and inheritance or retention of successful strategies and innovations (Kallis & Norgaard, 2010).

MULTI-LEVEL SYSTEMS AND INDIVIDUAL PREFERENCES AND BEHAVIOR

The variation, selection and retention of problem and solution ideas take place within a system of heterogeneous actors, of which the heterogeneous character is pivotal to the ongoing process of variation, selection and retention (cf. Van den Bergh & Gowdy, 2000; Foster & Hölzl, 2004). Two issues must be dealt with here: the properties of such systems and the relation between different (types of) systems. To start with the first issue: systems where selection takes place are, in fact, systems of systems, i.e., multilevel systems (Potts, 2000). One of the more enduring debates in evolutionary biology is about the level at which selection mechanisms operate: genes, organisms, or groups. Biologists have made progress in this area by finding evidence for group selection to take place. One of the conclusions is that selection at the genetic level is not necessarily dominant. Also, they have analyzed how groups may evolve into single organisms over time.

Although there is not a singular equivalent to the gene in economics, a similar discussion can be found in evolutionary economics, i.e., to determine the level that is most important for selection in systems (Dopfer, 2005; Worden, 2010). Perhaps this debate about the most important selection level is intractable as micro-level behavior of individual actors influences macro-level behavior and vice versa. Thus, when talking about systems, one should focus on the rules that govern the interaction between actors and which processes and interactions influence the evolution of the system over time (Foster & Hölzl, 2004). Van den Bergh & Gowdy (2000: 38) attempt to solve the issue by stating that the evolutionary process of choice "consists of three basic principles, that is on micro-level diversity (variation), heredity and individual selection, as well as macro-level trends and shocks ('large-scale accidents')."

Looking at the individual level, it appears that actors attempt to make rational, calculated decisions but that this rationality is not only bounded by the classic boundaries as identified by Simon and Lindblom. Individuals act based on the knowledge they have about the systems they are in. This knowledge is partial at best. Systems at a macro level (as opposed to the individual level) constrain the possible choices and actions that actors can make. The knowledge that these actors have of the system at a macro level and the way they learn (mostly through trial and error, i.e., a type of variation) about the system shapes their behavior. Since all actors learn differently and create different knowledge about the system and the (possible) behavior of others all actors coevolve in different manners with the system and other actors. Consequently, the problem-solution space evolves, too. "The selection process results from the success or failure of different behaviors and strategies in the competitive and cooperative dynamical game that is running" (Allen, 2001: 346-347).

Individual actors are also bounded by institutions that guide how individuals behave (Safarzynska & Van den Bergh, 2010). A single actor can hardly ever alter existing structures and institutions. A group of actors an influence such structures and institutions, e.g., lobby groups may be alter institutions or stop them from changing (Olson, 1965). There is therefore an intricate relationship between what happens at the individual level and the level of institutions and structure. Accepting into or keeping collective actors out of collective decision-making alters the evolution of the problem-and-solution space (Gerrits, 2008, 2011)[1]. "Consequently, in group selection, power acts at the lower level as a mechanism of social transmission. At a higher level, group dynamics involve assorting, migration between groups, group re-structuring, survival, reproduction, and conflict. Due to the interactions between groups, certain behaviors are enforced and ultimately fixed. Evolved policies may then act as enabling or discouraging certain individual strategies" (Safarzynska & Van den Bergh, 2010: 744).

Focusing on the macro-level, then, shows that the decision-making scales up from the individual level to aggregates of individuals to fields of organizations and institutions. Together they form nested systems, their systemic character stemming from the persistence of relations over time and the temporarily sharing of goals. Not only are actors connected to other actors through dynamic boundaries, they are also connected to other physical systems (cf. Norgaard, 1984, 1994; Gerrits, 2008). From this perspective, the environment is not constant and exogenous to the individual species' evolution, but influenced by it.

1. It should be noted that we use the word collective to denote decision-making processes that involve multiple actors. This is different from the way it is often used in relation to Olson's work, where it signifies the acts of specific political actors (like trade unions).

Material Feedback and Chance Events

The connection between seemingly incompatible systems and mutual influence on each system's evolution has been conceptualized by Norgaard (1984, 1994, 1995) and applied to collective decision-making by, e.g., Gerrits (2010, 2011), Moreno-Peñaranda and Kallis (2010) under the header of coevolution. This perspective shows how connected systems, such as social and physical systems, can exert selection pressures on each other. This influences the variation and with that selection at a later stage. The consequence is a continuous pattern of feedback loops between the systems. Whereas some social feedback loops between actors can be regarded as communication, it takes the form of material feedback when it concerns the physical world.

Gual and Norgaard (2010) show in several examples how there are complex relationships between the social and physical realm that are often ignored analytically. Cultural evolution can influence the natural evolution and the dynamics of physical systems. The changed physical conditions affect how selection forces operate in influencing one or several populations. Actors can then choose to respond, or not. For example, the genetic composition of pest populations is in a reciprocal relation with agriculture practices. Farmers that are in a competitive market usually change to a single crop, which makes them more vulnerable to pests. Farmers thus create a pesticide killing most of the pests. The highly adaptive pests in turn become increasingly immune. Hence a positive feedback loop that exerts selection pressures (Norgaard, 1994). The focus here is somewhat holistic, as it is not about a single technology or product but "on the mutual adaptation of technological and policy paradigms." (Kallis & Norgaard, 2010: 694).

Apart from such feedback loops, variation and selection are also influenced by chance events. Chance events are discrete events that come together by chance and whose results influence the systems' state (Sibeon, 1999). These events do have an impact although they are effectively detached from actors' direct influence instead of being (complex) connected to them

Direction and Improvement

Some accounts on evolution suggest that the best survives, e.g., the best idea, but that is not necessarily so. Which problem and solution ideas survive is not just a matter of survival of the fittest. "Selection should not be simplified as 'survival of the fittest', but rather as the survival of the sufficiently adapted unit in a changing environment." (Van den Bergh & Kallis, 2009: 4) Ideas that are relatively well-adapted in the selection environment can become increasingly adopted by other actors who subscribe to what becomes an increasingly dominant idea, either absolutely or relatively. This may be regarded as selective adaptation (Van den Bergh & Gowdy, 2000). It is now common to discard social evolution as 'progress' or a process of global optimization (ibid.). Selection is a local search process, which at best leads to a local optimum. However, confusion about the direction of evolution, i.e., whether it constitutes an improvement, is something that needs discussing in the light of Public Administration because collective decision-making relates individual behavior to benefits for the collective.

Sementelli (2007) notes that a number of accounts on evolution in collective decision-making appear to relate evolution to progress, but progress implies a certain directional tendency towards an improvement that is absent in evolutionary theories. A direction as a property of the time dimension is a necessary ingredient for an evolutionary theory but direction does not necessarily equate progress or improvement. It is confusion over direction vs. improvement that has led to diverging, contradictory and sometimes intersecting interpretations of the nature of evolution in collective decision-making (Kerr, 2002; Porter, 2006; Gerrits, 2008). Perhaps such ideas about evolution as improvement through the decisions of actors towards a distant goal are inspired by Darwinism but can be traced back to a materialist ontology that rules out the fact that what constitutes 'improvement' depends on the perspective of a certain actor (Norgaard, 1994, 1995; Sementelli, 2007). Also, given the reciprocal nature of evolutionary mechanisms, a lasting improvement towards a better fit is therefore purely a theoretical possibility (Porter, 2006; Sementelli, 2007).

Likewise, evolution does not imply working (together) to create the most optimal collective outcome. The point of departure remains a heterogeneous population with a high variety of ideas and this diversity creates the continuous disequilibrium mentioned before. Personal convictions may lead to benevolent behavior in actors but such considerations and actions are bounded within the full problem-solution space and as such not the necessarily logical outcome in the long run (Trivers, 1971; Manner & Gowdy, 2010). Economists have treated other-regarding behavior and cooperation as exceptional results that need to be explained in each case they appear (Worden, 2010). In biological and social systems collective operation is the exception to the rule, free-riding and the tragedy of the commons are prevalent. Governmental institutions are methods to coordinate the avoidance of such matters but it is not necessarily so that the optimal solution presents itself. As discussed before, the problem-solution space is the result of the competition within local diversity. Still, other-regarding behavior is often in certain circumstances better than acting selfish (Manner & Gowdy, 2010).

Waring (2010) proposes a model that combines evolutionary biology, economics and psychology, gene-culture evolution and mimetics that helps analyzing how other-regarding behavior can evolve. From this perspective, selfish behavior is selected at the level of individual actors. Other-regarding behavior and cooperation is favored at the group-level. Manner & Gowdy (2010) argue that other-regarding behavior is likely to emerge over time as group selection benefits group-beneficial attributes over individually-beneficial—but group-maladaptive—attributes. The evaluation of the literature on this subject boils down to this: evolution of the problem-and-solution space takes place regardless of whether this evolution is favorable to some actor or not. Actors attempt to influence the direction of the evolution but their influence is limited because of the considerations and actions of others that may contradict their intents. Despite these limitations, there are indications that other-regarding behavior at the individual level could benefit the group as a whole.

CONCLUSIONS AND A RESEARCH AGENDA

Central to collective decision-making processes is that populations of ideas, consisting of problem framing and solution framing evolve. Ideas are selected and become connected to one or more solution ideas. Actions and perceptions of actors within networks, material changes induced by decisions, and chance events drive the evolution in decision-making. The creation and selection of problem ideas and solution ideas within populations are intertwined and are part of the same feedback loop. The variation, selection and retention of problem and solution ideas take place within a system of heterogeneous actors and this takes place in multilevel systems.

Individual actors attempt to make (bounded) rational, calculated decisions based on the (partial) knowledge they have about the systems they are in. Systems constrain the possible choices and actions that actors can make.

All actors learn differently about the system and adapt accordingly. Hence, the problem-solution space evolves. Not only are actors connected to other actors through dynamic boundaries, they are also connected to other physical systems, which is not constant and exogenous to the individual species' evolution, but influenced by it. Social and physical systems can exert selection pressures on each other, influencing the variation and connected selections in a continuous process. This continuous process is not necessarily progress, but at best a local optimum or selective adaptation. Heterogeneous populations with a variety of problem and solution ideas and the reciprocal nature of the evolutionary selection mechanisms optimality is an academic exercise.

The conceptualization of the evolutionary framework is helpful in understanding collective decision-making processes. By making use of the evolutionary framework collective decision-making we can derive a number of generic research questions.

A first question deals with the fact that collective decision-making covers a wider area than intentional and purposeful activity (policymaking) with traceable consequences (successful implementation). In social systems, actors may deal with collective problems in alternative ways: they converge on certain problem-solution combinations through self-organization, where the boundary conditions are created by the evolutionary mechanisms of variation, selection and retention. Here it is important to note that each of the mechanisms can be more or less important in shaping the process: sometimes the available variation explains most of the collective decision-making, in other cases selection may be totally absent, but convergence is the result of successful transmission of a single idea, etc. The specific combination of these mechanisms is thus an empirical question.

Q1: How do variation, selection and retention create the boundary conditions for a process of self-organisation where actors converge on a specific problem-solution combination?

A second question concerns the relationship between the content of collective decision-making (the population of problem and solution ideas) as well as the process (the mechanisms through which ideas emerge, are transmitted, and selected).While these mechanisms may have physical components, to a large extent they will be social constructions: education, training, as well as workshops and visioning events are the ways through which variation is produced; ideas are transmitted along the connections provided by social networks, and selection can be the result of the decisions of autonomous actors (consumers) as well as formal rules (which specify a goal rather than the specific action to be performed; Boons, 2009). Our conceptualization thus provides an alternative way to look at the interaction between the cognitive, cultural component of collective decision making, and the institutional structures in which such decision-making takes place. For both dimensions there is a large body of literature, ranging from the role of frames in policy processes to the effects of specific network structures on policy outcomes. Thus, a second question is:

Q2: How does our knowledge of the cognitive and structural dimensions of collective decision-making expand by translating it into an evolutionary perspective?

A third question centers on the crucial point that according to an evolutionary perspective, the fact that some problem-solution combinations persist is not due to their truth, rationality, or high moral standing. It means that those ideas survive because they suit a particular conjunction of circumstances. It may also be the case that their survival is due to characteristics of the ideas themselves (Blackmore, 2000). One question is how this affects the system state of the social system in which collective decision-making takes place: stable, dynamic, chaotic or inert (Boons *et al.*, 2009).

Q3: What are the properties of collective decision-making and the results it produces in different system states?

A final question concerns the prescriptive part of public administration. If we adopt the evolutionary perspective, including the fact that 'fitness' does not equal rationality, truth, or high moral standing, and if we accept that collective decision-making to some extent goes on 'behind the back' if individual actors, what are the possibilities for a normative theory of governance? This question also relates to a central quality of human beings: their ability to reflect on the world, their actions and change actions accordingly. As actors in collective decision-making become aware of the way in which evolutionary mechanisms operate, they may decide to influence these mechanisms, or design collaborative processes that streamline the evolutionary process to some extent.

Q4: How can individuals in social systems shape processes of collective decision-making that are shaped by evolutionary mechanisms?

We end with a note on method. Analyzing the operation of evolutionary mechanisms requires a methodology that is able to capture changes over time. Thus, methods such as event history analysis and time series analysis may be useful (see Poole *et al.*, 2000 for an overview). The time period that needs to be delineated for analysis will need to be established empirically. In some policy fields, the dynamics in populations of problem and solutions happens quickly, while in others it moves slowly. Theoretically, there is no way to set a specific timeframe.

REFERENCES

Allen, P.M. (2001). Knowledge, ignorance and the evolution of complex systems. In: Foster, J and Stanley Metcalf, J.. Frontiers of evolutionary economics. Competition, self-organization and innovation policy, Cheltham, Edward Elgar, pp. 313-350.

Arthur, W.B., Durlauf, S.N. and Lane, D.A. (1997). The Economy as an Evolving Complex System II, Proceedings volume XXVII in the Santa Fe Institute studies in the sciences of complexity. Reading, Massachusetts: Addison-Wesley.

Axelrod, R. (1984). The Evolution of Cooperation, Basic Books.

Baumgartner, F. R. and Jones, B.D. (1993). Agendas and Instability in American Politics. Chicago: The University of Chicago Press.

Bergh, J.C.J.M. v. d. and Gowdy, J.M. (2000). Evolutionary Theories in Environmental and Resource Economics: Approaches and Applications. *Environmental and Resource Economics*, 17: 37-57.

Bergh, J.C.J.M. (2004). Evolutionary Thinking in Environmental Economics: Retrospect and Prospect. In: J. Foster & W. Hölzl (Eds.), Applied Evolutionary Economics and Complex Systems (pp. 293). Cheltenham Edward Elgar.

Bergh, J.C.J.M. (2007). Evolutionary Thinking in Environmental Economics. *Evolutionary Economics*, 17(5): 521-549.

Berry, F.S., and Berry, W.D. (1990). "State Lottery Adoptions as Policy Innovations: An Event History Analysis." *American Political Science Review,* 84(2): 395-415.

Blackmore, S. (1999). The meme machine. Oxford University Press.

Boons, F.A.A. (2009). Creating ecological value. An evolutionary approach to business strategies and the natural environment. Cheltenham: Edward Elgar.

Boudon, R. (1986). Theories of social change: A critical appraisal. Berkeley: University of California Press.

Braybrooke, D. and Lindblom, C.E. (1963). A Strategy of Decision. New York: The Free Press.

Campbell, D. (1969). Variation and selective retention in socio-cultural evolution. *General Systems*, 16: 69-85.

Chettiparamb, A. (2006). Metaphors in Complexity Theory and Planning', *Planning Theory*, 5(1): 71-91.

Cohen, M., March, J.G. and Olsen, J.P. (1972). A Garbage Can Model of Decision-making. *Administrative Science Quarterly*, 17: 1-25.

Corning, P.A. (1983). The Synergism Hypothesis: a Theory of Progressive Evolution. New York: McGraw-Hill, 1983.

Daneke, G. A. (1990). A Science of Public Administration? *Public Administration Review*, 50: 383-392.

Darwin, C. (1859). The origin of species. London: John Murray.

Dawkins R. (1976): The Selfish Gene. New York: Oxford University Press.

Dennard, L. (1996). The New Paradigm in Science and Public Administration. *Public Administration Review*, 56 (459-499).

Donk, W.B.H.J. van den (1998). Beyond Incrementalism? Redistributive Policy Making, Information Systems and the Revival of Synopticism. In: Snellen, I.T.M. & Donk, W.B.H.J. Public Administration in an Information Age. Amsterdam: IOS Press.

Dopfer, K. (2005). The evolutionary foundations of economics. Cambridge/New York: Cambridge University Press.

Elster, J. (2007). Explaining social behavior. Cambridge: Cambridge University Press.

Foster, J. and Hölzl, W. (Eds.). (2004). Applied Evolutionary Economics and Complex Systems. Cheltenham UK, Northampton USA: Edward Elgar Publishing.

Geddes, P. (1915). Cities in Evolution. London: William and Norgate.

Gerrits, L. (2008). The Gentle Art of Coevolution: a complexity theory perspective on decision-making over estuaries in Germany, Belgium and the Netherlands. Rotterdam: Erasmus University.

Gerrits, L. and Marks, P. (2008). Complex bounded rationality in dyke construction: Path-dependency, lock-in in the emergence of the geometry of the Zeeland delta. *Land Use Policy*, 25(3): 330-337.

Gerrits, L. (2010). Public Decision-Making as Coevolution. *Emergence: Complexity & Organization*, 12(1): 19-28.

Gerrits, L. (2011). A Coevolutionary Revision Of Decision Making Processes: An Analysis of Port Extensions in Germany, Belgium and The Netherlands. *Public Administration Quarterly*, 35(3): 309-329.

Gual, M.A. and Norgaard, R.B. (2010). Bridging Ecological and Social Systems Coevolution: A Review and Proposal. *Ecological Economics*, 69(4): 707-717.

Hird, M.J. (2010). Coevolution, Symbiosis and Sociology. *Ecological Economics*, 69(4): 737-742.

Hodgson, G.M. (2010). Darwinian Coevolution of Organizations and the Environment. *Ecological Economics*, 69(4): 700-706.

Hogwood, B. (1987) From Crisis to Complacency? Shaping Public Policy in Britain. Oxford: Oxford University Press.

Hrebiniak, L.G. and Joyce, W.F. (1985). Organizational adaptation: Strategic choice and environmental determinism. *Administrative Science Quarterly*, 30: 336-349.

Huxley, J. (1942). Evolution: The Modern Synthesis – The Definitive Edition, with a new forward by Massimo Pigliucci and Gerd B. Müller. MIT Press.

Kallis, G. and Norgaard, R.B. (2010). Coevolutionary Ecological Economics. *Ecological Economics*, 69(4): 690-699.

Kaufman, H. (1991). Time, Chance and Organizations: Natural Selection in a Perilous Environment (2nd ed.). Chatham: Chatham House.

Kerr, P. (2002). Saved from Extinction: Evolutionary Theorising, Politics and the State. *British Journal of Politics and International Relations*, 4(2): 330-358.

Kingdon, J.W. (1984). Agenda's, Alternatives and Public Policies. Boston: Harper Collin Publishing.

Lindblom, C.E. (1959). The Science of 'Muddling Through'. *Public Administration Review*, 19(2): 79-88.

Manner, M. and Gowdy, J. (2010). The Evolution of Social and Moral Behavior: Evolutionary Insights for Public Policy. *Ecological Economics*, 69(4): 737-761.

March, J.G. and Olsen (eds.) (1976). Ambiguity and Choice in Organizations. Oslo: Universitetsforlaget.

Marshall, S. (2009). Cities, Design & Evolution. London, New York: Routledge.

Mitleton-Kelly, E. (2003). Ten Principles of Complexity and Enabling Infrastructures. In: Mitleton-Kelly, E, (ed.) Complex Systems and Evolutionary Perspectives on Organisations: the Application of Complexity Theory to Organizations. Oxford: Elsevier Science, pp. 3-20.

Morçöl, G. (2010). Reconceptualizing Public Policy from the Perspective of Complexity Theory. *Emergence: Organizations and Management*, 12(1): 52-60.

Moreno-Peñaranda, R. and Kallis, G. (2010). A Coevolutionary Understanding of Agroenvironmental Change: A case-study of a rural community in Brazil. *Ecological Economics*, 69(4): 770-778.

Nelson, R. (2006). Evolutionary Social Science and Universal Darwinism. *Evolutionary Economics*, 16(5): 491-510.

Neuman, J. von and Morgenstern, O. (1944). Theory of Games and Economic Behavior, Princeton: Princeton University Press.

Norgaard, R.B. (1984). Coevolutionary Development Potential. *Land Economics* 60(2): 160-173.

Norgaard, R.B. (1994). Development Betrayed: The End of Progress and a Coevolutionary Revisioning of the Future. London, New York: Routledge.

Norgaard, R.B. (1995). Beyond Materialism: a Coevolutionary Reinterpration. *Review of Social Economy,* 53(4): 475-486.

Olson, M. (1965) The logic of collective action, Cambridge: Harvard University Press.

Parsons, T. (1966). Societies: Evolutionary and Comparative Perspectives. Englewood Cliffs: Prentice Hall.

Parsons, W. (1995). Public Policy; an Introduction to the Theory and Practice of Policy Analysis. Cheltenahm: Edward Elgar Publishing Limited.

Pierson, P. (2000). Increasing Returns, Path Dependence and the Study of Politics, *American Political Science Review*, 94(2): 251-267.

Pigliucci, M. and Müller, G. (eds.) (2010). Evolution: The Extended Synthesis. MIT Press, Cambridge, MA, USA.

Pollit, C. (2010). Envisioning Public Administration as a Scholarly Field in 2020. *Public Administration Review*, 70(6s): s292-s294.

Porter, T.B. (2006). Coevolution as a Research Framework for Organizations and the Natural Environment. *Organization and Environment*, 19(4): 479-504.

Potts, J. (2000). The new evolutionary microeconomics. Complexity, competence and adaptive behavior, Cheltenham, Edward Elgar.

Ross, L. (1977). The intuitive psychologist and his shortcomings: Distortions in the attribution process. In L. Berkowitz (Ed.), Advances in Experimental Social Psychology, 10: 173-220. New York: Academic Press.

Ross, L. and Nisbett, R.E. (1991). The Person and the Situation: Perspectives of Social Psychology. Philadelphia: Temple University Press.

Safarzyńska, K. and Bergh, J.C.J.M. (2004). Evolving Power and Environmental Policy: Explaining Institutional Change with Group Selection. *Ecological Economics*, 69(4): 743-752.

Sanderson, S.K. (1990). Social Evolutionism. A Critical History. Cambridge, M.A.: Blackwell.

Scott, R.J. (2010). The Science of Muddling Through Revisited. *Emergence, Complexity & Organization*, 12(1): 5-18.

Sementelli, A. (2007). Distortions of Progress: Evolutionary Theories and Public Administration. *Administration and Society*, 39(6): 740-760.

Sibeon, R. (1999). Anti-reductionist Sociology. *Sociology,* 33(2): 317-334.

Silva, S.T. and Teixeira, A.A.C. (2009). On the Divergence of Evolutionary Research Paths in the Past 50 Years: a Comprehensive Bibliometric Account. Evolutionary Economics. 19(5): 605-642.

Sober, E. (1993). Philosophy of Biology. Boulder: Westview Press.

Sotarauta M. and Srinivas S. (2006). Co-evolutionary policy processes: understanding innovative economies and future resilience, *Futures*, 38: 312-336.

Teisman, G.R., Van Buuren, A., and Gerrits, L. (2009). Managing Complex Governance Systems; Dynamics, Self-Organization and Coevolution in Public Investments. London, New York: Routledge.

Toulmin, S. (1972). Human understanding. Princeton: Princeton University Press.

Trivers, R. (1971). Evolution of reciprocal altruism. *Quarterly Review of Biology*, 46: 35.

Ven, A. van de, and Poole, M. (1995). Explaining Development and Change in Organizations, *Academy of Management Review*, 20(3): 510-540.

Volberda, H.W. & Lewin, A.Y. (2003). Co-evolutionary Dynamics Within and Between Firms: From Evolution to Co-evolution. *Journal of Management Studies*, 40(8): 2111-2136.

Waring, T. M. (2010). New evolutionary foundations: Theoretical requirements for a science of sustainability. *Ecological Economics*, 69(4): 718-730.

Wilson, E.O. (1975). Sociobiology: The New Synthesis. Cambridge: Belknap Press.

Worden, L. (2010). Notes from the greenhouse world: A study in coevolution, planetary sustainability, and community structure. *Ecological Economics*, 69(4): 762-769.

Dr. **Frank Boons** currently works as researcher and teacher at the Department of Public Administration of Erasmus University Rotterdam, The Netherlands. As an organizational & economic sociologist, his interest is in the dynamics of interorganisational processes that relate to sustainable development. He has published widely on the ecological strategies of firms in global and regional networks of firms and policy actors and issues of complex governance.

Lasse Gerrits, Ph.D. studied Public Administration and Urban Planning at the Erasmus University Rotterdam and now works as associate professor at the research group Governance of Complex Systems. His research interests concern the complexity of public decision-making and the development of the urban. He also teaches courses in complexity and systems.

Dr. **Peter Marks** works at the department of Public Administration at the Erasmus University Rotterdam. Having studied economics and philosophy his research mainly focuses on complex decision making processes from an evolutionary (economics) perspective. The target domain in his research is public safety policy issues. He also teaches courses in systems and evolutionary economics.

www.ingramcontent.com/pod-product-compliance
Lightning Source LLC
LaVergne TN
LVHW081940200926
848484LV00009B/2481
* 9 7 8 1 9 3 8 1 5 8 0 1 8 *